T0245423

# CAMBRIDGE LIBRARY COLLECTION

*Books of enduring scholarly value*

## Botany and Horticulture

Until the nineteenth century, the investigation of natural phenomena, plants and animals was considered either the preserve of elite scholars or a pastime for the leisured upper classes. As increasing academic rigour and systematisation was brought to the study of 'natural history', its subdisciplines were adopted into university curricula, and learned societies (such as the Royal Horticultural Society, founded in 1804) were established to support research in these areas. A related development was strong enthusiasm for exotic garden plants, which resulted in plant collecting expeditions to every corner of the globe, sometimes with tragic consequences. This series includes accounts of some of those expeditions, detailed reference works on the flora of different regions, and practical advice for amateur and professional gardeners.

## Hortus Kewensis

Trained as a gardener in his native Scotland, William Aiton (1731–93) had worked in the Chelsea Physic Garden prior to coming to Kew in 1759. He met Joseph Banks in 1764, and the pair worked together to develop the scientific and horticultural status of the gardens. Aiton had become superintendent of the entire Kew estate by 1783. This important three-volume work, first published in 1789, took as its starting point the plant catalogue begun in 1773. In its compilation, Aiton was greatly assisted with the identification and scientific description of species, according to the Linnaean system, by the botanists Daniel Solander and Jonas Dryander (the latter contributed most of the third volume). Aiton added dates of introduction and horticultural information. An important historical resource, it covers some 5,600 species and features a selection of engravings. Listing the printed works consulted, Volume 1 provides plant descriptions from Monandria to Heptandria.

Cambridge University Press has long been a pioneer in the reissuing of out-of-print titles from its own backlist, producing digital reprints of books that are still sought after by scholars and students but could not be reprinted economically using traditional technology. The Cambridge Library Collection extends this activity to a wider range of books which are still of importance to researchers and professionals, either for the source material they contain, or as landmarks in the history of their academic discipline.

Drawing from the world-renowned collections in the Cambridge University Library and other partner libraries, and guided by the advice of experts in each subject area, Cambridge University Press is using state-of-the-art scanning machines in its own Printing House to capture the content of each book selected for inclusion. The files are processed to give a consistently clear, crisp image, and the books finished to the high quality standard for which the Press is recognised around the world. The latest print-on-demand technology ensures that the books will remain available indefinitely, and that orders for single or multiple copies can quickly be supplied.

The Cambridge Library Collection brings back to life books of enduring scholarly value (including out-of-copyright works originally issued by other publishers) across a wide range of disciplines in the humanities and social sciences and in science and technology.

# Hortus Kewensis

*Or, a Catalogue of the Plants Cultivated in the Royal Botanic Garden at Kew*

VOLUME 1:
MONANDRIA TO HEPTANDRIA

WILLIAM AITON

CAMBRIDGE
UNIVERSITY PRESS

# CAMBRIDGE
## UNIVERSITY PRESS

University Printing House, Cambridge, CB2 8BS, United Kingdom

Published in the United States of America by Cambridge University Press, New York

Cambridge University Press is part of the University of Cambridge.
It furthers the University's mission by disseminating knowledge in the pursuit of
education, learning and research at the highest international levels of excellence.

www.cambridge.org
Information on this title: www.cambridge.org/9781108069670

© in this compilation Cambridge University Press 2014

This edition first published 1789
This digitally printed version 2014

ISBN 978-1-108-06967-0 Paperback

# Selected botanical reference works available in the
# CAMBRIDGE LIBRARY COLLECTION

al-Shirazi, Noureddeen Mohammed Abdullah (compiler), translated by
Francis Gladwin: *Ulfáz Udwiyeh, or the Materia Medica* (1793)
[ISBN 9781108056090]

Arber, Agnes: *Herbals: Their Origin and Evolution* (1938)
[ISBN 9781108016711]

Arber, Agnes: *Monocotyledons* (1925) [ISBN 9781108013208]

Arber, Agnes: *The Gramineae* (1934) [ISBN 9781108017312]

Arber, Agnes: *Water Plants* (1920) [ISBN 9781108017329]

Bower, F.O.: *The Ferns (Filicales)* (3 vols., 1923–8) [ISBN 9781108013192]

Candolle, Augustin Pyramus de, and Sprengel, Kurt: *Elements of the Philosophy
of Plants* (1821) [ISBN 9781108037464]

Cheeseman, Thomas Frederick: *Manual of the New Zealand Flora*
(2 vols., 1906) [ISBN 9781108037525]

Cockayne, Leonard: *The Vegetation of New Zealand* (1928)
[ISBN 9781108032384]

Cunningham, Robert O.: *Notes on the Natural History of the Strait of Magellan
and West Coast of Patagonia* (1871) [ISBN 9781108041850]

Gwynne-Vaughan, Helen: *Fungi* (1922) [ISBN 9781108013215]

Henslow, John Stevens: *A Catalogue of British Plants Arranged According to
the Natural System* (1829) [ISBN 9781108061728]

Henslow, John Stevens: *A Dictionary of Botanical Terms* (1856)
[ISBN 9781108001311]

Henslow, John Stevens: *Flora of Suffolk* (1860) [ISBN 9781108055673]

Henslow, John Stevens: *The Principles of Descriptive and Physiological Botany*
(1835) [ISBN 9781108001861]

Hogg, Robert: *The British Pomology* (1851) [ISBN 9781108039444]

Hooker, Joseph Dalton, and Thomson, Thomas: *Flora Indica* (1855)
[ISBN 9781108037495]

Hooker, Joseph Dalton: *Handbook of the New Zealand Flora* (2 vols., 1864–7) [ISBN 9781108030410]

Hooker, William Jackson: *Icones Plantarum* (10 vols., 1837–54) [ISBN 9781108039314]

Hooker, William Jackson: *Kew Gardens* (1858) [ISBN 9781108065450]

Jussieu, Adrien de, edited by J.H. Wilson: *The Elements of Botany* (1849) [ISBN 9781108037310]

Lindley, John: *Flora Medica* (1838) [ISBN 9781108038454]

Müller, Ferdinand von, edited by William Woolls: *Plants of New South Wales* (1885) [ISBN 9781108021050]

Oliver, Daniel: *First Book of Indian Botany* (1869) [ISBN 9781108055628]

Pearson, H.H.W., edited by A.C. Seward: *Gnetales* (1929) [ISBN 9781108013987]

Perring, Franklyn Hugh et al.: *A Flora of Cambridgeshire* (1964) [ISBN 9781108002400]

Sachs, Julius, edited and translated by Alfred Bennett, assisted by W.T. Thiselton Dyer: *A Text-Book of Botany* (1875) [ISBN 9781108038324]

Seward, A.C.: *Fossil Plants* (4 vols., 1898–1919) [ISBN 9781108015998]

Tansley, A.G.: *Types of British Vegetation* (1911) [ISBN 9781108045063]

Traill, Catherine Parr Strickland, illustrated by Agnes FitzGibbon Chamberlin: *Studies of Plant Life in Canada* (1885) [ISBN 9781108033756]

Tristram, Henry Baker: *The Fauna and Flora of Palestine* (1884) [ISBN 9781108042048]

Vogel, Theodore, edited by William Jackson Hooker: *Niger Flora* (1849) [ISBN 9781108030380]

West, G.S.: *Algae* (1916) [ISBN 9781108013222]

Woods, Joseph: *The Tourist's Flora* (1850) [ISBN 9781108062466]

For a complete list of titles in the Cambridge Library Collection please visit:
**www.cambridge.org/features/CambridgeLibraryCollection/books.htm**

# *HORTUS KEWENSIS*;

## OR, A

# CATALOGUE

### OF THE

# PLANTS

### CULTIVATED IN THE

## ROYAL BOTANIC GARDEN AT KEW.

### BY *WILLIAM AITON*,

#### GARDENER TO

## HIS MAJESTY.

### IN THREE VOLUMES.

---

## VOL. I.

### MONANDRIA — HEPTANDRIA.

---

## LONDON:

### PRINTED FOR GEORGE NICOL, BOOKSELLER
### TO HIS MAJESTY, PALL MALL.

#### M.DCC.LXXXIX.

TO THE

# K I N G.

PERMIT, Sir, a fervant rendered happy
by Your Majefty's benevolence, to obey
the impulfe of gratitude, which urges him
to lay at Your Majefty's feet, this attempt
to make public the prefent ftate of the Royal
Botanic Garden at Kew.

Small as the book appears, the compofition
of it has coft him a large portion of the leifure
allowed by the daily duties of his ftation, during more than fixteen years: in all that time
it has been thought worthy the affiftance
of men more learned than himfelf; a cir-
cumftance, which flatters him with the hope,
that it may not be found utterly unworthy
of Your Majefty's infpection.

As throughout his life he has unremit-
tingly attended to the ftudies of Horticulture

A 2                                    and

and Botany, he may, without prefumption, expect to have the fruits of his labours received with candour, by the cultivators of thofe valuable fciences; fhould they be approved, his gratification will be extreme; but his ambition has ftill a higher aim : the approbation of a Royal Mafter, to whom he is attached by every tie that gratitude, as well as duty, can impofe, is what he feeks to deferve; and, fhould Your Majefty be gracioufly pleafed to crown his labours by a fingle fentiment of applaufe, it will fatisfy, to the utmoft, the moft elevated wifhes of his heart; who is,

with unfeigned humility,

and uninterrupted attachment,

Your Majefty's moft devoted fervant,

WILLIAM AITON.

# PREFACE.

IT has been thought proper, in order that this work may not be fwelled with unneceffary fynonyms, that the plants defcribed in the Species Plantarum and the two Mantiffa's of the elder Linnæus, fhould fimply be referred to thofe books, with the addition only of references to the works of Jacquin, Curtis, L'Heritier, and the Flora Roffica.

The Supplementum Plantarum of the younger Linnæus was not publifhed till great progrefs had been made in compiling this Catalogue; plants that were inferted in it before the time of that publication, will therefore frequently be found to have the fame fynonyms as are given in it; all that have been added fince are referred to it in the fame manner as others are to the works of the Father; but errors which too frequently occur in the Son's publication, are, wherever they have been difcovered, corrected.

All

All plants that are not defcribed by either of the Linnæus's have every fynonym of careful writers, that could be difcovered, annexed to them.

When the younger Linnæus was in England, in 1781 and 1782, he compofed a treatife on the Palms and Liliaceous Plants, extracts of which, as far as was thought likely to be ufeful to this Catalogue, he communicated to the Author; this Manufcript is quoted under the abbreviation of *Linn. fil.* His death, which happened foon after his return to Sweden, prevented its publication; but it is in the poffeffion of Dr. James Edward Smith, of London, as are alfo the libraries and collections of both the Father and the Son.

References are frequently made to the works of M. L'Heritier, under Plants of which he has not yet publifhed either defcriptions or figures; thefe are taken from communications this gentleman frequently made, during the courfe of printing, of every thing he had prepared for the prefs. But, as the public will in due time be put into poffeffion of the whole, little need be faid on this fubject.

Throughout the whole of this Catalogue, an attempt is made to trace back, as far as poffible, how long each Plant has been cultivated in the Britifh Gardens, and to fix, with as much precifion as the nature of the fubject would allow, the epoch of its introduction.

introduction. This, like all new undertakings, will be liable to many corrections. For the use, therefore, of those who may undertake to improve it, a chronological list of the printed authorities that have been made use of is annexed.

1551. Turner's herbal, part 1.
1562. — — — part 2.
1568. — — — part 3.
1570. Lobelii adversaria.
1596. Catalogus horti Johannis Gerardi.
1597. Gerard's herbal.
1605. Lobelii adversariorum pars altera.
1629. Parkinson's paradisus.
1633. Gerard's herbal, enlarged by Johnson.
1640. Parkinson's herbal.
1648. Catalogus horti Oxoniensis.
1656. Catalogus horti Johannis Tradescanti, in museo Tradescantiano.
1658. Catalogus horti Oxoniensis, cura Ph. Stephani et Gul. Brownei.
1680. Morisoni historia Plantarum Oxoniensis, pars 2.
1683. Sutherland's hortus Edinburgensis.
1686. Raii historia plantarum, tom 1.
1687. — — — — — tom. 2.
1691. Plukenetii phytographia, pars 1 & 2.
1692. — — — — — pars 3.
1696. — — — — — pars 4, seu Almagestum.
1699. Historia plantarum Oxoniensis, pars 3, absoluta a J. Bobartio.
1700. Plukenetii mantissa almagesti.
1704. Raii historia plantarum, tom. 3.
1705. Plukenetii amaltheum.

1711—1715. Petiver's botanicum hortense, in the Philoso-
phical Transactions, vol. 27, 28, 29.

1716. Bradley's history of succulent plants, 1 decade.

1717. — — — — — — — 2 decade.

1724. Catalogue of trees fold by Robert Furber, in Miller's Gardener's and Florift's dictionary.

1725. Bradley's history of succulent plants, 3 decade.

1727. — — — — — — — 4 and 5 decade.

1728. Martyn plantarum rariorum decas 1.

1729. — — — — — — decas 2.

1730. — — — — — — decas 3.
Catalogue of plants propagated for sale in the gardens near London.

1731. Miller's gardener's dictionary, 1ft edition.

1732. Dillenii hortus Elthamensis.
Martyn plantarum rariorum decas 4.

1736. — — — — — — decas 5.

1739. Rand horti Chelseiani index.
Miller's gardener's dictionary, vol. 2.

1743. ——— 4th edition.

1748. ——— 5th edition.

1752. ——— 6th edition.

1755—1760. Miller's figures of plants described in the
gardener's dictionary.

1759. Miller's gardener's dictionary, 7th edition.

1768. — — — — — — 8th edition.

Whether Miller's dictionary, and especially the
second volume of the edition of 1739, can be considered as sufficient authority for concluding the
plants mentioned in it to have been actually cultivated in England at the time of its publication,
may be a matter of doubt. Lyte's Herbal is an
actual

actual tranflation of Dodonæus; and Parkinfon's
Paradifus terreftris little better than a compilation
from other books. Miller's dictionary is certainly
a more original work than either of thefe; it is
feldom, however, if ever, that the author has
quoted either of thefe books as authority, without
having been induced by fome additional reafon, to
believe the plants alluded to were actually cultivated here at the time ftated.

Several Manufcripts preferved in the Sloanean
Collection at the Britifh Mufeum have been
made ufe of in this part of the Work, particularly
N° 3370, intitled, Horti Regii Hamptonienfis
exoticarum Plantarum Catalogus; to which an-
other hand-writing has added, by Dr. Gray. On a
blank page in this book is the following memo-
randum : " This Catalogue I took from one
" which the Intendant of the garden they were in
" at Hampton-Court, lent to me upon the place,
" with liberty fufficient to infpect the Plants : they
" were brought from Soefdyke, a houfe belonging
" to Mr. Bentink, afterwards Earl of Portland,
" about the year 1690, and given by him to King
" William."

The abbreviation *Br. Muf. H. S.* fignifies the
Sloanean Hortus ficcus, kept in the Britifh Mufeum; from whence much information, principally
concerning the Plants cultivated by the Dutchefs

of

of Beaufort, has been obtained. *R. S.* means the
specimens of Plants annually, in obedience to Sir
Hans Sloane's will, presented by the Company of
Apothecaries to the Royal Society, part of which
are deposited in the library belonging to that body,
in Somerset Place, and the remainder in the British
Museum.

On the authority of various letters and other
papers communicated by Michael Collinson, Esq;
many Plants are said to have been introduced by
his father, Mr. Peter Collinson, of Mill Hill.

Mr. Knowlton, formerly Gardener to James Sherard, M. D. at Eltham, gave a variety of useful
information, to which his name is always annexed.
He died in 1782, at the age of 90.

Mr. James Lee, nurseryman at the vineyard,
Hammersmith, who remembers the gardens of
Archibald duke of Argyle, at Whitton, near
Hounflow, cultivated with much care and liberal
expence, has furnished the Author with a lift of the
Trees that were introduced by his Grace.

From his own memory the Author states several
Plants to have been cultivated by Mr. Ph. Miller,
in the Physick Garden at Chelsea, though no reference is made to them in his Gardener's Dictionary.

Some Plants are by tradition known to have
been

been introduced by Robert James Lord Petre, but the times when are utterly forgot; to remedy, as much as poffible, this inconvenience, they are always ftated as having been introduced before 1742, the time of his Lordfhip's death.

Mr. Miller, in his Dictionary, often mentions Plants as having been fent to him by Dr. Houftoun, but he frequently omits the time when he received them; thefe, therefore, are in like manner ftated as having been introduced before the period of the Doctor's deceafe, which happened in 1733.

As the figure of Calceolaria Fothergillii differs a little from the defcription, it is neceffary to mention, that the defcription was made in 1779, from a weak Plant, the firft that flowered in England; and that the plate is from a drawing, taken fince the letter-prefs was printed, from a ftrong healthy plant: hence it is alfo, that no reference is made in the text to this figure.

BOOKS

# BOOKS QUOTED.

*Aĉt. ac. nat. curiof.* Aĉta phyfico-medica Academiæ Cæfareæ Naturæ Curioforum. Voll. X. Norimbergæ, 1727—1754. 4to.

*Aĉt. acad. patav.* Saggi fcientifici e letterarii dell' Academia di Padova. Tom. 1. Padova, 1786. 4to.

*Aĉt. gothob.* Kongl. Götheborgfka Wetenfkaps och Witterhets Samhällets Handlingar. Wetenfkaps Afdelningen. Götheborg, 1778, feqq. 8vo.

*Aĉt. harlem.* Verhandelingen uitgegeeven door de Hollandfe Maatfchappye der Weetenfchappen te Haarlem. Haarlem, 1754, feqq. 8vo.

*Aĉt. helvet.* Aĉta Helvetica phyfico, mathematico, botanico-medica. Bafileæ, 1751, feqq. 4to.

*Aĉt. laufann.* Memoires de la Societé des Sciences phyfiques de Laufanne. Tome 1. Laufanne 1784. 4to.

*Aĉt. lund.* Phyfiographifka Sälfkapets Handlingar. Stockholm, 1776, feqq. 8vo.

*Aĉt. palat.* Hiftoria et Commentationes Academiæ Electoralis Scientiarum et Elegantiorum Literarum Theodoro-Palatinæ. Mannhemii, 1766, feqq. 4to.

*Aĉt. parif.* Hiftoire de l'Academie Royale des Sciences, avec les Memoires de Mathematique et de Phyfique. Paris, 1702, feqq. 4to.

*Aĉt. petrop.* Aĉta Academiæ Scientiarum Imperialis Petropolitanæ. Petropoli, 1778, feqq. 4to.

*Aĉt. ftockholm.* Kongl. Vetenfkaps Academiens Handlingar. XL Tomer. Stockholm, 1739—1779. 8vo.
Kongl. Vetenfkaps Academiens nya Handlingar. Stockholm, 1780, feqq. 8vo.

*Aĉt.*

*Act. tolof.* Hiftoire et Memoires de l'Academie Royale des Sciences, Infcriptions et Belles Lettres de Tou- loufe. Touloufe, 1782, feqq. 4to.

*Aldini hort. farnef.* Tob. Aldini rariorum plantarum quæ continentur in Horto Farnefiano defcriptio. Romæ, 1625. fol.

*Allion. pedem.* Car. Allioni Flora Pedemontana. Tomi III. Auguftæ Taurinorum, 1785. fol.

*Alp. 'exot.* Profp. Alpini de plantis exoticis libri II. Ve- netiis, 1629. 4to.

*Arduin. animadv.* Petr. Arduini animadverfionum botani- carum fpecimen I. Patavii 1759. Specimen II. Ve- netiis, 1764. 4to.

*Aublet guian.* Hiftoire des plantes de la Guiane Françoife, par M. Fufée Aublet. Tomes IV. Paris, 1775. 4to.

*Bauh. hift.* Hiftoria plantarum univerfalis Joh. Bauhini et. Joh. Henr. Cherleri. Tomi III. Ebroduni, 1650. 1651. fol.

*Bauh. pin.* Cafp. Bauhini Pinax theatri botanici. Bafileæ, 1671. 4to.

*Berg. cap.* Petr. Jon. Bergii defcriptiones plantarum ex Capite bonæ fpei. Stockholmiæ, 1767. 8vo.

*Bergius om läckerheter.* Tal om läckerheter hållct för K. Vetenfkaps Academien af Bengt Bergius. Stock- holm, 1785. 8vo.

*Beft. eyft.* Baf. Befleri Hortus Eyftettenfis. 1613. fol.

*Blair's bot. effays.* In Patr. Blair's botanick eflays, London, 1728, 8vo; a plate infcribed: Thefe, and many other fucculent plants, are to be fold by Mr. Thomas Fair- child at Hoxton.

*Bocc. muf.* Mufeo di piante rare, di Don Paulo Boccone. Venetia, 1697. 4to.

*Bocc. fic.* Icones et defcriptiones rariorum plantarum Si- ciliæ, Melitæ, Galliæ & Italiæ; auctore Paulo Boc- cone. Oxonii, 1674. 4to.

*Boerh.*

*Boerh. lugdb.* Herm. Boerhaave Index alter plantarum
quæ in horto Academiæ Lugduno Batavæ aluntur.
Tomi II. Lugd. Batav. 1720. 4to.

*Boym fl. finenf.* Mich. Boym Flora Sinenfis. Viennæ,
1656. fol.

*Bradl. fucc.* Ric. Bradley hiftoria plantarum fucculenta-
rum. Decades V. Londini, 1716—1727. 4to.

*Breyn. ic.* Jac. Breynii Icones rariorum et exoticarum
plantarum. Gedani, 1739. 4to.

*Brown jam.* The civil and natural hiftory of Jamaica, by
Patr. Browne. London, 1756. fol.

*Burm. afr.* Jo. Burmanni rariorum Africanarum plantarum Decades X. Amftelædami, 1738. 1739. 4to.

*Burm. fl. cap.* Nic. Laur. Burmanni prodromus Floræ
Capenfis; in Flora Indica. Lugd. Batav. 1768. 4to.

*Burm. ger.* Nic. Laur. Burmanni de Geraniis Differtatio. Lugd. Batav. 1759. 4to.

*Burm. ind.* Nic. Laur. Burmanni Flora Indica. Lugd.
Bat. 1768. 4to.

*Burm. zeyl.* Jo. Burmanni thefaurus Zeylanicus. Amftelædami, 1737. 4to.

*Buxb. cent.* Joh. Chr. Buxbaumii plantarum minus cognitarum Centuriæ V. Petropoli, 1728—1740. 4to.

*Cam. epit.* De plantis epitome Petr. Andr. Matthioli, aucta
a Joach. Camerario. Francof. ad Mœn. 1586. 4to.

*Camel. luz.* Herbarum in Infula Luzonum nafcentium, a
Ge. Jof. Camello obfervatarum fyllabus; in Tomo
III. Hiftoriæ plantarum Raji.

*Camer. hort.* Joach. Camerarii hortus medicus et philofo-
phicus. Francofurti, 1588. 4to.

*Catefb. carol.* The natural hiftory of Carolina, Florida, and
the Bahama Iflands, by Mark Catefby. II Vols.
London, 1731. 1743. fol.

*Cavan. diff.* Ant. Jof. Cavanilles Differtationes botanicæ.
Paris, 1785, feqq. 4to.

Cluf.

*Cluf. cur. poft.* Car. Clufii curæ pofteriores. Antverpiæ, 1611. fol.

*Cluf. hift.* Car. Clufii rariorum plantarum hiftoria. Antverpiæ, 1601. fol.

*Col. ecphr.* Fab. Columnæ minus cognitarum ftirpium εκφρασις. Romæ, 1616. 4to.

*Comm. amftel.* ⎫ Horti Medici Amftelodamenfis rariorum
*Comm. hort,* ⎬ plantarum defcriptio et icones, auctore Jo. Commelino. Amftelodami 1697. Pars altera, auctore Cafp. Commelino. 1701. fol. ⎭

*Comm. petrop.* Commentarii Academiæ Scientiarum Imperialis Petropolitanæ. Tomi XIV. Petropoli, 1728— 1751. 4to.

*Comm. præl.* Cafp. Commelini præludia botanica. Lugd. Batav. 1703. 4to.

*Comm. rar.* Cafp. Commelini plantæ rariores et exoticæ horti medici Amftelodamenfis. Lugd. Batav. 1706. 4to.

*Commentat. gotting.* Commentationes Societatis Regiæ Scientiarum Gottingenfis. Gottingæ, 1779 feqq. 4to.

*Cook's voyage.* A voyage towards the South Pole, and round the World, in the years 1772—1775, by James Cook. II Vols. London, 1777. 4to.

*Corn. canad.* Jac. Cornuti Canadenfium plantarum hiftoria. Parifiis, 1635. 4to.

*Curtis lond.* Flora Londinenfis, by William Curtis. London. fol. (quoted as far as N° 60.)

*Curtis magax.* The Botanical magazine. London, 1787, feqq. 8vo. (quoted as far as N° 29, publifhed June 1. 1789.)

*Cyrill. neapol.* Dom. Cyrilli plantarum rariorum regni Neapolitani fafciculus 1. Neapoli, 1788. fol.

*Dahl obf. bot.* And. Dahl Obfervationes botanicæ circa fyftema vegetabilium divi a Linné Gottingæ, 1784 editum. Havniæ, 1787. 8vo.

*Dalech.*

*Dalech. hiſt.* Hiſtoria generalis plantarum. Tomi II.
   Lugduni, 1586. fol.

*Dalib. pariſ.* Floræ Pariſienſis prodromus, par M. Dali-
   bard. Paris, 1749. 12mo.

*D'Aſſo arag.* (Ign. de Aſſo) Synopſis ſtirpium indigenarum Aragoniæ. Maſſiliæ, 1779. 4to.

*De Lamarck encycl.* Encyclopedie methodique : Botanique, par M. le Chevalier, de Lamarck. Tome I. II.
   Paris, 1783. 1786. 4to.

*De la Roche pl. nov.* Dan. de la Roche deſcriptiones plantarum aliquot novarum. Lugd. Batav. 1766. 4to.

*Dill. elth.* Joh. Jac. Dillenii hortus Elthamenſis. Voll. II.
   Londini, 1732. fol.

*Dod. mem.* Memoires pour ſervir à l'hiſtoire des plantes,
   par M. Dodart. Paris, 1676. fol.

*Douglas deſcr. of the Guernſey Lily.* A deſcription of the
   Guernſey Lilly, by James Douglas. London, 1737.
   fol.

*Douglas's hiſtory of the Coffee-tree.* A deſcription and hiſtory of the Coffee-tree, by James Douglas. London
   1727. fol.

*Ducheſne fraiſ.* Hiſtoire naturelle des Fraiſiers, par M.
   Ducheſne fils. Paris, 1766. 12mo.

*Du Halde chin.* Deſcription de l'Empire de la Chine et de
   la Tartarie Chinoiſe, par le P. J. B. du Halde. Tomes
   IV. Paris, 1735. fol.

*Du Hamel arb.* Traité des Arbres et Arbuſtes, qui ſe
   cultivent en France en pleine terre, par M. Du Hamel Du Monceau. Tomes II. Paris, 1755. 4to.

*Durande bourg.* Flore de Bourgogne (par Durande.)
   II parties. Dijon, 1782. 8vo.

*Du Roi hort. harbecc.* Die Harbkeſche wilde Baumzucht,
   von Joh. Phil. du Roi. II Bände. Braunſchweig,
   1771. 1772. 8vo.

                                                          *Du*

*Du Roi obf. bot.* Joh. Phil. du Roi diſſertatio ſiſtens ob-
ſervationes botanicas. Helmſtadii, 1771. 4to.

*Edwards' birds.* A natural hiſtory of Birds, by Ge. Edwards, 2d part. London, 1747. 4to.

*Ehret pict.* Plantæ et papiliones rariores, depictæ et æri
inciſæ a Ge. Dion. Ehret. Tabulæ XV. Londini,
1748—1759. fol.

*Evelyn's kalend. hort.* Kalendarium hortenſe, by John
Evelyn. Printed with the following:

*Evelyn's ſylva.* Sylva, or a diſcourſe of Foreſt Trees, by
John Evelyn. London, 1664. fol.

*Fabricii it. norveg.* Joh. Chr. Fabricius Reiſe nach Nor-
wegen. Hamburg, 1779. 8vo.

*Feuillée it.* Journal des obſervations faites ſur les côtes
orientales de l'Amerique meridionale, par le R. P.
Louis Feuillée. Tomes III. Paris, 1714, 1725. 4to.

*Fl. dan.* Icones plantarum ſponte naſcentium in regnis
Daniæ et Norvegiæ, editæ a Ge. Chr. Oeder, Oth.
Frid. Müller et Mart. Vahl. Hafniæ, 1761, ſeqq. fol.

*Fl. lapp.* Car. Linnæi Flora Lapponica. Amſtelædami,
1737. 8vo.

*Fl. ſuec. ed.* 1. Car. Linnæi Flora Svecica. Holmiæ, 1745.
8vo.

*Forſk. deſcr.* Flora Ægyptiaco-Arabica, ſive deſcriptiones
plantarum quas per Ægyptum inferiorem et Arabiam
Felicem detexit Pet. Forſkäl. Havniæ, 1775. 4to.

*Forſt. fl. auſtr.* Ge. Forſter Florulæ Inſularum Auſtra-
lium prodromus. Gottingæ, 1786. 8vo.

*Forſt. gen.* Characteres generum plantarum, quas in iti-
nere ad Inſulas Maris Auſtralis collegerunt Joh. Reinh.
Forſter et Ge. Forſter. Londini, 1776. fol.

*Forſt. pl. eſcul.* Ge. Forſter de plantis eſculentis Inſularum
Oceani Auſtralis commentatio. Berolini, 1786. 8vo.

*Fuchſ. hiſt.* Leon. Fuchſii de hiſtoria Stirpium commentarii. Baſileæ, 1542. fol.

*Furber's catal.* A catalogue of plants fold by Mr. Robert
Furber; in the 2d volume of the Gardener's and
Florift's dictionary, by Ph. Miller. London, 1724.
8vo.

*Gærtn. fem.* Jof. Gærtner de fructibus et feminibus plantarum. Stutgardiæ, 1788. 4to.

*Garid. prov.* Hiftoire des plantes qui naiffent en Pro-
vence, et principalement aux environs d'Aix, par Mr.
Garidel. Paris, 1719. fol.

*Gen. pl.* Car. v. Linné Genera plantarum. Editio fexta.
Holmiæ, 1764. 8vo.

*Ger. emac.* Joh. Gerard's Herball, enlarged by Th. Johnfon. London, 1633. fol.

*Ger. herb.* Joh. Gerard's Herball. London, 1597. fol.

*Ger. prov.* Lud. Gerard Flora Gallo-Provincialis. Parifiis, 1761. 8vo.

*Gloxin obf. bot.* Benj. Petr. Gloxin Obfervationes botanicæ. Argentorati, 1785. 4to.

*Gmel. fib.* Jo. Ge. Gmelin Flora Sibirica. Tomi IV.
Petropoli 1747—1769. 4to.

*Gouan illuftr.* Ant. Gouan illuftrationes et obfervationes
botanicæ. Tiguri, 1773. fol.

*Gron. virg.* Joh. Fred. Gronovii Flora Virginica. Lugd.
Batav. 1762. 4to.

*Gron. virg. ed. 1.* ——————— Partes II. Lugd. Batav.
1743. 8vo.

*Gunn. norv.* Jo. Ern. Gunneri Flora Norvegica. Pars
Prior. Nidrofiæ, 1766. Pars Pofterior. Hafniæ,
1772. fol.

*Hall. hift.* Alb. v. Haller hiftoria ftirpium indigenarum
Helvetiæ inchoata. Tomi III. Bernæ, 1768. fol.

*Hannov. magaz.* Hannoverifches Magazin, 1783. 4to.

*Heifteri Brunfvigia.* Laur. Heifteri defcriptio Brunfvigiæ,
novi generis plantæ. Brunfvigæ, 1753. fol.

*Herm.*

*Herm. lugdb.* Pauli Hermanni catalogus Horti Academici Lugduno-Batavi. Lugd. Batav. 1687. 8vo.

*Herm. par.* P. Hermanni Paradifus Batavus. Lugd. Batav. 1705. 4to.

*Hort. angl.* A catalogue of Trees, Shrubs, Plants, and Flowers, which are propagated for fale in the Gardens near London. London, 1730. fol.

*Hort. cliff.* Car. Linnæi Hortus Cliffortianus. Amftelodami, 1737. fol.

*Hort. Ger.* Catalogus arborum, fruticum ac plantarum, tam indigenarum, quam exoticarum, in horto Johannis Gerardi nafcentium. Londini, 1596. fol.

*Hort. oxon. edit.* 1. Catalogus plantarum Horti Medici Oxonienfis. Oxonii, 1648. 8vo.

*Hort. oxon. edit.* 2. Catalogus Horti Botanici Oxonienfis, cura Phil. Stephani et Gul. Brovnei. Oxonii, 1658. 8vo.

*Hort. upf.* Car. Linnæi Hortus Upfalienfis. Holmiæ, 1748. 8vo.

*Houtt. nat. hift.* Natuurlyke hiftorie of uitvoerige befchryving der Dieren, Planten en Mineraalen, volgens het Samenftel van den Heer Linnæus (door Mart. Houttuyn.) Tweede Deel. XIV Vols. Amfterdam, 1773—1783. 8vo.

*Hudf. angl.* Gul. Hudfoni Flora Anglica. Tomi II. Londini, 1778. 8vo.

*Hudf. angl. edit.* 1. ———————— Londini, 1768. 8vo.

*Jacqu. auftr.* Nic. Jof. Jacquin Flora Auftriaca. Voll. V. Viennæ, 1773–1778. fol.

*Jacqu. collect.* N. J. Jacquin Collectanea ad Botanicam, Chemiam et Hiftoriam Naturalem fpectantia. Vol. I. Vindobonæ, 1786. 4to.

*Jacqu. hift.* N. J. Jacquin felectarum ftirpium Américanarum hiftoria. Viennæ, 1763. fol.

*Jacqu.*

*Jacqu. hort.* N. J. Jacquin Hortus Botanicus Vindobonen-
　fis. Tomi III. Viennæ, 1770—1776. fol.

*Jacqu. ic.* Icones plantarum rariorum, editæ a N. J. Jac-
　quin. Vindobonæ, 1781, feqq. fol. (quoted as far as
　Fafc. 2dus Vol. 2di.)

*Jacqu. mifcell.* N. J. Jacquin Mifcellanea Auftriaca ad
　Botanicam, Chemiam et Hiftoriam Naturalem fpec-
　tantia. Voll. II. Vindobonæ, 1778, 1781. 4to.

*Jacqu. obf.* N. J. Jacquin Obfervationum Botanicarum
　Partes IV. Viennæ, 1764—1771. fol.

*Jacqu. vind.* N. J. Jacquin enumeratio ftirpium quæ
　fponte crefcunt in agro Vindobonenfi. Viennæ,
　1762. 8vo.

*Ic. Kæmpfer.* Icones felectæ plantarum quas in Japonia
　collegit et delineavit Eng. Kæmpfer. (not yet pub-
　lifhed.)

*Johnf. it. cant.* (Thom. Johnfon) Defcriptio itineris plan-
　tarum inveftigationis ergo fufcepti in agrum Cantia-
　num. (Londini) 1632. 8vo.

*Journ. de Rozier.* Obfervations fur la Phyfique, fur
　l'Hiftoire naturelie et fur les Arts, par M. l'Abbé
　Rozier. Paris, 1771, feqq. 4to.

*Kæmpf. amœn.* Eng. Kæmpfer amœnitatum exoticarum
　fafciculi V. Lemgoviæ, 1712. 4to.

*Kretzfchmar monogr.* Sam. Kretzfchmar Befchreibung der
　Martyniæ annuæ villofæ. Friedrichftadt, (1764) 4to.

*Lauth Acer.* Th. Lauth Differtatio de Acere. Argen-
　torati, 1781. 4to.

*Le Brun it. perf.* Voyages de Corn. Le Brun par la Mof-
　covie, en Perfe, et aux Indes Orientales. Tomes II.
　Amfterdam, 1718. fol.

*Le Comte mem. de la Chine.* Nouveaux memoires fur l'etat
　prefent de la Chine, par le R. Pere Louis le Comte.
　Tomes II. Amfterdam, 1697. 12mo.

*Leyf. hal.* Frid. Wilh. a Leyffer Flora Halenfis. Halæ,
　1783. 8vo.

*L'Herit.*

*L'Herit.* (in Geraniis) Car. Lud. L'Heritier monographia de Geranio. (not yet publifhed.)

*L'Herit. corn.* Car. Lud. L'Heritier Cornus. Parifiis, 1788. fol.

*L'Herit. monogr.* Michauxia, fol.

Buchozia. Car. Lud. L'Heritier, fol.

*L'Herit. fert. angl.* Car. Lud. L'Heritier Sertum Anglicum. Parifiis, 1788. fol. (publifhed as far as pag. 36, and tab. 2.)

*L'Herit. folan.* Car. Lud. L'Heritier Solana aliquot rariora. (not yet publifhed.)

*L'Herit. ftirp. nov.* Stirpes novæ defcriptionibus et iconibus illuftratæ a Car. Lud. L'Heritier. Fafcic. 1, 2. Parifiis, 1784. Fafc. 3. 1785. Fafc. 4. 1785. (publifhed 1788.) Fafc. 5, 1785. (publifhed 1789.) Fafc. 6, 1785. (not yet publifhed.)

*Lightf. fcot.* Flora Scotica, by John Lightfoot, 2 vols. London, 1777. 8vo.

*Linn. fil. diff. de Lavandula.* Differtatio de Lavandula Præfide Car. a Linné (filio). Upfaliæ, 1780. 4to.

*Linn. mant.* Car. a Linné Mantiffa Plantarum Generum editionis VI. et Specierum editionis II. Holmiæ, 1767. 8vo.

Mantiffa plantarum altera. Holmiæ, 1771. 8vo.

*Linn. fuppl.* Supplementum plantarum Syftematis vegetabilium editionis XIII. Generum plantarum editionis VI. et Specierum plantarum editionis II. editum a Car. a Linne (filio). Brunfvigæ, 1781. 8vo.

*Linn. zeyl.* Car. Linnæi Flora Zeylanica. Holmiæ, 1747. 8vo.

*Lob. ic.* Plantarum feu ftirpium icones. Antverpiæ, 1591. 4to obl.

*Lobel. adv.* Stirpium adverfaria nova, authoribus Petro Pena et Matthia de Lobel. Londini, 1571. fol.

a 3         Matthiæ

Matthiæ de Lobel adverfariorum altera pars. Londini, 1605. fol.

*Lœfl. it.* Petri Lœfling Refa til Spanfka länderna uti Europa och America, utgifven of Carl Linnæus. Stockholm, 1758. 8vo.

*Lœf. pruff.* Joh. Lœfelii Flora Pruffica, auxit et edidit Joh. Gottfched. Regiomonti, 1703. 4to.

*Ludolf. æthiop.* Jobi Ludolfi hiftoria Æthiopica. Francof. ad Mœn. 1681. fol.

*Ludolf. comment.* Jobi Ludolfi ad fuam hiftoriam Æthiopicam commentarius. Francof. ad Mœn. 1691. fol.

*Magn. hort.* Petri Magnol Hortus Regius Monfpelienfis. Monfpelii, 1697. 8vo.

*Marcgr. braf.* (Guil. Pifonis et Ge. Marcgravii) Hiftoria naturalis Brafiliæ. Lugd. Bat. et Amftelod. 1648. fol.

*Marfh. arb. amer.* Arbuftrum Americanum, by Humphry Marfhall. Philadelphia, 1785. 8vo.

*Mart. dec.* Jo. Martyn hiftoria plantarum rariorum. (Decades V.) Londini, 1728. (feqq.) fol.

*Martyn hort. cantabr.* Thom. Martyn catalogus Horti Botanici Cantabrigienfis. Cantabrigiæ, 1771. 8vo.

*Mattufchka enum. ftirp. filef.* Henr. Godefr. Comitis de Mattufchka enumeratio ftirpium in Silefia fponte crefcentium. Vratiflaviæ, 1779. 8vo.

*Medic. bot. beobacht.* Botanifche Beobachtungen des jahres 1783, von Friedr. Kafim. Medikus. Mannheim, 1784. 8vo.

*Medic. Theodora.* Theodora fpeciofa, ein neues Pflanzen gefchlecht, von F. K. Medikus. Mannheim, 1786. 8vo.

*Meerb. ic.* Afbeeldingen van zeldzaame Gewaffen, door Nic. Meerburgh. Leyden, 1775. (feqq.) fol.

*Mem. pour l'hift. nat. de la Suiffe.* Memoires pour fervir à l'Hiftoire phyfique et naturelle de la Suiffe, redigés par

par M. Reynier et par M. Struve. Tome I. Laufanne, 1788. 8vo.

*Merian furin.* Maria Sibylla Merian de generatione et metamorphofibus Infectorum Surinamenfium. Hagæ Comitum, 1726. fol.

*Merr. pin.* Chriftoph. Merrett pinax rerum naturalium Britannicorum. Londini, 1667. 8vo.

*Mill. dict.* Phil. Miller's Gardener's Dictionary. The eighth edition. London, 1768. fol.

*Mill. dict. edit.* 1. ———— (The firft edition) Lond. 1731. fol.
       4. ———— The fourth edition. Lond. 1743. fol.
       5. ———— The fifth edition. Lond. 1748. fol.
       6. ———— The fixth edition. Lond. 1752. fol.
       7. ———— The feventh edition. Lond. 1759. fol.

*Mill. dict. vol.* 2. The fecond volume of the Gardener's Dictionary, by Phil. Miller. London, 1739. fol.

*Mill. ic.* Figures of plants defcribed in the Gardener's Dictionary, by Phil. Miller. II Vols. London, 1760. fol.

*J. Mill. ic.* 7 coloured plates of Plants, by John Miller, 1780. fol.

*J. Miller illuftr.* Joh. Miller illuftratio fyftematis fexualis Linnæi. Londini, 1777. fol.

*J. Fr. Mill. ic.* 36 coloured plates of plants and animals, by John Fred. Miller, 1776—1782. fol.

*Mifcell. taurinens.* Melanges de Philofophie et de Mathematique de la Societé Royale de Turin. Tome 3 me. Turin, 1766. 4to.

*Mœnch. hort. weiffenft.* Verzeichnifs aufländifcher Bäume

und Stauden des Luftchloſſes Weiſſenſtein bey Caſſel, von Conr. Mönch. Frankf. und Leipz. 1785. 8vo.

*Moriſ. bleſ.* Rob. Moriſon Hortus Regius Bleſenſis auctus. Londini, 1669. 8vo.

*Moriſ. hiſt.* Rob. Moriſon hiſtoria plantarum univerſalıs Oxonienſis. Pars II. Oxonii, 1680. Pars III. abſoluta a Jac. Bobartio, 1699. fol.

*Muſ. Trad.* Muſæum Tradeſcantium. London, 1656. 8vo.

*Nov. act. nat. cur.* Nova Acta Phyſico-Medica Academiæ Cæſareæ Naturæ Curioſorum. Norimbergæ, 1757, ſeqq. 4to.

*Nov. act. upſal.* Nova Acta Regiæ Societatis Scientiarum Upſalienſis. Upſaliæ, 1773, ſeqq. 4to.

*Nov. comm. gotting.* Novi Commentarii Societatis Regiæ Scientiarum Gottingenſis. Tomi VIII. Gottingæ et Gothæ, 1771—1778. 4to.

*Nov. comm. petrop.* Novi Commentarii Academiæ Scientiarum Imperialis Petropolitanæ. Tomi XX. Petropoli, 1750—1776. 4to.

*Oſb. it.* Pehr Oſbecks Dagbok öfver en Oſtindiſk reſa. Stockholm, 1757. 8vo.

*Pallas hort. Demidov.* Petr. Sim. Pallas enumeratio plantarum, quæ in horto Procopii a Demidof Moſcuæ vigent. Petropoli, 1781. 8vo.

*Pallas it.* P. S. Pallas Reiſe durch verſchiedene Provintzen des Ruſſiſchen Reichs. III. Theile. Peterſburg, 1771—1776. 4to.

*Park parad.* Joh. Parkinſon Paradiſi in Sole Paradiſus terreſtris. London, 1629. fol.

*Park. theat.* Joh. Parkinſon Theatrum Botanicum. London,.1640. fol.

*Pet. gaz.* Jac. Petiver Gazophylacium naturæ et artis. fol.

*Pet.*

*Pet. muf.* Mufei Petiveriani Centuriæ X. Londini, 1695
—1703. 8vo.

*Philofoph. tranfaƈt.* Philofophical Tranfaƈtions of the Royal
Society of London. London, 1665. feqq. 4to.

*Piccivoli hort. panciat.* Hortus Panciaticus, o fia catalogo
delle piante efotiche, e dei fiori efiftenti nel Giardino
della Villa detta la Loggia, preffo a Firenze, di pro‡
prietà dell' ill. Sig. Marchefe Nicc. Fanciatichi, de-
fcritto da Giuf. Piccivoli. Firenze, 1783. 4to.

*Plat's garden of Eden.* The Garden of Eden, by Sir Hugh
Plat, Knight. London, 1660. 8vo.

*Pluk. alm.* Leon. Plukenett Almageftum Botanicum.
Londini, 1696. 4to.

*Pluk. amalth.* Leon. Plukenett Amaltheum Botanicum.
Londini, 1705. 4to.

*Pluk. mant.* Leon. Plukenett Mantiffa Almagefti Botanici. Londini, 1700. 4to.

*Pluk. phyt.* Leon. Plukenett Phytographia. Londini,
1691, 1692. 4to.

*Plum. ie.* Plantarum Americananum Fafciculi X, conti-
nentes plantas quas olim Car. Plumierius detexit et
depinxit, edidit Jo. Burmannus. Amftelædami, 1755
—1760. fol.

*Pollich palat.* Joh. Ad. Pollich Hiftoria Plantarum in Pa-
latinatu Eleƈtorali fponte nafcentium. Tomi III.
Mannhemii 1776, 1777. 8vo.

*Raj. hift.* Jo. Raji hiftoria Plantarum. Tomi III. Londini, 1686—1704. fol.

*Raj. fyn.* Jo. Raji Synopfis methodica Stirpium Britan-
nicarum. Londini, 1724. 8vo.

*Rand. chel.* If. Rand Horti Medici Chelfejani index com-
pendiarius. Londini, 1739. 8vo.

*Ray's letters.* Philofophical letters between the late Mr.
Ray and feveral of his correfpondents, publifhed by
W. Derham. London, 1718. 8vo.

*Rea's*

*Rea's flora.* Flora, or a complete Florilege, by John Rea.
     London, 1665. fol.

*Regn. bot.* La Botanique mife à la portée de tout le
     monde, ou collection des plantes d'ufage dans la
     Medecine, dans les Alimens et dans les Arts; par les
     Sr. et De Regnault. Paris, 1774. fol.

*Relhan cantabr.* Rich. Relhan Flora Cantabrigienfis.
     Cantabrigiæ, 1785. 8vo.

*Retzii obf.* And. Jah. Retzii obfervationum botanicarum
     Fafcic. I—IV. Lipfiæ, 1779—1786. fol.

*Retzii fcand.* A. J. Retzii Floræ Scandinaviæ prodro-
     mus. Holmiæ, 1779. 8vo.

*Rheed. mal.* Hortus Indicus Malabaricus, adornatus per
     Henr. van Rheede van Draakenftein. Tomi XII.
     Amftelædami, 1678—1703. fol.

*Richard hift. du Tonquin.* Hiftoire naturelle, civile et po-
     litique du Tonquin, par M. l'Abbé Richard. Tomes II.
     Paris 1778. 12mo.

*Riv. mon.* Aug. Quir. Rivini Ordo plantarum, quæ funt
     flore irregulari monopetalo. Lipfiæ, 1691. fol.

*Riv. pent.* A. Q. Rivini Ordo plantarum, quæ funt flore
     irregulari pentapetalo. Lipfiæ, 1699. fol.

*Rob. ic.* 319 plates of plants engraved by Nic. Robert,
     A. Boffe et Lud. de Chaftillon. fol.

*Roths beytr.* Albr. Wilh. Roths beyträge zur Botanik.
     1, 2. Theil. Bremen, 1782, 1783. 8vo.

*Roy. lugdb.* Adr. van Royen Floræ Leydenfis prodro-
     mus, exhibens plantas quæ in Horto Academico
     Lugduno-Batavo aluntur. Ludg. Batav. 1740. 8vo.

*Rumph. amb.* Ge. Everh. Rumphii Herbarium Amboi-
     nenfe. Partes VI. et Auctarium. Amftelædami,
     1750—1755. fol.

*Rupp. jen.* Henr. Bernh. Ruppii Flora Jenenfis. Fran-
     cof. et Lipf. 1726. 8vo.

*Salmon's herb.* Will. Salmon's Englifh herbal. II Vols.
London, 1710, 1711. fol.

*Sauv. monfp.* F. B. de Sauvages methodus foliorum, feu
plantæ Floræ Monfpelienfis, juxta foliorum ordinem
digeftæ. La Haye, 1751. 8vo.

*Schmid. ic.* Caf. Chriftoph. Schmidel Icones plantarum.
(Norimbergæ) 1762. fol.

*Schreb. gram.* Joh. Chr. Dan. Schrebers befchreibung der
Gräfer. Leipzig, 1769, feqq. fol.

*Scop. carn.* Jo. Ant. Scopoli Flora Carniolica. Tomi II.
Viennæ, 1772. 8vo.

*Scop. infubr.* J. A. Scopoli deliciæ Floræ et Faunæ Infu-
bricæ. Partes III. Ticini, 1786—1788. fol.

*Sloan. hift.* ⎰ A voyage to the Iflands Madera, Barbados,
⎱ Nieves, S. Chriftopher's, and Jamaica,
*Sloan. jam.* ⎰ with the natural hiftory of the laft of thofe
⎱ Iflands. II Vols. London, 1707, 1725. fol.

*Sonnerat it. ind.* Voyage aux Indes Orientales et à la
Chine, par M. Sonnerat. Tomes II. Paris, 1782.
4to.

*Sonnerat iter nov. guin.* Voyage à la Nouvelle Guinée,
par M. Sonnerat. Paris, 1776. 4to.

*Sp. pl.* Car. Linnæi Species plantarum. Tomi II. Holmiæ,
1762, 1763. 8vo.

*Sp. pl. ed.* 1. ———————————— Tomi II. Holmiæ,
1753. 8vo.

*Sutherl. hort. edin.* Hortus Medicus Edinburgenfis, or a
catalogue of the plants in the Phyfical Garden at
Edinburgh, by James Sutherland. Edinburgh, 1683.
8vo.

*Swartz prodr.* Nova genera et fpecies plantarum, feu
prodromus defcriptionum vegetabilium, quæ fub iti-
nere in Indiam Occidentalem annis 1783—87 digeffit
Olof Swartz. Holmiæ, Upf. et Ab. 1788. 8vo.

*Syft.*

*Syſt. nat. ed.* 10.   Car. Linnæi Syſtema Naturæ.   Editio
Decima.  Tom. II.  Holmiæ, 1759. 8vo.

*Syſt. nat. ed.* 12.   Car. a Linné Syſtema Naturæ.   Editio
Duodecima.  Tom. II.  Holmiæ, 1767. 8vo.

*Syſt. nat. vol.* 3. ———————— Tom. III.  Holmiæ, 1768.
8vo.

*Syſt. veget.*   Car. a Linné Syſtema Vegetabilium.   Editio
Decima quarta, curante Jo. Andr. Murray.  Gottingæ,
1784. 8vo.

*Syſt. veget. ed.* 13. ———— Editio Decima tertia.   Gottingæ et Gothæ, 1774. 8vo.

*Tabern. hiſt.*  Jac. Theod. Tabernæmontani Kräuterbuch.
3 Theile.  Baſel, 1664. fol.

*Thunb. Aloe.*  Car. Petr. Thunberg Præſide Diſſertatio de
Aloe.  Upſaliæ, 1785. 4to.

*Thunb. Erica.*  C. P. Thunberg Præſ. Diſſ. de Erica.
Upſaliæ, 1785. 4to.

*Thunb. Ficus.*  C. P. Thunberg Præſ. Diſſ. Ficus genus.
Upſ. 1786. 4to.

*Thunb. Gardenia.*  C. P. Thunberg Diff. de Gardenia.
Upſ. 1780. 4to.

*Thunb. Gladiolus.*  C. P. Thunberg Diſſ. Gladiolus.  Upſ.
1784. 4to.

*Thunb. japon.*  C. P. Thunberg Flora Japonica.  Lipſiæ,
1784. 8vo.

*Thunb. Iris.*  C. P. Thunberg Diſſ. Iris.  Upſ. 1782.
4to.

*Thunb. Ixia.*  C. P. Thunberg Diſſ. Ixia.  Upſ. 1783.
4to.

*Thunb. nov. gen.*  C. P. Thunberg Diſſ. Nova Plantarum Genera.  Pars I—V.  Upſ. 1781 — 1784.
4to.

*Thunb. Oxalis.*  C. P. Thunberg Diſſ. Oxalis.  Upſ.
1781. 4to.

*Thunb.*

# BOOKS QUOTED. xxix

*Thunb. Protea.* C. P. Thunberg Diff. de Protea. Upf. 1781. 4to.

*Till. pif.* Mich. Aug. Tilli catalogus plantarum Horti Pifani. Florentiæ, 1723. fol.

*Tourn. cor.* Jof. Pitton Tournefort Corollarium inftitutionum rei herbariæ. Parifiis, 1703. 4to.

*Tourn. inft.* J. P. Tournefort inftitutiones rei herbariæ. Tomi III. Lugduni 1719. 4to.

*Tourn. it.* Relation d'un voyage du Levant, par M. Pitton de Tournefort. Tomes II. Paris, 1717. 4to.

*Trew ehret.* Plantæ feleftæ, quarum imagines pinxit Ge. Dion. Ehret, collegit et illuftravit Chriftoph. Jac. Trew. Norimbergæ, 1750—1773. fol.

*Trew Seligm.* Hortus nitidiffimus, five amœniffimorum florum imagines, quas collegit Chr. Jac. Trew, in æs incifas vivifque coloribus piftas edidit Joh. Mich. Seligmann. Norimb. 1750, feqq. fol.

*Turn. herb.* Will. Turner's herball. London, 1551. fol.

*Turn. herb. part* 1, *2d eait.* —— The firft parte. Collen, 1568. fol.

     *part* 2. ——— The feconde parte. Collen, 1562. fol.

     *part* 3. ——— The thirde parte. Collen, 1568. fol.

*Vailli parif.* Seb. Vaillant Botanicum Parifienfe. Leide et Amfterdam, 1727. fol.

*Wale. brit.* Flora Britannica indigena, or plates of the indigenous plants of Great Britain, by John Walcott. Bath, 1778. 8vo.

*Weber dec.* Ge. Henr. Weber Præfide Diff. Plantarum minus cognitarum decuria. Kiloniæ, 1784. 4to.

*Weinm. phytanth.* Joh. Wilh. Weinmann Phytanthozai-
                     conogra-

conographia. IV. Bände. Regenſburg, 1737—1742.
fol.

*Zanon. biſt.* Jac. Zanonii rariorum ſtirpium hiſtoria.
Bononiæ, 1742. fol.

*Zuccagni monogr.* Diſſertazione concernente l'iſtoria di una
pianta panizzabile dell' Abiſſinia, conoſciuta da quei
popoli ſotto il nome di Tef, dal Dott. Attilio Zuccagni. Firenze, 1775. 8vo.


# A B B R E V I A T I O N S.


| | | | |
|---|---|---|---|
| D. S. | — | — | Dry Stove. |
| G. H. | — | — | Green Houſe. |
| H. | — | — | Hardy. |
| S. | — | — | Stove. |
| ⊙. | — | — | Annual. |
| ♂. | — | — | Biennial. |
| ♃. | — | — | Perennial. |
| ♄. | — | — | Shrubby. |

*Claſſis*

# Classis I.

# MONANDRIA

## MONOGYNIA.

### CANNA. *Gen. pl.* 1.

*Corolla* 6-partita, erecta : labio bipartito, revoluto.
*Stylus* lanceolatus, corollæ adnatus. *Calyx* 3-phyllus.

1. C. foliis ovatis utrinque acuminatis nervosis. *Sp.*     *indica.*
   *pl.* 1.

α flore toto rubescente, foliis elliptico-ovatis.     *rubra.*
   Common Indian Reed, or Shot.

β petalis interioribus erectis luteis, nectarii lacinia re-     *lutea.*
   voluta lineolis rubicundis irrorata, foliis ovato-ellipticis.
   Yellow Indian Reed.

γ petalis interioribus erectis coccineis, nectarii lacinia     *coccinea.*
   revoluta lutea lineolis rubicundis irrorata, foliis ovato-ellipticis.
   Canna coccinea. *Mill. dict.*
   Scarlet Indian Reed.

δ petalis interioribus reflexis coccineis, nectarii lacinia     *patens.*
   revoluta lutea lineolis rubris irrorata, foliis lanceolato-oblongis.
   Spreading-flowered Indian Reed.
   *Nat.* of both Indies.

B            *Cult.*

*Cult.* 1596, by Mr. John Gerard.  *Hort. Ger.*
*Fl.* moſt part of the Year.                    S. ♃.

*glauca.*   2. C. foliis petiolatis lanceolatis enervibus.  *Sp. pl.* 1.
Glaucous-leav'd Indian Reed.
*Nat.* of America.
*Cult.* 1732, by James Sherard, M. D.  *Dill. elth.* 69.
*t.* 59. *f.* 69.
*Fl.* July.                                     S. ♃.

A M O M U M.  *Gen. pl.* 2.
*Cor.* 4. fida : lacinia prima patente.

*Zingiber.*  1. A. ſcapo nudo, ſpica ovata.  *Sp. pl.* 1. *Jacqu. hort.* 1.
*p.* 31. *t.* 75.
Narrow-leav'd Ginger.
*Nat.* of the Eaſt Indies.
*Cult.* 1731, by Mr. Philip Miller.  *Mill. dict. edit.* 1.
Zingiber 1.
*Fl.* September.                                S. ♃.

*Zerumbet.*  2. A. ſcapo nudo, ſpica oblonga obtuſa.  *Sp. pl.* 1.
*Jacqu. hort.* 3. *p.* 30. *t.* 54.
Broad-leav'd Ginger.
*Nat.* of the Eaſt Indies.
*Cult.* 1690, in the royal garden at Hampton-court.
*Catal. mſcr.*
*Fl.* September—November.                       S. ♃.

*Granum*   3. A. ſcapo ramoſo breviſſimo.  *Sp. pl.* 2.
*paradiſi.*  Grains of paradiſe.
*Nat.* of Guinea.
*Introd.* 1785, by Meſſrs. Lee and Kennedy.
*Fl.*                                           S. ♃.

COSTUS.

## C O S T U S. *Gen. pl.* 3.

*Cor.* interior inflata, ringens : labio inferiore trifido.

1. Costus. *Sp. pl.* 2.                                 *arabicus.*
   Arabian Coftus.
   *Nat.* of both Indies.
   *Cult.* 1752, by Mr. Ph. Miller. *Mill. dict. edit.* 6.
   *Fl.* Auguft.                                        S. ♃.

## M A R A N T A. *Gen. pl.* 5.

*Cor.* ringens, 5-fida : laciniis 2 alternis patentibus.

1. M. culmo ramofo. *Sp. pl.* 2.                        *arundi-*
   Indian Arrow-root.                                   *nacea.*
   *Nat.* of South America.
   *Introd.* before 1732, by William Houftoun, M. D.
      *Martyn dec.* 4. *p.* 39.
   *Fl.* July and Auguft.                               S. ♃.

## C U R C U M A. *Gen. pl.* 6.

*Stamina* 4 fterilia, quinto fertili.

1. C. foliis lanceolatis : nervis lateralibus numerofiffi-  *longa.*
   mis. *Sp. pl.* 3.
   Amomum Curcuma. *Jacqu. hort.* 3. *p.* 5. *t.* 4.
   Long-rooted Turmerick.
   *Nat.* of the Eaft Indies.
   *Cult.* 1759, by Mr. Philip Miller. *Mill. dict. edit.* 7.
   *Fl.* Auguft.                                        S. ♃.

## K Æ M P F E R I A. *Gen. pl.* 7.

*Cor.* 6-partita : laciniis 3 majoribus patulis, unica
   bipartita. *Stigma* bilamellatum.

1. K. foliis ovatis feffilibus. *Sp. pl.* 3.            *Galanga.*
   Galangale.

*Nat.*

*Nat.* of the Eaſt Indies.

*Introd.* 1724, by Charles Dubois, Eſq.   *Mill. ic.* 18.

*Fl.* June—September,                S. ♃.

## BOERHAVIA.   *Gen. pl.* 9.

*Cal.* o.   *Cor.* 1-petala, campanulata, plicata. *Sem.* 1. nudum, inferum. *(Stam.* 1. ſ. 2.)

*erecta.*    1. B. caule erecto glabro, floribus diandris.   *Syſt. veget.* 52.   *Jacqu. hort.* 1. *p.* 2. *t.* 5, 6.

Upright Hogweed.

*Nat.* of both Indies.

*Introd.* before 1733, by William Houſtoun, M. D. *Mill. dict. edit.* 8.

*Fl.* July—September.                S. ♂.

*diffuſa.*    2. B. caule lævi diffuſo, foliis ovatis.   *Syſt. veget.* 52.

Spreading Hogweed.

*Nat.* of both Indies.

*Cult.* 1690, in the royal garden at Hampton-court. *Catal. mſcr.*

*Fl.* Auguſt and September.          S. ♃.

*ſcandens.*    3. B. caule erecto, floribus diandris, foliis cordatis acutis.   *Syſt. veget.* 52.   *Jacqu. hort.* 1. *p.* 2. *t.* 4.

Climbing Hogweed.

*Nat.* of Jamaica.

*Cult.* 1691, in the royal garden at Hampton-court. *Pluk. phyt. t.* 113. *f.* 7.

*Fl.* April——September             S. ♄.

## SALICORNIA.   *Gen. pl.* 10.

*Cal.* ventriculoſus, integer.   *Petala* o. *Sem.* 1.

*herbacea.*    1. S. herbacea patula, articulis apice compreſſis emarginato-bifidis.   *Syſt. veget.* 52.

Marſh

Marſh jointed Glaſs-wort, or Salt-wort.
*Nat.* of Britain.
*Fl.* Auguſt and September.  H. ☉.

2. S. articulis obtuſis baſi incraſſatis, ſpicis ovatis. *Sp.*  *arabica.*
   *pl.* 5.
   Arabian jointed Glaſs-wort.
   *Nat.* of Arabia.
   *Cult.* 1758, by Mr. Philip Miller.
   *Fl.*  S. ♄.

### HIPPURIS. *Gen. pl.* 11.

*Cal.* 0. *Petala* 0.  *Stigma* ſimplex. *Sem.* 1.

1. HIPPURIS. *Sp. pl.* 6.  *Curtis lond.*  *vulgaris.*
   Mare's-tail.
   *Nat.* of Britain.
   *Fl.* May.  H. ♃.

### POLLICHIA.

*Cal.* 1-phyllus, 5-dentatus.  *Cor.* 0.  *Sem.* 1.  *Rec.*
   ſquamæ baccatæ, fructus includentes.

1. POLLICHIA.  *campeſ-*
   Whorl-leav'd Pollichia.  *tris*
   *Nat.* of the Cape of Good Hope.  Mr. *William Pa-*
   *terſon.*
   *Introd.* 1780, by the Counteſs of Strathmore.
   *Fl.* September.  G. H. ♂.
   DESCR. *Caules* teretes, tomento raro pubeſcentes, ra-
   moſi.  *Folia* verticillata, verſus unum latus vergen-
   tia, lineari-lanceolata, acuta, integerrima, glabra,
   vix uncialia.  *Stipulæ* plurimæ, membranaceæ, lanceolatæ, perſiſtentes, inæquales, pleræque ſeſquili-
   neares.  *Flores* ſeſſiles, congeſti in capitulum, axil-
   las trium vel quatuor foliorum occupans.
   B 3  DIGYNIA.

# DIGYNIA.

## CORISPERMUM. *Gen. pl.* 12.

*Cal.* o.   *Petala* 2.   *Sem.* 1. ovale, nudum.

*hyſſopifo-*
*lium.*
1. C. floribus lateralibus. *Sp. pl.* 6.
Hyſſop-leav'd Tick-feed.
*Nat.* of the South of France and Ruſſia.
*Cult.* 1739, by Mr. Philip Miller. *Rand. chelſ.*
*Fl.* July.        H. ☉.

*ſquarro-*
*ſum.*
2. C. ſpicis ſquarroſis. *Sp. pl.* 6.
Rough-ſpiked Tick-feed.
*Nat.* of Ruſſia.
*Cult.* 1759, by Mr. Philip Miller. *Mill. dict. edit.* 7.
*Fl.* Auguſt and September.     H. ☉.

## CALLITRICHE. *Gen. pl.* 13.

*Cal.* o.   *Petala* 2.   *Capſula* 2-locularis, 4-ſperma.

*verna.*
1. C. foliis ſuperioribus ovalibus, floribus androgynis.
    *Sp. pl.* 6.
Vernal ſtar-headed Chickweed.
*Nat.* of Britain.
*Fl.* April——July.     H. ☉.

*autumna-*
*lis.*
2. C. foliis omnibus linearibus apice bifidis, floribus
    hermaphroditis. *Sp. pl.* 6.
Autumnal ſtar-headed Chickweed.
*Nat.* of Britain.
*Fl.* Auguſt and September.     H. ☉.

BLITUM.

BLITUM. *Gen. pl.* 14.

*Cal.* 3-fidus. *Petala* 0. *Sem.* 1. calyce baccato.

1. B. capitellis fpicatis terminalibus. *Sp. pl.* 6.      *capita-*
Berry-headed Strawberry Blite.                      *tum.*
*Nat.* of Auftria.
*Cult.* 1633, by Mr. John Parkinfon. *Ger. emac.* 326.
*f.* 8.
*Fl.* May——Auguft                H. ⊙.

2. B. capitellis fparfis lateralibus. *Sp. pl.* 7.      *virga-*
Slender-branch'd Strawberry Blite.                *tum.*
*Nat.* of France, Spain, and Tartary.
*Cult.* 1759, by Mr. Philip Miller. *Mill. dict. edit.* 7.
*Fl.* May——September.             H. ⊙.

# Classis II.

# DIANDRIA

## MONOGYNIA.

NYCTANTHES. (Scabrita. *Linn. mant.* 3.)

*Corolla* hypocrateriformis : *laciniæ* truncatæ. *Capsula* bilocularis, marginata. *Semina* folitaria.

*Arbor triftis.*

1. NYCTANTHES. *Sp. pl.* 8.
Scabrita triflora. *Linn. mant.* 37.
Scabrita fcabra. *Syft. nat. ed.* 12. *p.* 115.
Square-ftalked Nyctanthes.
*Nat.* of the Eaft Indies.
*Introd.* 1781, by Sir Jof. Banks, Bart.
*Fl.*                                          S. ♄.

JASMINUM. *Gen. pl.* 17.

*Corolla* hypocrateriformis. *Bacca* dicocca. *Semina* folitaria, arillata.

*Sambac.*

1. J. foliis oppofitis fimplicibus ellipticis ovatis fubcordatifque membranaceis opacis, ramulis petiolifque pubefcentibus, laciniis calycinis fubulatis.
Nyctanthes Sambac. *Sp. pl.* 8.
α flore fimplici.
Single-flower'd Arabian Jafmine.
β floribus multiplicatis, laciniis oblongis acutis tubo brevioribus.
Common double Arabian Jafmine.
γ floribus plenis, laciniis fubrotundis tubo longioribus.
Kudda-mulla. *Rheed. mal.* 6. *p.* 89. *t.* 51.
Great flower'd double Arabian, or Tufcan Jafmine.
*Nat.* of the Eaft Indies.

*Cult.*

*Cult.* 1730. *Hort. angl. tab.* 7.
*Fl.* moft part of the fummer. S. ♄.

2. J. foliis oppofitis fimplicibus lanceolatis nitidis, laciniis *glaucum.*
    calycinis fubulatis.
Nyctanthes glauca. *Linn. fuppl.* 82.
Jafminum Africanum foliis folitariis, floribus vulga-
    tiori fimilibus. *Comm. rar.* 5. *tab.* 5.
Glaucous-leav'd Jafmine.
*Nat.* of the Cape of Good Hope. Mr. *Francis*
    *Maffon.*
*Introd.* 1774.
*Fl.* Auguft. G. H. ♄.

3. J. foliis oppofitis ternatis, foliolis ovatis fubcordatif- *azoricum.*
    que undulatis, ramulis glabris teretibus, corollæ la-
    ciniis tubo æqualibus.
Jafminum azoricum. *Sp. pl.* 9.
Azorian Jafmine.
*Nat.* of Madeira.
*Cult.* 1731, in the royal garden at Hampton-court.
    *Mill. dict. edit.* 1. *n.* 9.
*Fl.* May——November. G. H. ♄.

4. J. foliis alternis ternatis: foliolis obovatis cuneiformi- *fruticans.*
    bufque obtufis, ramis angulatis, laciniis calycinis
    fubulatis.
Jafminum fruticans. *Sp. pl.* 9.
Common yellow Jafmine.
*Nat.* of the South of Europe, and the Levant.
*Cult.* 1597, by Mr. John Gerard. *Ger. herb.* 1129.
*Fl.* May——October. H. ♄.

5. J. foliis alternis acutis ternatis pinnatifque, ramis an- *humile.*
    gulatis, laciniis calycinis breviffimis.
                            Jafminum

Jaſminum humile. . *Sp. pl.* 9.
Italian yellow Jaſmine.
*Nat.*
*Cult.* 1730.   *Hort. angl. t.* 6.
*Fl.* July——September.                          H. ♄ .

*odoratiſſi-*   6. J. foliis alternis obtuſiuſculis ternatis pinnatiſque, ra-
*mum.*              mis teretibus, laciniis calycinis breviſſimis.
Jaſminum odoratiſſimum.   *Sp. pl.* 10.
Yellow Indian Jaſmine.
*Nat.* of Madeira.
*Cult.* 1730.   *Mill. dict. edit.* 1.
*Fl.* May——November.                        G. H. ♄ :

*officinale.*   7. J. foliis oppoſitis pinnatis : foliolis acuminatis, gem-
                       mulis erectiuſculis.
Jaſminum officinale.   *Sp. pl.* 9.
Common white Jaſmine.
*Nat.*
*Cult.* 1597.   *Ger. herb.* 745. *f.* 1.
*Fl.* June——October.                          H. ♄ .

*grandi-*    8. J. foliis oppoſitis pinnatis : foliolis obtuſiuſculis, gem-
*florum.*            mulis horizontalibus.
Jaſminum grandiflorum.   *Sp. pl.* 9.
Spaniſh, or Catalonian Jaſmine.
*Nat.* of the Eaſt Indies.
*Cult.* 1629.   *Park. parad.* 406. *n.* 2.
*Fl.* June——October.                         G. H. ♄ .

## L I G U S T R U M.  *Gen. pl.* 18.

*Cor.* 4-fida.   *Bacca* tetraſperma.

*vulgare.*   1. LIGUSTRUM.  *Sp. pl.* 10.  *Curtis lond.*
Common Privet.

                                                        *Nat.*

*Nat.* of Britain.

*Fl.* June and July.                                                      H. ♄.


## PHILLYREA. *Gen. pl.* 19.

*Cor.* 4-fida. *Bacca* monofperma.

1. P. foliis oblongo-lanceolatis integris ferratifque.          *media.*
   Phillyrea media. *Sp. pl.* 10.

α foliis oblongo-lanceolatis.                                   liguftri-
   Phillyrea liguftrifolia. *Mill. dict.*                       folia.
   Phillyrea III. *Cluf. hift.* 1. *p.* 52. *fig.*
   Privet-leav'd Phillyrea.

β foliis lanceolatis, ramis erectis virgatis.                  virgata.
   Long-branched Phillyrea.

γ foliis lanceolatis, ramis divaricato-pendulis.               pendula.
   Drooping Phillyrea.

δ foliis oblongo-lanceolatis, ramis fuberectis.                oleæfo-
   Phillyrea Oleæ Ephefiacæ folio. *Pluk. alm.* 295. *t.*      lia.
      310. *f.* 5.
   Olive-leav'd Phillyrea.

ε foliis ovali-oblongis obtufiufculis.                         buxifolia.
   Box-leav'd Phillyrea.
   *Nat.* of the South of Europe.
   *Cult.* 1597, by the Earl of Effex. *Ger. herb.* 1209. *f.* 2.
   *Fl.* May and June.                                          H. ♄.


2. P. foliis lineari-lanceolatis integerrimis. *Sp. pl.* 10.   *anguftifo-*
α foliis lanceolatis, ramis rectis.                            *lia.*
   Phillyrea IV. *Cluf. hift.* 1. *p.* 52. *fig.*              lanceola-
   Common narrow-leav'd Phillyrea.                              ta.

β foliis lanceolato-fubulatis elongatis, ramis rectis.         rofmari-
   Phillyrea rofmarinifolia. *Mill. dict.*                     nifolia.
   Phillyrea V. *Cluf. hift.* 1. *p.* 52.
   Rofemary-leav'd Phillyrea.

                                                      γ foliis

brachia-
ta.
γ foliis oblongo-lanceolatis brevioribus, ramis divaricatis.
Dwarf Phillyrea.
*Nat.* of the South of Europe.
*Cult.* 1597, by the Earl of Effex.  *Ger. herb.* 1209.
*f.* 1.
*Fl.* May and June.                                    H. ♄.

*latifolia.*   3. P. foliis ovato-oblongis fubcordatis ferratis.
Phillyrea latifolia.  *Sp. pl.* 10.
lævis.   α foliis ovatis planis obfolete ferratis.
Phillyrea latifolia.  *Mill. dict.*
Phillyrea arbor.  *Lob. ic.* 2. *p.* 132.  *Bauh. hift.* 1. *p.*
540. *fig.*
Smooth broad-leav'd Phillyrea.
fpinofa.   β foliis ovato-oblongis acutis argute ferratis planis.
Phillyrea fpinofa.  *Mill. dict.*
Phillyrea latifolia fpinofa triphyllos.  *Pluk. alm.* 295.
*t.* 310. *f.* 4.
Phillyrea I.  *Cluf. hift.* 1. *p.* 51. *fig.*
Prickly broad-leav'd Phillyrea.
obliqua.   γ foliis lanceolato-oblongis acutis ferratis oblique flexis.
Phillyrea II.  *Cluf. hift.* 1. *p.* 52. *fig.*
Ilex-leav'd Phillyrea.
*Nat.* of the South of Europe.
*Cult.* 1597, by the Earl of Effex.  *Ger. herb.* 1210.
*f.* 3.
*Fl.* May and June.                                    H. ♄.

O L E A.  *Gen. pl.* 20.

*Cor.* 4-fida : laciniis fubovatis.  *Drupa* monofperma.

*europæa.*   1. O. foliis lanceolatis integerrimis, racemis axillaribus
coarctatis.

Olea

Olea europæa. *Sp. pl.* 11.

α foliis lanceolatis planis fubtus incanis.      commu-
    Common European Olive.      nis.
β foliis lineari-lanceolatis planis fubtus argenteis.    longifo-
    Long leav'd European Olive.      lia.
γ foliis oblongis planis fubtus incanis.      latifolia.
    Broad-leav'd European Olive.
δ foliis lanceolatis fubtus ferrugineis.      ferrugi-
    Iron-coloured European Olive.      nea.
ε foliis oblongis oblique flexis fubtus pallidis.    obliqua.
    Twifted-leav'd European Olive.
ζ foliis oblongo-ovalibus, ramis patentibus divaricatis.   buxifolia.
    Box-leav'd European Olive.
    *Nat.* of the South of Europe.
    *Cult.* 1648, in Oxford garden. *Hort. oxon. edit.* 1.
     p. 37.
    *Fl.* June——Auguft.   α. H. ♄. β. γ. δ. ε. ζ. G. H. ♄.

2. O. foliis ovatis integerrimis, racemis paniculæformi-   *capenfis.*
     bus divaricatis.
    Olea capenfis. *Sp. pl.* 11.
α foliis ovato-oblongis rigidis planis, petiolis rubris.   coriacea.
    Sideroxylum foliis fubrotundis integris. *Burm. afr.*
     234. *t.* 81. *f.* 2.
    Leathery-leav'd cape Olive.
β foliis ellipticis undatis, petiolis viridibus.     undulata.
    Sideroxylum foliis oblongis integris. *Burm. afr.* 233.
     *t.* 81. *f.* 1.
    Wave-leav'd cape Olive.
    *Nat.* of the Cape of Good Hope.
    *Cult.* 1732, by James Sherard, M.D. *Dill. elth.* 193.
     *t.* 160. *f.* 194.
    *Fl.* June——September.     G. H. ♄.

3. O.

*america-*
*na.*

3. O. foliis lanceolato-ellipticis integerrimis, racemis anguſtatis, braĉteis omnibus perſiſtentibus connatis
parvis.

Olea americana. *Linn. mant.* 24.

American Olive.

*Nat.* of Carolina and Florida.

*Cult.* 1758, by Mr. Philip Miller.

*Fl.* June.                                      G. H. ♄.

*excelſa.*

4. O. foliis ellipticis integerrimis, racemis anguſtatis,
braĉteis perfoliatis : baſeos cyathiformibus perſiſtentibus ; ſuperioribus deciduis magnis foliaceis.

Laurel-leav'd Olive.

*Nat.* of Madeira.   Mr. *Francis Maſſon.*

*Introd.* 1784.

*Fl.*                                           G. H. ♄.

*fragrans.*

5. O. foliis lanceolatis ſerratis, pedunculis lateralibus aggregatis unifloris.   *Thunb. japon.* 18. *tab.* 2. *Syſt.*
*veget.* 57.

Quaj-fa.  *Oſb. it.* 250.

Sweet-ſcented Olive.

*Nat.* of Cochinchina, China, and Japan.

*Introd.* 1771, by Benjamin Torin, Eſq.

*Fl.* July and Auguſt.                          G. H. ♄.

## CHIONANTHUS. *Gen. pl.* 21.

*Cor.* 4-fida : laciniis longiſſimis.  *Drupæ* nucleus
ſtriatus.

*virginica.*
latifolia.

1. C. pedunculis trifidis trifloris.  *Sp. pl.* 11.

α foliis ovato-ellipticis.

Broad-leav'd Fringe-tree.

*anguſti-*
folia.

β foliis lanceolatis.

Narrow-leav'd Fringe-tree.

*Nat.*

*Nat.* of North America.
*Introd.* 1736, by Peter Collinſon, Eſq. *Coll. mſcr.*
*Fl.* June and July.                    H. ♄.


### SYRINGA. *Gen. pl.* 22.

*Cor.* 4-fida.    *Capſula* bilocularis.

1. S. foliis ovato-cordatis.  *Sp. pl.* 11.          *vulgaris.*
α floribus cæruleis.                           cærulea.
   Common blue Lilac.
β floribus violaceis.                          violacea.
   Common purple Lilac.
γ floribus albis.                              alba.
ι. Common white Lilac.
   *Nat.* of Perſia.
   *Cult.* 1597, by Mr. John Gerard.  *Ger. herb.* 1213.
   *f.* 2.
   *Fl.* April and May.                  H. ♄.


2. S. foliis lanceolatis.  *Sp. pl.* 11.          *perſica.*
α foliis ſimplicibus, floribus cæruleis.       cærulea.
   Blue Perſian Lilac.
β foliis ſimplicibus, floribus albis.          alba.
   White Perſian Lilac.
γ foliis pinnatifidis.                         laciniata.
   Cut-leav'd·Perſian Lilac.
   *Nat.* of Perſia.
   *Cult.* 1658.  *Hort. oxon. ed.* 2. *p.* 83.
   *Fl.* May.                             H. ♄.


### ANCISTRUM. *Linn. ſuppl.* 10.

*Cal.* 4-phyllus. *Cor.* o. *Stigma* multipartitum. *Drupa* exſucca, hiſpida, 1-locularis.

1. A. caulibus ſubdemerſis, pedunculis ſcapiformibus,   *lucidum.*
   ſpicis ovatis, foliolis oblongis integerrimis acutis
   ſubfaſciculatis.

                                      Shining

Shining Anciſtrum.

*Nat.* of Falkland Iſlands.

*Introd.* 1777, by John Fothergill, M. D.

*Fl.* May and June.                                    G. H. ♃.

DESCR. *Caules* breviſſimi, decumbentes. *Folia* plurima, ſubradicalia, approximata, impari pinnata, patula, bi- vel triuncialia : *foliola* plerumque octojuga,
ſeſſilia, ſupra nitida, ſubtus albida, carinata : carina
apiceque piloſo, bilinearia, patentia. *Petioli* baſi
dilatati, vaginantes, ciliati. *Pedunculi* communes
plerumque ſolitarii, longitudine foliorum, teretes,
villoſi, interdum folio uno alterove inſtructi. *Spica*
multiflora, vix ſemuncialis. *Bracteæ* longitudine
germinum : inferiores apice fiſſæ ; ſuperiores integerrimæ. *Calycis* foliola extus et apice piloſa, viridia, vix lineam longa. *Stamina* longitudine calycis, atro-rubentia. *Germen* ſupra medium denticulis
quatuor obſoletis notatum, piloſum. *Stigmata* faſ-
ciculata, atro-ſanguinea. *Drupa* ſuperne inæqualiter
cornuta.

*latebro-*    2. A. caulibus demerſis, pedunculis ſcapiformibus, ſpicis
*ſum.*              elongatis, foliolis oblongis inciſis villoſis, fructibus
                    undique armatis.

Agrimonia decumbens.    *Linn. ſuppl.* 251.

Hairy Anciſtrum.

*Nat.* of the Cape of Good Hope.    Mr. *Fr. Maſſon.*

*Introd.* 1774.

*Fl.*                                                       G. H. ♃.

DESCR. *Calycis* foliola extus villoſiuſcula, bilinearia.
*Filamenta* calyce paulo breviora. *Antheræ* bifidæ.
*Germen* villoſum, undique glochidibus obſitum.
*Stigma* multipartitum : lacinulis ſubulatis, brevibus,
patentiſſimis. *Drupa* quadrilinearis, tomentoſa,
undique glochidibus pedicellatis armata. *Nux* la-
cunoſa, per tegumentum drupæ glochides exſerens.

CIRCAEA.

CIRCAEA. *Gen. pl.* 24.

*Cor.* dipetala. *Cal.* diphyllus, fuperus. *Sem.* 1. biloculare.

1. C. caule erecto, racemis pluribus, foliis ovatis. *Syft.* *lutetiana.*
   *veget.* 58. *Curtis lond.*
   Enchanter's Nightfhade.
   *Nat.* of Britain.
   *Fl.* June——Auguft.                                   H. ♃.

2. C. caule proftrato, racemo unico, foliis cordatis. *Syft.* *alpina.*
   *veget.* 58.
   Mountain Enchanter's Nightfhade.
   *Nat.* of Britain.
   *Fl.* June——September.                                H. ♃.

VERONICA. *Gen. pl.* 25.

*Cor.* Limbo 4-partito : lacinia infima anguftiore. *Capfula* bilocularis.

\* *Spicatæ.*

1. V. fpicis terminalibus, foliis feptenis verticillatis, caule *fibirica.*
   fubhirto. *Sp. pl.* 12.
   Siberian Speedwell.
   *Nat.* of Siberia.
   *Introd.* 1779, by Chevalier Thunberg.
   *Fl.* July and Auguft.                                 H. ♃.

2. V. fpicis terminalibus, foliis quaternis quinifque. *Sp.* *virgini-*
   *pl.* 13.                                              *ca.*
   α floribus albis.
   Virginian white-flower'd Speedwell.
   β floribus incarnatis.
   Virginian blufh-flower'd Speedwell.
   *Nat.* of Virginia.

                    C                        *Cult.*

*Cult.* 1714, by the Duchess of Beaufort. *Br. Muf.*
H. S. 336. *fol.* 42.
*Fl.* July—September.                                    H. ♃.

*fpuria.*    3. V. fpicis terminalibus, foliis ternis æqualiter ferratis.
*Sp. pl.* 13.
Baftard Speedwell.
*Nat.* of Siberia and Germany.
*Cult.* 1731, by Mr. Philip Miller. *Mill. dict. edit.* 1.
*no.* 2.
*Fl.* May and June.                                    H. ♃.

*mariti-*    4. V. fpicis terminalibus, foliis ternis inæqualiter ferratis.
*ma.*            *Sp. pl.* 13.
α flore cæruleo.
Blue-flower'd Sea Speedwell.
β flore albo.
White-flower'd Sea Speedwell.
γ flore incarnato.
Flefh-coloured Sea Speedwell.
*Nat.* of the European Sea-coafts.
*Cult,* 1570, by Mr. Hugh Morgan. *Lobel. adv.* 145.
*Fl.* June——Auguft.                                    H. ♃.

*longifolia.*    5. V. fpicis terminalibus, foliis oppofitis lanceolatis fer-
ratis acuminatis. *Sp. pl.* 13.
Long-leav'd Speedwell.
*Nat.* of Ruffia and Auftria.
*Cult.* 1731, by Mr. Philip Miller. *Mill. dict. edit* 1.
*no.* 3.
*Fl.* July——September.                                    H. ♃.

*incana.*    6. V. fpicis terminalibus, foliis oppofitis crenatis obtufis,
caule erecto tomentofo. *Sp. pl.* 14.
Hoary Speedwell.

*Nat.*

*Nat.* of Ruffia.

*Cult.* 1759, by Mr. Philip Miller. *Mill. dict. edit.* 7. *no.* 14.

*Fl.* July——September. H. ♃.

7. V. fpica terminali, foliis oppofitis crenatis obtufis, caule *fpicata.*
adfcendente fimpliciffimo. *Sp. pl.* 14.
Upright fpiked male Speedwell.
*Nat.* of England.
*Fl.* June——Auguft. H. ♃.

8. V. fpicis terminalibus, foliis oppofitis obtufe ferratis *hybrida.*
fcabris, caule erecto. *Sp. pl.* 14.
Welfh Speedwell.
*Nat.* of Wales.
*Fl.* June—Auguft. H. ♃.

9. V. fpica terminali, foliis linearibus pinnatifidis fub- *pinnata.*
fafciculatis: laciniis filiformibus divaricatis.
Veronica pinnata. *Linn. mant.* 24.
Winged-leav'd Speedwell.
*Nat.* of Siberia.
*Introd.* 1776, by Chevalier Murray.
*Fl.* July. H. ♃.

10. V. racemo fubfpicato terminali, foliis pinnatifidis laci- *laciniata.*
niatis.
Jagged-leav'd Speedwell.
*Nat.* of Siberia.
*Introd.* 1780, by William Pitcairn, M. D.
*Fl.* June and July. H. ♃.

11. V. fpicis terminalibus, foliis lanceolatis incifo-pinnati- *incifa.*
fidis glabris.
Cut-leav'd Speedwell.
*Nat.* of Siberia.

C 2 *Introd.*

*Introd.* 1779, by Meſſrs. Kennedy and Lee.
*Fl.* July and Auguſt.                   H. ♃.

*officinalis.*  12. V. ſpicis lateralibus pedunculatis, foliis oppoſitis,
          caule procumbente.  *Sp. pl.* 14.  *Curtis lond.*
          Officinal Male Speedwell.
          *Nat.* of Britain.
          *Fl.* May—July.                   H. ♃.

*decuſſata.*  13. V. ſpicis terminalibus paniculatis, foliis oblongis in-
          tegerrimis lævigatis coriaceis, caule fruticoſo.
          Croſs-leav'd Speedwell.
          *Nat.* of Falkland Iſlands.
          *Introd.* 1776, by John Fothergill, M. D.
          *Fl.* July and Auguſt.             G. H. ♄.
          Descr.  *Frutex* ramoſiſſimus.  *Rami* teretes, gla-
          bri, cicatricibus petiolorum ſubannulati.  *Folia*
          oppoſita, acuta, margine cartilaginea, pollice paulo
          breviora.  *Petioli* breviſſimi.  *Racemi* ſpiciformes
          e ſupremis axillis oppoſiti, brachiati, numeroſi,
          pauciflori, foliis breviores.  *Braĉteæ* ad baſin pe-
          dicellorum ovatæ, patentes, ciliatæ, ſeſquilineares.
          *Braĉteolæ* duæ ·oppoſitæ, ovatæ, minutæ, in me-
          dio ſinguli pedicelli.  *Calyx* fere tetraphyllus, ſeſ-
          quilinearis.  *Corolla* infundibuliformis, alba.  *Fila-*
          *menta* corolla breviora, alba.  *Antheræ* cordatæ.
          *Germen* ovato-oblongum, glabrum.  *Stylus* ſubu-
          latus, brevis.

                * * *Corymboſo-racemoſæ.*

*aphylla.*  14. V. corymbo terminali, ſcapo nudo.  *Sp. pl.* 14.
          Naked-ſtalk'd Speedwell.
          *Nat.* of Switzerland, and Italy.
          *Introd.* 1775, by the Doĉtors Pitcairn and Fothergill.
          *Fl.* May.                       H. ♃.

                                    15. V.

15. V. corymbo terminali, caule adfcendente diphyllo, *bellidioi-*
foliis obtufis crenatis, calycibus hirfutis. *Syft.* *des.*
*veget.* 59.
Daify leav'd Speedwell.
*Nat.* of the Alps of Switzerland.
*Introd.* 1775, by the Doctors Pitcairn and Fothergill.
*Fl.* June and July. H. ♃.

16. V. corymbo terminali, foliis lanceolatis obtufiufculis *fruticu-*
crenatis, caulibus fruticulofis. *Syft. veget.* 59. *lofa.*
Purple-ftalk'd evergreen Speedwell.
*Nat.* of the Alps of Switzerland and Auftria.
*Cult.* 1748, by Mr. Philip Miller. *Mill. dict. edit.* 5.
*no.* 14.
*Fl.* June and July. H. ♄.

17. V. corymbo terminali, foliis oppofitis, calycibus *alpina.*
hifpidis. *Sp. pl.* 15.
Alpine Speedwell.
*Nat.* of Scotland.
*Fl.* May. H. ♃.

18. V. corymbo terminali, foliis oppofitis glabriufculis, pe- *faxatilis.*
dunculis folia floralia fuperantibus. *Linn. fuppl.* 83.
Veronica fruticans. *Jacqu. vind.* 200.
Veronica alpina fruticans Serpylli majoris folio lon-
giore. *Pluk. alm. p.* 384. *t.* 232. *f.* 5.
Veronica. *Hall. hift. n.* 545. β.
Rock Speedwell.
*Nat.* of Auftria and Switzerland.
*Introd.* 1768, by Profeffor de Sauffure.
*Fl.* June. H. ♃.

19. V. racemo terminali fubfpicato, foliis ovatis glabris *ferpyllifo-*
crenatis. *Sp. pl.* 15. *Curtis lond.* *lia.*
C 3 Smooth

Smooth Speedwell, or Paul's Betony.
*Nat.* of Britain.
*Fl.* May—July.                                   H. ♃.

*Becca-*
*bunga.*    20. V. racemis lateralibus, foliis ovatis planis, caule re-
pente. *Sp. pl.* 16. *Curtis lond.*
Brooklime Speedwell.
*Nat.* of Britain.
*Fl.* May and June.                               H. ♃.

*Anagal-*
*lis.*    21. V. racemis lateralibus, foliis lanceolatis ferratis,
caule erecto. *Sp. pl.* 16. *Curtis lond.*
Long-leav'd water Speedwell.
*Nat.* of Britain.
*Fl.* July.                                        H. ☉.

*ſcutellata.*    22. V. racemis lateralibus alternis: pedicellis pendulis,
foliis linearibus integerrimis. *Syſt. veget.* 59. *Cur-*
*tis lond.*
Narrow-leav'd water Speedwell.
*Nat.* of Britain.
*Fl.* June and July.                              H. ♃.

*Teucri-*
*um.*    23. V. racemis lateralibus longiſſimis, foliis ovatis rugoſis
dentatis. obtuſiuſculis, caulibus procumbentibus.
*Syſt. veget.* 59.
Hungarian Speedwell.
*Nat.* of Germany and Hungary.
*Cult.* 1596, by Mr. John Gerard. *Hort. Ger.*
*Fl.* July and Auguſt.                            H. ♃.

*proſtrata.*    24. V. racemis lateralibus, foliis ovato-oblongis ferratis,
caulibus proſtratis. *Sp. pl.* 17.
Trailing Speedwell.
*Nat.* of Germany and Italy.

*Introd.*

*Introd.* 1774, by Chevalier Murray.
*Fl.* May and June. H. ♃.

25. V. racemis lateralibus paucifloris, calycibus hirfutis, *montana.*
foliis ovatis rugofis crenatis petiolatis, caule debili.
*Sp. pl.* 17. *Curtis lond. Jacqu. auftr.* 2. *p.* 6.
*t.* 109.
Mountain Speedwell.
*Nat.* of Britain.
*Fl.* May and June. H. ♃.

26. V. racemis lateralibus, foliis ovatis feffilibus rugofis *Chamæ-*
dentatis, caule bifariam pilofo. *Syft. veget.* 60. *drys.*
*Curtis lond.*
Wild Germander, or Speedwell.
*Nat.* of Britain.
*Fl.* May and June. H. ♃.

27. V. racemis lateralibus, foliis pinnatifidis glabris acutis *orientalis.*
bafi attenuatis, calycibus inæqualibus, pedicellis
capillaribus braćtea longioribus.
Veronica orientalis. *Mill. dić̆t.*
Veronica auftriaca, β. *Sp. pl.* 17.
Oriental Speedwell.
*Nat.* of the Levant.
*Cult.* 1759, by Mr. Philip Miller. *Mill. dić̆t. edit.* 7.
*no.* 12.
*Fl.* July and Auguft. H. ♃.

28. V. racemis lateralibus, foliis multipartitis pilofiufcu- *multifida.*
lis : laciniis linearibus : bafeos divaricatis, calyci-
bus inæqualibus, pedicellis longitudine braćtearum.
Veronica multifida. *Sp. pl.* 17. (exclufo fynonymo Buxbaumii.) *Jacqu. auftr.* 4. *p.* 15. *t.* 329.
Veronica auftriaca, x. *Sp. pl.* 17.

Multifid

Multifid Speedwell.

*Nat.* of Auſtria.

*Cult.* 1748, by Mr. Philip Miller.  *Mill. dict. edit.* 5. *no.* 17:

*Fl.* June—Auguſt.                                      H. ♃.

*urticæ-*     29. V. racemis lateralibus, foliis cordatis ſeſſilibus argute
*folia.*             ſerratis acuminatis, caule ſtricto, foliolis calycinis
                    quaternis.

            Veronica urticæfolia.  *Jacqu. auſtr.* 1. *p.* 37. *t.* 59. *Linn. ſuppl.* 83.

            Veronica foliis hirſutis nervoſis ſeſſilibus cordatis
                 lanceolatis ſerratis, floribus racemoſis longe petio-
                 latis.  *Hall. hiſt.* 535.

            Nettle-leav'd Speedwell.

            *Nat.* of Auſtria and Switzerland.

            *Introd.* 1776, by Joſeph Nicholas de Jacquin, M. D.

            *Fl.* June and July.                          H. ♃.

*latifolia.*  30. V. racemis lateralibus, foliis cordatis ſeſſilibus rugoſis
                    obtuſe ſerratis, caule ſtricto, foliolis calycinis quinis.

            Veronica latifolia.  *Sp. pl.* 18.

            Veronica pſeudo-chamædrys.  *Jacqu. auſtr.* 1. *p.* 37. *t.* 60.

            Broad-leav'd Speedwell.

            *Nat.* of Auſtria and Switzerland.

            *Introd.* 1775, by Joſeph Nicholas de Jacquin, M. D.

            *Fl.* May and June.                           H. ♃.

                     *** *Pedunculis unifloris.*

*agreſtis.*   31. V. floribus ſolitariis, foliis cordatis inciſis pedunculo
                    brevioribus.  *Sp. pl.* 18.  *Curtis lond.*

            Germander-Speedwell.

            *Nat.* of Britain.

            *Fl.* May and June.                           H. ☉.

                                                    32. V.

32. V. floribus folitariis, foliis cordatis incifis pedunculo *arvenfis.*
longioribus. *Sp, pl.* 18. *Curtis lond.*
Corn Speedwell.
*Nat.* of Britain.
*Fl.* May.        H. ☉.

33. V. floribus folitariis, foliis cordatis planis quinquelo- *hederifo-*
bis. *Sp. pl.* 19. *Curtis lond.*      *lia.*
Ivy-leav'd Speedwell.
*Nat.* of Britain.
*Fl.* April—June.       H. ☉.

34. V. floribus folitariis, foliis digitato-partitis, pedun- *triphyllos.*
culis calyce longioribus. *Syft. veget.* 60.
Trifid Speedwell.
*Nat.* of Britain.
*Fl.* April and May.       H. ☉.

35. V. floribus folitariis, foliis digitato-partitis, pedun- *verna.*
culis calyce brevioribus. *Syft. veget.* 60.
Spring Speedwell.
*Nat.* of England.
*Fl.* April and May.       H. ☉.

36. V. floribus folitariis fubfeffilibus, foliis oblongis fub- *romana.*
dentatis, caule erecto. *Sp. pl.* 19.
Roman Speedwell.
*Nat.* of the South of Europe.
*Introd.* 1775, by Monf. Thouin.
*Fl.* May and June.       H. ☉.

37. V. floribus folitariis feffilibus, foliis lanceolato-linea- *peregri-*
ribus glabris obtufis integerrimis, caule erecto.   *na.*
*Sp. pl.* 20.

                                   Knot-

Knot-grafs-leav'd Speedwell.
*Nat.* of the North of Europe.
*Cult.* 1680.   *Morif. hift.* 2. *p.* 322. *no.* 19.
*Fl.* May and June.                                H. ☉.


JUSTICIA.   *conf.*  *Gen. pl.* 27.

*Cor.* 1-petala, irregularis.  *Cal.* fimplex.  *Capf.* ungue
   elaftico diffiliens: *Diffepimentum* contrarium, adnatum.

*fexangu-*      1. J. herbacea, corollæ labiis integris, bracteis cunei-
*laris.*            formibus, foliis ellipticis, ramis fexangularibus,
               antheris parallelis.
             Jufticia fexangularis.   *Sp. pl.* 23.
             Chick-weed-leav'd Jufticia.
             *Nat.* of South America.
             *Cult.* 1739, by Mr. Philip Miller.  *Mill. dict. vol.* 2.
             *no.* 1.
             *Fl.* July.                                S. ☉.


*coccinea.*     2. J. fruticofa, corollis bilabiatis: labio fuperiori indi-
               vifo, foliis bracteifque fpicarum ellipticis acumi-
               natis, antheris parallelis.
             Jufticia coccinea.   *Aublet guian.* 10. *t.* 3.
             Scarlet-flower'd Jufticia.
             *Nat.* of South America.
             *Introd.* about 1770.
             *Fl.* February.                            S. ♄.


*Ecboli-*       3. J. fruticofa, corollis bilabiatis: labio fuperiori lineari,
*um.*              fpicarum bracteis ovalibus cufpidatis ciliatis, an-
               theris parallelis.
             Jufticia Ecbolium.   *Sp. pl.* 20.
             Long-fpiked Jufticia.
             *Nat.* of the Eaft Indies.
                                                    *Cult.*

*Cult.* 1759, by Mr. Philip Miller.  *Mill. dict. edit.* 7.
*no.* 8.
*Fl.* March——Auguft.                                S. ♄.

4. J. herbacea, corollis bilabiatis divifis, foliis ovatis acu-   *malaba-*
    minatis, paniculis axillaribus, bractea extima li-   *rica.*
    neari duplo longiore, antheris divaricatis.
  Dianthera malabarica.   *Linn. fuppl.* 85. (exclufo fy-
    nonymo Rheedii.)
  Malabar Jufticia.
  *Nat.* of the Eaft Indies.
  *Introd.* 1785, by Sir Jofeph Banks, Bart.
  *Fl.* Auguft.                                          S. ☉.

5. J. herbacea, corollis bilabiatis: labio fuperiori indi-   *pectoralis.*
    vifo, foliis lanceolatis petiolatis, fpicis paniculatis,
    bracteis minutis, antheris binis.
  Jufticia pectoralis.   *Jacqu. hift.* 3. *t.* 3.
  Dianthera pectoralis.   *Syft. veget.* 64.
  Forked Jufticia.
  *Nat.* of the Weft Indies.
  *Introd.* 1787, by Mr. Alex. Anderfon.
  *Fl.*                                                   S.

6. J. herbacea hirta, corollis bilabiatis divifis, foliis lan-   *ciliaris.*
    ceolatis obtufiufculis, floribus axillaribus fpicifque
    foliofis, antheris parallelis appendiculatis.
  Jufticia ciliaris.   *Linn. fuppl.* 84. (exclufis fynonymis
    Burmanni et Hermanni.)
  Jufticia ciliata.   *Jacqu. hort.* 2. *p.* 47. *t.* 104.
  Ciliated Jufticia.
  *Nat.*
  *Introd.* 1780, by Monf. Thouin.
  *Fl.* June—Auguft.                                      S. ☉.

                                                          7. J.

*Adhato-*
*da.*

7. J. arborea, corollis ringentibus, foliis ovato-lanceolatis
    acuminatis, braɩeis ovato-ellipticis foliaceis, an-
    theris parallelis.
Jufticia Adhatoda.  *Sp. pl.* 20.
Malabar Nut.
*Nat.* of Ceylon.
*Cult.* 1699, by the Duchefs of Beaufort.  *Br. muſ.*
    *Sloan. mſs.* 525 and 3329.
*Fl.* May and July.             G. H. ♄.

*orchioi-*
*des.*

8. J. fruticofa, corollis ringentibus, pedunculis axillari-
    bus folitariis unifloris, braɩeis calyce brevioribus,
    foliis lanceolatis feffilibus, antheris binatis appendi-
    culatis.
Jufticia orchioides.  *Linn. ſuppl.* 85.
Broom-leav'd Jufticia.
*Nat.* of the Cape of Good Hope.  Mr. *Fr. Maſſon.*
*Introd.* 1774.
*Fl.* Auguft and September.        G. H. ♄.

*hyſſopifo-*
*lia.*

9. J. fruticofa, corollis ringentibus, foliis lanceolatis,
    braɩeis calyce brevioribus, laciniis calycinis ob-
    longis, antheris binatis appendiculatis.
Jufticia hyffopifolia.  *Sp. pl.* 21.
Hyffop-leav'd Jufticia, or Snap-tree.
*Nat.* of the Canary Iflands.
*Cult.* 1690, in the royal garden at Hampton-court.
    *Mill. ic.* 9. *t.* 13.
*Fl.* March——Auguft.          G. H. ♄.

G R A T I O L A.  *Gen. pl.* 29.
*Cor.* irregularis, refupinata.  *Stamina* 2 fterilia.  *Capſ.*
2-locularis.  *Cal.* 7-phyllus : 2 exterioribus patulis.

1. G.

1. G. floribus pedunculatis, foliis lanceolatis ferratis. *officinalis.*
    *Sp. pl.* 24.
    Hedge-hyffop.
    *Nat.* of the South of Europe.
    *Cult.* 1568. *Turn. herb. part.* 3. *fol.* 33.
    *Fl.* June—Auguft. H. ♃.

2. G. foliis ovali-oblongis, pedunculis unifloris, caule *Monnie-*
    repente. *Sp. pl.* 24. *ria.*
    Thyme-leav'd Gratiola.
    *Nat.* of both Indies and the South Sea Iflands.
    *Introd.* 1772, by Monf. Richard.
    *Fl.* July—September. S. ♃.

SCHWENKIA. *Gen. pl. p.* 577.

*Cor.* fubæqualis, fauce plicata glandulofa. *Stam.* 3
    fterilia. *Capf.* 2-locularis, polyfperma.

1. SCHWENKIA. *Syft. veget.* 64. *america-*
    Guinea Schwenkia. *na.*
    *Nat.* of Guinea.
    *Introd.* 1781, by Meffrs. Kennedy and Lee.
    *Fl.* Auguft and September. S. ♂.

CALCEOLARIA. *Linn. mant.* 143.

*Cor.* ringens, inflata. *Capf.* 2-locularis, 2-valvis.
    *Cal.* 4-partitus, æqualis.

1. C. foliis pinnatis. *Linn. mant.* 171. *pinnata.*
    Winged-leav'd Calceolaria.
    *Nat.* of Peru.
    *Introd.* 1773, by Sir Jofeph Banks, Bart.
    *Fl.* July——Octtober. S. ☉.

2. C.

*Fother-*
*gillii.*

2. C. foliis fpathulatis integerrimis, pedunculis fcapifor-
mibus unifloris.
Spatula-leav'd Calceolaria.
*Nat.* of Falkland Iflands.
*Introd.* 1777, by John Fothergill, M. D.
*Fl.* May——Auguft.                                    G. H. ♂.
*Caules* vix unciales, prope radicem fubdivifi. *Folia*
oppofita, petiolata, obtufa, fupra pilofa, vix uncialia.
*Pedunculi* terminales, folitarii vel gemini, teretes,
foliis duplo longiores. *Calycis* laciniæ acutæ, apice
inflexæ, extus pilofæ, trilineares. *Corollæ* labium
fuperius reniformi-fubrotundum, erectum, fornica-
tum, flavum, çalyce paulo brevius ; labium inferius
defcendens, fuperiori quater longius, antice dilata-
tum, inflatum, fubtus dilute flavicans, fupra ad la-
tera rubicundum, antice luteum maculis rubris,
prope palatum flavum : Faux magna, aperta, tetra-
gono-ovata. *Filamenta* corollæ lateribus ad bafin
faucis inferta, fubulata. *Antheræ* fubrotundæ,
magnæ. *Stylus* craffus, longitudine ftaminum.
*Stigma* incraffatum, planum.


PINGUICULA. *Gen. pl.* 30.

*Cor.* ringens, calcarata. *Cal.* bilabiatus, 5-fidus.
*Capf.* unilocularis.


*vulgaris.*    1. P. nectario cylindraceo longitudine petali. *Sp. pl.* 25.
Common Butter-wort.
*Nat.* of Britain.
*Fl.* May.                                                H. ⚃.


UTRICULARIA.

Tab.1.Vol.1.Page.30.

Calceolaria Fothergillii.

Sowerby. del.

M<sup>c</sup>Kenzie sc.

## UTRICULARIA. *Gen. pl.* 31.

*Cor.* ringens, calcarata. *Cal.* 2-phyllus, æqualis.
*Capf.* unilocularis.

1. U. nectario conico, fcapo paucifloro. *Sp. pl.* 26.　　*vulgaris.*
Common Hooded-Milfoil.
*Nat.* of Britain.
*Fl.* June and July.　　　　　　　　　H. ♃.

2. U. nectario carinato. *Sp. pl.* 26.　　　　　*minor.*
Leffer Hooded-Milfoil.
*Nat.* of Britain.
*Fl.* June and July.　　　　　　　　　H. ♃.

## VERBENA. *Gen. pl.* 32.

*Cor.* infundibulif. fubæqualis, curva. *Calycis* unico
dente truncato. *Semina* 2. f. 4. nuda. (*Stam.* 2.
f. 4.)

1. V. diandra, fpicis longiffimis carnofis nudis, foliis lan-　*indica.*
ceolato-ovatis oblique dentatis, caule lævi. *Sp.*
*pl.* 27.
Indian Vervain.
*Nat.* of Ceylon.
*Cult.* 1732, by Mr. Philip Miller, *R. S. no.* 547.
*Fl.* Auguft and September.　　　　　　　S. ☉.

2. V. diandra, fpicis longiffimis carnofis nudis, foliis　*jamai-*
fpathulato-ovatis ferratis, caule hirto. *Sp. pl.* 27.　*cenfis.*
Jamaica Vervain.
*Nat.* of the Weft Indies.
　　　　　　　　　　　　　　　　*Cult.*

*Cult.* 1714, by the Duchefs of Beaufort.  *Br. muf.*
H. S. 136. *fol.* 45.
*Fl.* June——September.                               S. ♂.

mexica-
na.

3. V. diandra, fpicis laxis, calycibus fructus reflexis ro-
tundato-didymis hifpidis.  *Sp. pl.* 28.
Mexican Vervain.
*Nat.* of Mexico.
*Cult.* 1726, by James Sherard, M. D. *Dill. elth.* 407.
*t.* 302. *f.* 389.
*Fl.* Auguft and September.                          S. ♃.

nodiflora.

4. V. tetrandra, fpicis capitato-conicis, foliis ferratis,
caule repente.  *Sp. pl.* 28.
Creeping Vervain.
*Nat.* of Jamaica.
*Introd.* before 1733, by William Houftoun, M. D.
*Mill. dict. edit.* 8.
*Fl.* moft part of the year.                         S. ♃.

bonarien-
fis

5. V. tetrandra, fpicis fafciculatis, foliis lanceolatis ,am-
plexicaulibus.  *Sp. pl.* 28.
Clufter-flower'd Vervain.
*Nat.* of Buenos Ayres.
*Cult.* 1732, by James Sherard, M.D.  *Dill. elth.* 406.
*t.* 300. *f.* 387.
*Fl.* July——October.                                H. ♂.

baftata.

6. V. tetrandra, fpicis longis acuminatis, foliis haftatis.
*Sp. pl.* 29.
Halberd-leav'd Vervain.
*Nat.* of Canada.
*Cult.* 1731, by Mr. Philip Miller.  *Mill. dict. edit.* 1.
*no.* 4.
*Fl.* June—Auguft.                                   H. ♃.

7. V.

7. V. tetrandra, ſpicis filiformibus, foliis indiviſis lan-　*carolina.*
　ceolatis ſerratis obtuſiuſculis ſubſeſſilibus. *Spec.*
　*pl.* 29.
Carolina Vervain.
*Nat.* of North America.
*Cult.* 1732, by James Sherard, M. D. *Dill. elth.*
　407. *t.* 301. *f.* 388.
*Fl.* June——September. 　　　　　　　　H. ♃.

8. V. tetrandra, ſpicis filiformibus paniculatis, foliis in-　*urticifo-*
　diviſis ovatis ſerratis acutis petiolatis. *Sp. pl.* 29.　*lia.*
Nettle-leav'd Vervain.
*Nat.* of North America.
*Cult.* 1731, by Mr. Philip Miller. *Mill. dict. edit.*
　1. *n.* 3.
*Fl.* July——September. 　　　　　　　　H. ♃.

9. V. tetrandra, ſpicis laxis ſolitariis, foliis trifidis in-　*Aubletia.*
　ciſis.
Verbena Aubletia. *Jacqu. hort.* 2. *p.* 82. *t.* 176.
　*Retzius act. ſtockh.* 1773. *p.* 144. *t.* 5. *Meditus*
　*act. palat. vol.* 3. *phyſ. p.* 194. *t.* 7. *Linn. ſuppl.* 86.
Obletia. *Journ. de Rozier, introd.* 1. *p.* 367. *t.* 2.
Buchnera canadenſis. *Linn. mant.* 88.
Cut-leav'd Roſe Vervain.
*Nat.* of America.
*Introd.* 1774, by Monſ. Richard.
*Fl.* June and July. 　　　　　　　　H. ♂.

10. V. tetrandra, ſpicis filiformibus paniculatis, foliis　*officinalis.*
　multifido-laciniatis, caule ſolitario. *Sp. pl.* 29.
　*Curtis lond.*
Officinal Vervain.
*Nat.* of Britain.
*Fl.* June——September. 　　　　　　　H. ♂.

　　　　　D 　　　　　　　　11. V.

*fupina.*    11. V. tetrandra, fpicis filiformibus folitariis, foliis bi-
pinnatifidis.  *Sp. pl.* 29.
Trailing Vervain.
*Nat.* of Spain and Portugal.
*Cult.* 1640.  *Park. theat.* 675. *f.* 2.
*Fl.* June and July.               H. ☉.

### L Y C O P U S.  *Gen. pl.* 33.

*Cor.* 4-fida : lacinia unica emarginata.  *Stamina* dif-
tantia.  *Semina* 4, retufa.

*europæus.*    1. L. foliis finuato-ferratis.  *Sp. pl.* 30.  *Curtis lond.*
European Lycopus, or Water-Horehound.
*Nat.* of Britain.
*Fl.* July——September.            H. ♃.

*virgini-*    2. L. foliis æqualiter ferratis.  *Sp. pl.* 30.
*cus.*        Virginian Lycopus.
*Nat.* of Virginia.
*Cult.* 1760, by Mr. James Gordon, fen.
*Fl.* Auguft and September.        H. ♃.

### A M E T H Y S T E A.  *Gen. pl.* 34.

*Cor.* 5-fida : lacinia infima patentiore.  *Stamina* approximata.  *Calyx* fubcampanulatus.  *Sem.* 4, gibba.

*cærulea.*    1. AMETHYSTEA.  *Sp. pl.* 30.
Blue Amethyft.
*Nat.* of Siberia.
*Cult.* 1759, by Mr. Philip Miller.  *Mill. dict. edit.* 7.
*Fl.* June and July.             H. ☉.

**CUNILA.**

CUNILA. *Gen. pl.* 35.

*Cor.* ringens: labio fuperiore erecto, plano. *Filamenta* caftrata duo. *Semina* 4.

1. C. foliis ovatis ferratis, corymbis terminalibus dicho- *mariana.*
tomis. *Sp. pl.* 30.
Mint-leav'd Cunila.
*Nat.* of North America.
*Cult.* 1760, by Mr. James Gordon.
*Fl.* July——September. H. ♃.

2. C. foliis oblongis bidentatis, floribus verticillatis. *Syft.* *pulegioi-*
*veget.* 67. *des.*
Pennyroyal-leav'd Cunila.
*Nat.* of North America.
*Introd.* 1777, by Monf. Thouin.
*Fl.* Auguft. H. ☉.

ZIZIPHORA. *Gen. pl.* 36.

*Cor.* ringens: labio fuperiore reflexo, integro. *Cal.* filiformis. *Semina* 4.

1. Z. capitulis terminalibus, foliis ovatis. *Sp. pl.* 31. *capitata.*
Oval-leav'd Ziziphora.
*Nat.* of Syria.
*Cult.* 1752, by Mr. Phil. Miller. *Mill. dict. ed.* 6. *n.* 1.
*Fl.* July and Auguft. H. ☉.

2. Z. floribus lateralibus, foliis lanceolatis. *Sp. pl.* 31. *tenuior.*
Spear-leav'd Ziziphora.
*Nat.* of the Levant.
*Cult.* 1752, by Mr. Ph. Miller. *Mill. dict. edit.* 6 *n.* 2.
*Fl.* June and July. H. ☉.

*acinoides.*    3. Z. floribus lateralibus, foliis ovatis.   *Sp. pl.* 31.
Thyme-leav'd Ziziphora.
*Nat.* of Siberia.
*Introd.* 1786, by William Pitcairn, M. D.
*Fl.* July and Auguft.        H. ♃.

### MONARDA.   *Gen. pl.* 37.

*Cor.* inæqualis : labio fuperiore lineari filamenta invol-
vente.   *Semina* 4.

*fiftulofa.*    1. M. foliis oblongo-lanceolatis cordatis villofis planis.
Monarda fiftulofa.   *Sp. pl.* 32.
Purple Monarda.
*Nat.* of Canada.
*Cult.* 1656, by Mr. John Tradefcant, jun.   *Muf.*
*Trad.* 148.
*Fl.* June——Auguft.        H. ♃.

*oblongata.*    2. M. foliis oblongo-lanceolatis bafi rotundato-attenuatis
villofis planis.
Long-leav'd Monarda.
*Nat.* of North America.
*Cult.* 1761, by Mr. James Gordon.
*Fl.* July——September.        H. ♃.

*didyma.*    3. M. floribus capitatis fubdidynamis, caule acutangulo.
*Sp. pl.* 32.
Scarlet Monarda, or Ofwego-tea.
*Nat.* of North America.
*Introd.* 1755, by Peter Collinfon, Efq.   *Coll. mfcr.*
*Fl.* June——Auguft.        H. ♃.

*rugofa.*    4. M. foliis ovato-lanceolatis cordatis glabris rugofis.
White Monarda.
*Nat.* of North America

                                 *Cult.*

*Cult.* 1761, by Mr. James Gordon.
*Fl.* July——September. H. ♃.

5. M. floribus verticillatis, corollis punctatis, bracteis co-  *punctata,*
   loratis. *Syst. veget.* 68.
   Spotted Monarda.
   *Nat.* of Maryland and Virginia.
   *Cult.* 1714, by Mr. Thomas Fairchild. *Philosoph.*
   *transf. n.* 346. *p.* 358. *no.* 109.
   *Fl.* June——October. H. ♂,

   R O S M A R I N U S. *Gen. pl.* 38.
   *Cor.* inæqualis: labio superiore bipartito. *Filamenta*
   longa, curva, simplicia cum dente.

1. ROSMARINUS. *Sp. pl.* 33.  *officinalis.*
   Common Rosemary.
   *Nat.* of the South of Europe and the Levant.
   *Cult.* 1596, by Mr. John Gerard. *Hort. Ger.*
   *Fl.* January——May. H. ♄.

   S A L V I A. *Gen. pl.* 39.
   *Cor.* inæqualis. *Filamenta* transverse pedicello affixa.

1. S. foliis lanceolatis denticulatis, floribus pedunculatis.  *ægyptia-*
   *Syst. veget.* 68. *Jacqu. hort.* 2. *p.* 49. *t.* 108.  *ca.*
   Egyptian Sage.
   *Nat.* of Egypt and the Canary Islands.
   *Introd.* 1770, by Monsf. Richard.
   *Fl.* June and July. H. ☉.

2. S. foliis lineari-oblongis dentato-pinnatifidis, verticillis  *dentata.*
   bifloris, laciniis calycinis obtusis.
   Tooth-leav'd Sage.
   *Nat.* of the Cape of Good Hope. Mr. *Fr. Masson.*
   D 3  *Introd.*

*Introd.* 1774.

*Fl.* December and January.                                 G.H. ♄.

*cretica.*   3. S. foliis lanceolatis, calycibus diphyllis.   *Sp. pl.* 33.
             Cretan Sage.
             *Nat.* of the Ifland of Candia.
             *Cult.* 1760, by Mr. James Gordon.
             *Fl.* June——Auguft.                              H. ♄.

*lyrata.*    4. S. foliis radicalibus lyratis dentatis, corollarum galea
                breviffima.  *Sp. pl.* 33.
             Horminum virginicum.  *Sp. pl.* 832.
             Lyre-leav'd Sage.
             *Nat.* of Virginia and Carolina.
             *Cult.* 1728, by James Sherard, M.D.   *Dill elth.* 219.
             *Fl.* June——Auguft.                              H. ♃.

*officinalis.*  5. S. foliis lanceolato-ovatis integris crenulatis, floribus
                fpicatis, calycibus acutis.  *Sp. pl.* 34.
             Garden Sage.
             *Nat.* of the South of Europe.
             *Cult.* 1597, by Mr. John Gerard.  *Ger. herb.* 623. *f.* 1.
             *Fl.* June and July.                             H. ♄.

*triloba.*   6. S. tomentofa, foliis petiolatis rugofiffimis trilobis : lo-
                bo intermedio producto oblongo ; lateralibus ovatis
                obtufis.  *Linn. fuppl.* 88.
             Three-lobed Sage.
             *Nat.* of the South of Europe.
             *Cult.* 1597, by Mr. John Gerard.  *Ger. herb.* 623. *f.* 2.
             *Fl.* June and July.                             H. ♄.

*pomifera.*  7. S. foliis lanceolato-ovatis integris crenulatis, floribus
                fpicatis, calycibus obtufis.  *Sp. pl.* 34.
             Apple-bearing Sage.
             *Nat.* of the Ifland of Candia.

                                                             *Cult.*

*Cult.* 1699, by the Hon. Charles Howard. *Morif. hift.*
3. *p.* 399. *no.* 4.
*Fl.* July and Auguſt.                    H. ♃.

8. S. foliis oblongis crenatis, corollarum galea ſemi-orbi-   *viridis.*
culata, calycibus fructiferis reflexis.   *Sp. pl.* 34.
*Jacqu. ic. miſcell.* 2. *p.* 366.
Green-topp'd Sage.
*Nat.* of Italy.
*Introd.* 1776, by Monſ. Thouin.
*Fl.* July and Auguſt.                    H. ☉.

9. S. foliis obtuſis crenatis, bracteis ſummis ſterilibus   *Hormi-*
majoribus coloratis.   *Sp. pl.* 34.                    *num.*
α coma violacea.
Purple-topp'd Sage.
β coma rubra.
Red-topp'd Sage.
*Nat.* of the South of Europe.
*Cult.* 1597, by Mr. John Gerard. *Ger. herb.* 628. *f.* 2.
*Fl.* June and July.                    H. ☉.

10. S. foliis oblongis cordatis rugoſis crenatis, pilis caulis   *virgata.*
calyciſque apice glanduloſis.
Salvia virgata. *Jacqu. hort.* 1. *p.* 14. *t.* 37. *Syſt. ve-*
*get.* 70.
Long-branch'd Sage.
*Nat.*
*Cult.* 1758, by Mr. Philip Miller.
*Fl.* July——November.                    H. ♃.

11. S. foliis cordatis rugoſis biſerratis, bracteis coloratis   *ſylveſtris.*
flore brevioribus acuminatis, pilis caulis calyciſque
ſimplicibus.
                    D 4                    Salvia

Salvia fylveftris.  *Sp. pl.* 34.  *Jacqu. auftr.* 3. *p.* 7.
  *t.* 212.
Spotted-ftalk'd Bohemian Sage.
*Nat.* of Auftria and Bohemia.
*Cult.* 1759, by Mr. Ph. Miller.  *Mill. dict. edit.* 7.
  Sclarea 8.
*Fl.* June——October.                          H. ♃.

*nemorofa.*  12. S. foliis cordato-lanceolatis ferratis planis, bracteis
  coloratis, corollæ labio infimo reflexo. *Sp. pl.* 35.
  Spear-leav'd Sage.
  *Nat.* of Auftria and Tartary.
  *Cult.* 1728, by Mr. Ph. Miller.  *R. S. no.* 342.
  *Fl.* June——October.                          H. ♃.

*fyriaca.*  13. S. foliis cordatis dentatis : inferioribus repandis,
  bracteis cordatis brevibus acutis, calycibus tomen-
  tofis. *Syft. veget.* 69.
  Syrian Sage.
  *Nat.* of the Levant.
  *Cult.* 1759, by Mr. Ph. Miller. *Mill. dict. edit.* 7.
  Sclarea 6.
  *Fl.* July.                                   G. H. ♄.

*vifcofa.*  14. S. foliis oblongis obtufis erofo-crenatis vifcidis, floribus
  verticillatis, bracteis cordatis acutis.  *Jacqu. ic.*
  *mifcell.* 2. *p.* 328.
  Clammy Sage.
  *Nat.* of Italy.
  *Introd.* 1773, by John Earl of Bute.
  *Fl.* May and June.                           H. ♃.

*pratenfis.*  15. S. foliis cordato-oblongis crenatis : fummis amplexicaulibus, verticillis fubnudis, corollis galea
  glutinofis. *Syft. veget.* 69.
  Meadow Sage, or Clary.

                                               *Nat.*

*Nat.* of England.
*Fl.* May——November.                    H. ♃.

16. S. foliis cordatis lateribus fublobatis; fummis feffi-   *indica.*
    libus, verticillis fubnudis remotiffimis. *Sp. pl.* 37.
    *mant.* 318.   *Jacqu. hort.* 1. *p.* 33. *t.* 78.
    Indian Sage.
    *Nat.* of India.
    *Cult.* 1731, by Mr. Ph. Miller. *Mill. dict. edit.* 1.
    Sclarea 5.
    *Fl.* May——July.                    H. ♃.

17. S. foliis ferratis finuatis læviufculis, corollis calyce   *verbena-*
    anguftioribus. *Sp. pl.* 35.                *ca,*
    Vervain Sage, or Clary.
    *Nat.* of Britain.
    *Fl.* June——October.                H. ♂.♃.

18. S. fcabra, foliis lyratis dentatis rugofis, caule pani-   *fcabra.*
    culato ramofo. *Linn. fuppl.* 89.
    Rough-leav'd Sage.
    *Nat.* of the Cape of Good Hope.   Mr. *Fr. Maſſon.*
    *Introd.* 1774.
    *Fl.* moft part of the Summer.           G. H. ♄.

19. S. foliis ferratis pinnatifidis rugofiſſimis, fpica obtufa,   *clandefti-*
    corollis calyce anguftioribus. *Sp. pl.* 36.         *na.*
    Cut-leav'd Sage.
    *Nat.* of Italy.
    *Cult.* 1-68, by Mr. Ph. Miller. *Mill. dict. edit.* 8.
    Horminum 2.
    *Fl.* May——July.                    H. ♂.

20. S. foliis ovatis cordatifque erofo-finuatis; radicalibus   *auftriaca.*
                                     petiolatis,

petiolatis, caule fubaphyllo, ftaminibus corolla duplo longioribus.

Salvia auftriaca. *Jacqu. auftr.* 2. *p.* 8. *t.* 112. *Syft.
veget.* 69.

Auftrian Sage.

*Nat.* of Auftria.

*Introd.* 1776, by Jofeph Nicholas de Jacquin, M.D.

*Fl.* June and July.                                    H. ♃.

*difermas.*   21. S. foliis cordato-oblongis erofis, ftaminibus corollam
æquantibus. *Sp. pl.* 36.

Long-fpiked Sage.

*Nat.* of Syria.

*Introd.* 1773, by Chevalier Murray.

*Fl.* July.                                             H. ♃.

*rugofa.*   22. S. foliis cordatis oblongo-lanceolatis erofo-crenatis
rugofis pilofiufculis, ftaminibus corolla brevioribus.

Wrinkle-leav'd Sage.

*Nat.* of the Cape of Good Hope. Mr. *Fr. Maffon.*

*Introd.* 1775.

*Fl.* July and Auguft.                                 G. H. ♄.

OBS.   Valde affinis Salviæ difermas; vix differt nifi
ftaminibus brevibus faucem tantummodo æquantibus. *Flores* albi.

*nubia.*   23. S. foliis oblongis .fubcordatis inæquilateralibus rugofis crenatis bafi fubauritis.

Salvia nubia. *Murray comm. gotting.* 1778. *pag.* 90.
*tab.* 3. *Syft. vegetab.* 70.

Nubian Sage.

*Nat.* of Africa.

*Introd.* 1784, by Monf. Thouin.

*Fl.* June and July.                                   G. H. ♃.

*mexicana.*   24. S. foliis ovatis utrinque acuminatis ferratis. *Sp. pl.* 37.

Mexican

Mexican Sage.
*Nat.* of Mexico.
*Cult.* 1724, by James Sherard, M. D. *Dill. elth.*
339. *t.* 254. *f.* 330.
*Fl.* May——July                    G. H. ♄.

25. S. foliis fubcordatis, corollarum galea barbata, caly-    *formofa.*
cibus trilobis, caule frutefcente. *L'Heritier ftirp.*
*nov. p.* 41. *t.* 21.
Salvia Leonuroides. *Gloxin obf. bot. p.* 15. *tab.* 2.
Shining-leav'd Sage.
*Nat.* of Peru.
*Introd.* 1783, by Monf. Thouin.
*Fl.* moft part of the Summer.              G. H. ♄.

26. S. foliis cordatis acutis tomentofis ferratis, corollis    *coccinea.*
calyce duplo longioribus anguftioribus.
Salvia coccinea. *Linn. fuppl.* 88. (exclufo loco natali.)
*Murray commentat. gotting.* 1778. *p.* 86. *tab.* 1.
Scarlet-flower'd Sage.
*Nat.* of Eaft Florida. Mr. *John Bartram.*
*Cult.* 1774.
*Fl.* moft part of the Year.                   S. ♄.

27. S. foliis ovatis, petiolis utrinque mucronatis, fpicis    *hifpanica.*
imbricatis, calycibus trifidis. *Sp. pl.* 37.
Spanifh Sage.
*Nat.* of Spain and Italy.
*Cult.* 1739, by Mr. Ph. Miller. *Rand. chel.* Horminum 1.
*Fl.* June and July.                     H. ☉.

28. S. foliis inferioribus lyratis; fummis cordatis, floribus    *abyffinica.*
verticillatis, calycibus mucronatis ciliatis. *Jacqu.*
*ic. coll.* 1. *p.* 132.
                                        Abyffinian

Abyffinian Sage.
*Nat.* of Africa.
*Introd.* 1775, by James Bruce, Efq.
*Fl.* June and July. G.H. ♃

*verticil-*
*lata.*

29. S. foliis cordatis crenato-dentatis, verticillis fubnudis,
ftylo corollæ labio inferiori incumbente. *Syft.*
*veget.* 70.
Whorl-flower'd Sage.
*Nat.* of Germany.
*Cult.* 1683, by Mr. James Sutherland. *Sutherl.*
*hort. edin.* 156.
*Fl.* June——November. H. ♃

*napifolia.*

30. S. foliis cordatis crenato-dentatis : inferioribus haf-
tatis lyratifque, verticillis fubnudis, labio fuperiore
breviore.
Salvia napifolia. *Jacqu. hort.* 2. *p.* 71. *tab.* 152.
*Syft. veget.* 70.
Rape-leav'd Sage.
*Nat.*
*Introd.* 1776, by Jofeph Nicholas de Jacquin, M. D.
*Fl.* June and July. H. ♃

*glutinofa.*

31. S. foliis cordato-fagittatis ferratis acutis. *Sp. pl.* 37.
Yellow Sage, or Clary.
*Nat.* of Germany and Italy.
*Cult.* 1596, by Mr. John Gerard. *Hort. Ger.*
*Fl.* June——November. H. ♃

*canarien-*
*fis.*

32. S. foliis haftato-triangularibus oblongis crenatis ob-
tufis. *Sp. pl.* 38.
Canary Sage.
*Nat.* of the Canary Iflands.

*Cult.*

*Cult.* 1697, by the Duchefs of Beaufort. *Br. Muf.*
*Sloan. mfs.* 3357. *fol.* 62.
*Fl.* June——September. G. H. ♄.

33. S. foliis fubrotundis ferratis: bafi truncatis dentatis. *africana.*
*Sp. pl.* 38.
Blue-flower'd African Sage.
*Nat.* of the Cape of Good Hope.
*Cult.* 1739, by Mr. Ph. Miller. *Rand. chel. n.* 11.
*Fl.* April——June. G. H. ♄.

34. S. foliis fubrotundis integerrimis: bafi truncatis den- *aurea.*
tatis. *Sp. pl.* 38.
Gold-flower'd African Sage.
*Nat.* of the Cape of Good Hope.
*Cult.* 1731, by Mr. Ph. Miller. *Mill. dict. edit.* 1. *n.* 13.
*Fl.* May——November. G. H. ♄.

35. S. foliis obovato-cuneiformibus denticulatis nudis, *panicula-*
caule frutefcente. *Linn. mant.* 25. *ta.*
Panicled Sage.
*Nat.* of the Cape of Good Hope.
*Cult.* 1758, by Mr. Ph. Miller. *Mill. ic.* 150. *t.*
225. *f.* 1.
*Fl.* June——September. G. H. ♄.

36. S. foliis rugofis cordatis oblongis villofis ferratis, *Sclarea.*
bracteis coloratis calyce longioribus concavis acu-
minatis. *Syft. veget.* 71.
Common Clary.
*Nat.* of Syria and Italy.
*Cult.* 1562. *Turn. herb. part* 2. *fol.* 70.
*Fl.* July——September. H. ♂.

37. S.

*cerato-*
*phylla.*

37. S. foliis rugofis pinnatifidis lanatis, verticillis fummis
sterilibus.  *Sp. pl.* 39.
Horn-leav'd Sage.
*Nat.* of Perfia.
*Cult.* before 1699, by Jacob Bobart.  *Morif. hift. 3.*
*p.* 393. *no.* 6. *f.* 11. *t.* 13. *f.* 6.
*Fl.* July and Auguft.                                     H. ♂.

*Æthiopis.*

38. S. foliis oblongis erofis lanatis, verticillis lanatis, co-
rollæ labio crenato, bracteis recurvatis fubfpinofis.
*Syft. veget.* 71.  *Jacqu. auftr.* 3. *p.* 7. *t.* 211.
Woolly Sage, or Clary.
*Nat.* of Auftria.
*Cult.* 1570.  *Lobel. adv.* 242.
*Fl.* May and June.                                        H. ♂.

*pinnata.*

39. S. foliis lyrato-pinnatis.  *Syft. veget.* 71.
Winged-leav'd Sage.
*Nat.* of the Levant.
*Cult.* 1739, in Chelfea garden.  *Rand. chelf.* 176. *n.* 9.
*Fl.* July.                                                H. ♂.

*argentea.*

40. S. foliis oblongis dentato-angulatis lanatis, verticillis
fummis sterilibus, bracteis concavis.  *Sp. pl.* 38.
Silvery-leav'd Sage, or Clary.
*Nat.* of the Ifland of Candia.
*Cult.* 1768, by Mr. Ph. Miller.  *Mill. dict. edit.* 8.
Sclarea 15.
*Fl.* May——Auguft.                                         H. ♂.

*cerato-*
*phylloides.*

41. S. foliis pinnatifidis rugofis villofis, caule paniculato
ramofiffimo.  *Linn. mant.* 26.
Branchy Sage.
*Nat.* of Egypt.

*Introd.*

*Introd.* 1771, by Monſ. Richard.
*Fl.* June——Auguſt.                                    H. ♂.

42. S. foliis cordatis inæqualiter baſi exciſis, caule nudo,   *nutans.*
   ſpicis ante floreſcentiam cernuis. *Syſt. veget.* 72.
   Nodding Sage.
   *Nat.* of Ruſſia.
   *Introd.* 1780, by Peter Simon Pallas, M.D.
   *Fl.* June——September.                           H. ♃.


   COLLINSONIA. *Gen. pl.* 40.

*Cor.* inæqualis : labio inferiore multifido, capillari.
            *Sem.* 1, perfectum.

1. C. foliis ovatis caulibuſque glabris.                  *canaden-*
   Collinſonia canadenſis. *Sp. pl.* 39.                   *ſis.*
   Nettle-leav'd Collinſonia.
   *Nat.* of North America.
   *Introd.* 1735, by Peter Collinſon, Eſq. *Coll. mſcr.*
   *Fl.* Auguſt——October.                            H. ♃.

2. C. foliis ovatis ſubcordatis piloſiuſculis, caule piloſiuſ-   *ſcabriuſ-*
   culo ſcabrido.                                         *cula.*
   Rough-ſtalk'd Collinſonia.
   *Nat.* of Eaſt Florida.   Mr. *John Bartram.*
   *Cult.* 1776, by John Fothergill, M. D.
   *Fl.*                                            G. H. ♃.


                    *DIGYNIA.*

# DIGYNIA.

## ANTHOXANTHUM. *Gen. pl.* 42.

*Cal.* Gluma 2-valvis, 1-flora. *Cor.* Gluma 2-valvis, acuminata, ariftata.

*odoratum.* 1. A. fpica ovato-oblonga, flofculis fubpedunculatis arif-
ta longioribus. *Sp. pl.* 40. *Curtis lond.*
Sweet-fcented Spring Grafs.
*Nat.* of Britain.
*Fl.* May. H. ♃.

## CRYPSIS.

*Cal.* Gluma 2-valvis, 1-flora. *Cor.* Gluma 2-valvis, mutica.

*aculeata.* 1. CRYPSIS.
*a* Schoenus aculeatus. *Sp. pl.* 63. *Schreb. gram.* 2. *p.* 62. *t.* 32.
Phleum fchoenoides. *Jacqu. auftr.* 5. *p.* 29. *t. app.* 7.
Anthoxanthum aculeatum. *Linn. fuppl.* 89.
*β* Phleum fchoenoides. *Sp. pl.* 88. *Jacqu. ic. collect.* 1. *p.* 111.
Prickly Crypfis.
*Nat.* of the South of Europe.
*Introd.* 1783, by Monf. Thouin.
*Fl.* Auguft. H. ☉.

*TRIGYNIA.*

# *TRIGYNIA.*

## PIPER. *Gen. pl.* 43.

*Cal.* o. *Cor.* o. *Bacca* monofperma.

1. P. foliis lanceolato-ovatis quinquenerviis rugofis. *Sp.*    *Amalago.*
   *pl.* 41.
   Rough-leav'd Pepper.
   *Nat.* of Jamaica.
   *Cult.* 1759, by Mr. Ph. Miller. *Mill. dict. edit.* 7. *n.* 3.
   *Fl.*                           S. ♄.

2. P. foliis cordatis petiolatis, caule herbaceo. *Sp. pl.* 42.    *pelluci-*
   Shining-leav'd Pepper.                            *dum.*
   *Nat.* of South America.
   *Cult.* 1759, by Mr. Philip Miller. *Mill. dict. edit.* 7.
   *n.* 2.
   *Fl.* April——September.             S. ☉.

3. P. foliis obovatis enerviis. *Sp. pl.* 42.        *obtufifi-*
   Blunt-leav'd Pepper.                         *lium.*
   *Nat.* of the Weft Indies.
   *Cult.* 1739, by Mr. Philip Miller. *Rand. chel.* Sau-
   rurus.
   *Fl.* April——September.             S. ♃.

4. P. foliis verticillatis rhombeo-ovatis integerrimis pe- *polyfta-*
   tiolatis trinerviis pubefcentibus.                *chyon.*
   Piper obtufifolium. *Jacqu. ic. collect.* 1. *p.* 141.
   Many-fpiked Pepper.
   *Nat.* of Jamaica.

E

Introd.

*Introd.* 1775, by John Fothergill, M. D.
*Fl.*                                           S. ♃.

*pulchel-*    5. P. foliis quaternis fubfeffilibus oblongis enerviis inte-
*lum.*              gerrimis, fpicis terminalibus.
                   Small-leav'd Pepper.
                   *Nat.* of Jamaica.   *Thomas Clark,* M. D.
                   *Introd.* 1778, by Mr. William Forfyth.
                   *Fl.* July——September.                S. ♃.

*Claſſis*

*Claſſis III.*

# T R I A N D R I A

## *MONOGYNIA.*

**VALERIANA.** *Gen. pl.* 44.

*Cal.* o. *Cor.* 1-petala, baſi hinc gibba, ſupera. *Sem.* 1.

1. V. floribus monandris caudatis, foliis lanceolatis in-    *rubra.*
    tegerrimis. *Sp. pl.* 44.
    Red Valerian.
    *Nat.* of France and Italy.
    *Cult.* 1597, by Mr. John Gerard. *Ger. herb.* 550. *f.* 1.
    *Fl.* June——October.                                         H. ♃.

2. V. floribus monandris, foliis pinnatifidis. *Sp. pl.* 44.    *calcitra-*
    Cut-leav'd Valerian.                                         *pa.*
    *Nat.* of Portugal.
    *Cult.* 1683, by Mr. James Sutherland. *Sutherl. hort.*
    *edin.* 348. *no.* 4.
    *Fl.* May——July.                                           H. ☉.

3. V. floribus diandris ringentibus, foliis ovatis ſeſſili-    *cornuco-*
    bus. *Sp. pl.* 44.                                           *piæ.*
    Purple Valerian.
    *Nat.* of Barbary, Spain, and Sicily.
    *Cult.* 1596, by Mr. John Gerard. *Hort. Ger.*
    *Fl.* May——Auguſt.                                         H. ☉.

4. V. floribus triandris dioicis, foliis pinnatis integerrimis.  *dioica.*
    *Sp. pl.* 44. *Curtis lond.*

E 2                              Marſh

Marſh Valerian.
*Nat.* of Britain.
*Fl.* May and June.                                      H. ♃.

*officinalis.*  5. V. floribus triandris, foliis omnibus pinnatis.  *Sp.*
*pl.* 45.
Officinal Valerian.
*Nat.* of Britain.
*Fl.* June——September.                              H. ♃.

*Phu.*  6. V. floribus triandris, foliis caulinis pinnatis; radicali-
bus indiviſis.  *Sp. pl.* 45.
Garden Valerian.
*Nat.* of Germany.
*Cult.* 1597.   *Ger. herb.* 917. *f.* 1.
*Fl.* May——July.                                      H. ♃.

*tripteris.*  7. V. floribus triandris, foliis dentatis : radicalibus cor-
datis, caulinis ternatis ovato-oblongis.  *Sp. pl.* 45.
*Jacqu. auſtr.* 3. *p.* 38. *t.* 268.
Three-leav'd Valerian.
*Nat.* of the Alps of Switzerland.
*Cult.* 1739, by Mr. Philip Miller.  *Mill. dict. vol.* 2.
*no.* 2.
*Fl.* March——May.                                    H. ♃.

*montana.*  8. V. floribus triandris, foliis ovato-oblongis ſubdenta-
tis, caule ſimplici.  *Sp. pl.* 45.  *Jacqu. auſtr.* 3.
*p.* 38. *t.* 269.
Mountain Valerian.
*Nat.* of Auſtria and Switzerland.
*Cult.* 1739, by Mr. P. Miller.  *Mill. dict. vol.* 2. *no.* 4.
*Fl.* June and July.                                   H. ♃.

*tuberoſa.*  9. V. floribus triandris, foliis radicalibus lanceolatis integerrimis; reliquis pinnatifidis.  *Syſt. veget.* 80.
Tuberous-

Tuberous-rooted Valerian.

*Nat.* of the South of Europe.

*Cult.* 1739, by Mr. Philip Miller. *Rand. chel. no.* 6.

*Fl.* May and June. H. ♃.

10. V. floribus triandris, foliis caulinis cordatis ferratis *pyrenaica.*
petiolatis: fummis ternatis. *Sp. pl.* 46.

Pyrenean Valerian.

*Nat.* of the Pyrenees.

*Cult.* 1692, by Charles Dubois, Efq. *Pluk. phyt.*
t. 232. f. 1.

*Fl.* May and June. H. ♃.

11. V. floribus triandris, caule dichotomo, foliis linea-  *Locufta.*
ribus. *Sp. pl.* 47. *Curtis lond.*

α Valeriana caule dichotomo, foliis lanceolatis integris  olitoria.
fructu fimplici. *Hort. cliff.* 16.

Corn Valerian, or Lamb's-Lettuce.

β Valeriana caule dichotomo, foliis lanceolatis ferratis,  vefica-
fructu inflato. *Hort. cliff.* 16.  ria.

Bladder-cupp'd Corn Valerian.

γ Valeriana caule dichotomo, foliis lanceolatis dentatis,  coronata.
fructu fexdentato. *Hort. cliff.* 16.

Coronate Corn Valerian.

*Nat.* of Britain, β. of the Ifland of Candia, γ. of Portugal.

*Fl.* April——June. H. ☉.

12. V. floribus tetrandris æqualibus, foliis pinnatifidis,  *fibirica.*
feminibus paleæ ovali adnatis. *Sp. pl.* 48.

Siberian Valerian.

*Nat.* of Siberia.

*Cult.* 1759, by Mr. Philip Miller. *Mill. dict. edit.* 7.
no. 8.

*Fl.* May and June. H. ☉.

E 3 TAMA-

## TAMARINDUS. *Gen. pl.* 46.

*Cal.* 4-partitus. *Petala* 3. *Nectarium* fetis 2 brevibus fub filamentis. *Legumen* pulpofum.

*indica.*     1. TAMARINDUS. *Sp. pl.* 48.
Tamarind-tree.
*Nat.* of Egypt, and both Indies.
*Cult.* before 1633, by Mr. Tuggy. *Ger. emac.* 1607.
*Fl.* June and July.       S. ♄.

## CNEORUM. *Gen. pl.* 48.

*Cal.* 3-dentatus. *Petala* 3, æqualia. *Bacca* 3-cocca.

*tricoccum.*     1. CNEORUM. *Sp. pl.* 49.
Widow-wail.
*Nat.* of the South of France and Spain.
*Cult.* 1596, by Mr. John Gerard. *Hort. Ger.*
*Fl.* May——September.       G. H. ♄.

## COMOCLADIA. *Gen. pl.* 49.

*Cal.* 3-partitus. *Cor.* 3-partita. *Drupa* oblonga: nucleo bilobo.

*integrifolia.*     1. C. foliolis integris. *Sp. pl.* 49.
Intire-leav'd Maiden Plumb.
*Nat.* of Jamaica.
*Introd.* 1778, by Mr. William Forfyth.
*Fl.*       S. ♄.

## MELOTHRIA. *Gen. pl.* 50.

*Cal.* 5-fidus. *Cor.* campanulata, 1-petala. *Bacca* 3-locularis, polyfperma.

*pendula.*     1. MELOTHRIA. *Sp. pl.* 49.
Pendulous Melothria.

      *Nat.*

*Nat.* of America.
*Cult.* 1759, by Mr. Philip Miller. *Mill. dict. edit.* 7.
*Fl.* June and July. S. ⊙.

## ORTEGIA. *Gen. pl.* 51.

*Cal.* 5-phyllus. *Cor.* o. *Capf.* 1-locularis. *Sem.* plurima.

1. O. floribus fubverticillatis, caule fimplici. *Syft.* *hifpanica.*
veget. 82.
Spanifh Ortegia.
*Nat.* of Spain.
*Cult.* 1768, by Mr. Philip Miller. *Mill. dict. edit.* 8.
*Fl.* June and July. H. ♃.

2. O. floribus folitariis axillaribus, caule dichotomo. *Syft.* *dichoto-*
veget. 82. *ma.*
Ortegia dichotoma. *Allioni in mifcell. taurinenf.* 3.
*p.* 176. *t.* 4. *f.* 1.
Fork'd Ortegia.
*Nat.* of Italy.
*Introd.* 1781, by Monf. Thouin.
*Fl.* Auguft and September. H. ♃.

## LOEFLINGIA. *Gen. pl.* 52.

*Cal.* 5-phyllus. *Cor.* 5-petala, minima. *Capf.* 1-lo-
cularis, 3-valvis.

1. LOEFLINGIA. *Sp. pl.* 50. *hifpanica.*
Spanifh Loeflingia.
*Nat.* of Spain.
*Introd.* 1770, by Monf. Richard.
*Fl.* June. H. ⊙.

E 4. POLYC-

POLYCNEMUM. *Gen. pl.* 53.

*Cal.* 3-phyllus.   *Petala* 5, calyciformia.   *Sem.* 1, subnudum.

*arvense.*    1. POLYCNEMUM.   *Sp. pl.* 50.   *Jacqu. austr.* 4. *p.* 34. *t.* 365.
Trailing Polycnenum.
*Nat.* of Italy, France, and Germany.
*Cult.* 1758, by Mr. Philip Miller.
*Fl.* July.                                             H. ☉.

C R O C U S.   *Gen. pl.* 55.

*Cor.* 6-partita, æqualis.   *Stigmata* convoluta.

*sativus.*    1. C. spatha univalvi radicali, corollæ tubo longissimo. *Sp. pl.* 50.

officina-   *a* Crocus autumnalis, foliis angustioribus margine re-
lis.              volutis. *Syst. veget.* 83.
Saffron Crocus.

vernus.    *β* Crocus vernalis, foliis latioribus margine patulo. *Syst. veget.* 83. *Curtis magaz.* 45. *Jacqu. austr.* 5. *p.* 47. *t. app.* 36.
Spring Crocus.
*Nat.* of England.
*Fl. a.* October, *β.* February.                       H. ♃.

I X I A.   *Gen. pl.* 56.

*Cor.* 6-partita, campanulata, regularis.   *Stigmata* 3.

*rosea.*    1. I. scapo unifloro aphyllo brevissimo. *Syst. veget. ed.* 13. *p.* 75.
Rose-colour'd Ixia.
*Nat.* of the Cape of Good Hope.
*Cult.* 1758, by Mr. P. Miller. *Mill. ic.* 160. *t.* 240.
*Fl.* May.                                             G. H. ♃.

*Bulboco-*   2. I. scapo unifloro brevissimo, foliis angulatis caulinis,
*dium.*          stigmatibus sextuplicibus. *Syst. veget. ed.* 13. *p.* 76.
Thunb. Ixia, *n.* 3.

Crocus-

Crocus-leav'd Ixia.
*Nat.* of the Alps of Italy.
*Cult.* 1739, by Mr. Philip Miller. *Rand. chel.* Bulbocodium.
*Fl.* March and April. H. ♃.

3. I. floribus racemosis, bracteis integris, foliis ensifor- *aulica.*
mibus planis nervosis lævibus.
Clufter-flower'd Ixia.
*Nat.* of the Cape of Good Hope. Mr. *Francis Maffon.*
*Introd.* 1774.
*Fl.* April. G. H. ♃.

4. I. foliis linearibus, axillis bulbiferis, floribus alternis, *bulbifera.*
ftaminibus lateralibus. *Sp. pl.* 51. *Thunb. Ixia,*
*n.* 17.
Bulb-bearing Ixia.
*Nat.* of the Cape of Good Hope.
*Cult.* 1758, by Mr. Philip Miller. *Mill. ic.* 158.
*t.* 236. *f.* 2.
*Fl.* May and June. G. H. ♃.

5. I. foliis ensiformibus glabris, floribus alternis feffili- *ariftate.*
bus, fpathis longitudine tubi laceris.
Ixia ariftata. *Thunb. Ixia, n.* 15.
α corollis purpureis.
Ixia uniflora. *Linn. mant.* 27.
Ixia grandiflora. *Houtt. hift. nat.* 12. *p.* 29. *tab.* 77.
*fig.* 3. *de la Roche pl. nov. p.* 23.
Purple-flower'd bearded Ixia.
β corollis violaceis: laciniis margine ftramineis.
Ixia foliis gladiolatis nervofis, fpatha lacera. *Mill.*
*ic.* 158. *t.* 237. *f.* 1. 2.
Violet-flower'd bearded Ixia.

Cult.

Cult. 1758, by Mr. Philip Miller. *Mill. ic. loc. cit.*
Fl. April. G. H. ♃

*villofa.* 6. I. foliis oblongo-lanceolatis acutis villofis fubplicatis diftichis, tubo fpathæ æquali.
Dark red Ixia.
*Nat.* of the Cape of Good Hope.
*Introd.* 1778, by Patrick Ruffell, M. D.
Fl. Auguft. G. H. ♃

*flexuofa.* 7. I. foliis linearibus, racemo flexuofo multifloro. *Sp. pl.* 51.
Bending-ftalk'd Ixia.
*Nat.* of the Cape of Good Hope.
*Cult.* 1757, by Mr. Philip Miller. *Mill. ic.* 104. *t.* 156. *f.* 2.
Fl. May. G. H. ♃

*polyfta-* 8. I. foliis linearibus, fcapo fpicis pluribus. *Sp. pl.* 51.
*chia.* Ixia erecta. *Thunb. Ixia, n.* 18.
Many-fpik'd Ixia.
*Nat.* of the Cape of Good Hope.
*Cult.* 1757, by Mr. Philip Miller. *Mill. ic.* 104. *t.* 155. *f.* 2.
*Fl.* May and June. G. H. ♃

*longiflo-* 9. I. foliis enfiformi-linearibus ftrictis, tubo filiformi
*ra.* longiffimo.
Ixia longiflora. *Berg. cap.* 7.
Ixia paniculata. *De la Roche pl. nov.* 26. *t.* 1.
Gladiolus longiflorus. *Linn. fuppl.* 96. *Thunb. Gla-diolus, n.* 22.
Long-flower'd Ixia.
*Nat.* of the Cape of Good Hope. Mr. *Francis Maffon. Introd.*

*Introd.* 1774.
*Fl.* April——June.                                     G. H. ♃.


10. I. foliis linearibus ſtriƈtis, ſpica diſticha imbricata.      *plantagi-*
   Gladiolus alopecuroides.   *Sp. pl.* 54.   *Thunb. Gla-*   *nea.*
      *diolus, n.* 14.
   Fox-tail Ixia.
   *Nat.* of the Cape of Good Hope.
   *Introd.* 1774, by Mr. Francis Maſſon.
   *Fl.* June and July.                               G. H. ♃.


11. I. foliis enſiformibus ſtriatis, ſpica elongata.  *Sp.*   *ſcillaris.*
      *pl.* 52.
   Squil-flower'd Ixia.
   *Nat.* of the Cape of Good Hope.
   *Introd.* 1787, by Mr. Francis Maſſon.
   *Fl.* January.                                     G. H. ♃.


12. I. polyſtachia, foliis enſiformibus nervoſis margine      *margina-*
      incraſſatis, ſpicis pluribus adpreſſis, tubo incurvo,   *ta.*
      ſtigmatibus bifidis.
   Gladiolus marginatus.  *Linn. ſuppl.* 95.  *Thunb. Gla-*
      *diolus, n.* 20.
   Broad-leav'd Ixia.
   *Nat.* of the Cape of Good Hope.  Mr. *Francis Maſſon.*
   *Introd.* 1774.
   *Fl.* June.                                        G. H. ♃.


13 I. foliis ſubenſiformibus glabris, racemo terminali,      *patens.*
      corollis campanulatis patulis : laciniis alternis an-
      guſtioribus, filamentis erectis.
   Spreading-flower'd Ixia.
   *Nat.* of the Cape of Good Hope.
   *Introd.* 1779, by William Pitcairn, M. D.
   *Fl.* April.                                       G. H. ♃.
                                                        14. I.

*maculata.* 14. I. foliis ensiformibus, floribus alternis, petalis basi obscuris. *Sp. pl.* 1664. *Thunb. Ixia, n.* 19.
Spotted Ixia.
*Nat.* of the Cape of Good Hope.
*Cult.* 1757, by Mr. Philip Miller. *Mill. ic.* 104. *t.* 156. *f.* 1.
*Fl.* May and June. G. H. ♃.

*deusta.* 15. I. foliis lanceolatis nervosis, floribus alternis seffilibus, tubo bracteis breviore, laminis obtusis : exterioribus basi maculatis carinatisque.
Copper-colour'd Ixia.
*Nat.* of the Cape of Good Hope. Mr. *Francis Maffon. Introd.* 1774.
*Fl.* May. G. H. ♃.
Descr. *Folia* lineari-lanceolata, acuta, integerrima, plana, glabra, nervosa, spithamæa. *Scapus* teres, glaber, simplex. *Flores* remoti. *Bracteæ* duæ ad basin singuli germinis, latæ, membranaceæ, fissæ, vix semunciales. *Corolla* monopetala, regularis, fulva : *Tubus* cylindraceo-campanulatus, angustus ; *Limbus* sexpartitus : *laciniæ* ovatæ, obtusæ, leviter emarginatæ, uncia longiores, inferne in ungues latos attenuatæ : tres *exteriores* inferne gibbæ (unde limbus basi e trigono urceolatus) in medio macula atro-rubente et carina intus elevata notatæ, superne patulæ ; tres *interiores* erectiusculæ, planæ, immaculatæ. *Faux* macula virescenti flava, stellata.

*crocata.* 16. I. foliis ensiformibus, floribus alternis, tubo longitudine bractearum, corollæ laminis ovatis integerrimis basi hyalinis.
α floribus croceo-rufescentibus.
Ixia crocata. *Sp. pl.* 52. *Thunb. Ixia, n.* 20.
Common

Common Crocus-flower'd Ixia.

β floribus læte rubris.

Red Crocus-flower'd Ixia.

*Nat.* of the Cape of Good Hope.

*Cult.* 1758, by Mr. Philip Miller. *Mill. ic.* 160. *t.* 239. *f.* 2.

*Fl.* May and June. G. H. ♃.

17. I. foliis lineari-lanceolatis, floribus alternis feffilibus, *squalida.* tubo bracteis longiore, laminis ovato-oblongis.

α laminis cuneiformi-oblongis obtufe emarginatis bafi patula. fubhyalinis.

Spreading fqualid Ixia.

β foliis ftrictis, laminis ovato-oblongis integerrimis bafi ftricta. concoloribus.

Upright fqualid Ixia.

*Nat.* of the Cape of Good Hope. Mr. *Francis Maffon.* *Introd.* 1774.

*Fl.* May. G. H. ♃.

Obs. Varietas α. Ixiæ crocatæ valde affinis, fed differt laminis corollæ anguftioribus, magis pellucidis, ideoque venis extantioribus ; leviter emarginatis. Color etiam corollæ in hac pallide rufefcens, feu fordide carneus, cum parvo flavedinis.

Descr. Varietatis β. *Folia* acuminata, ftricta, plana, glabra; vix fpithamæa. *Scapus* teres, glaber, foliis duplo longior. *Flores* remoti. *Bracteæ* duæ ad bafin finguli germinis, membranaceæ, apice fiffæ, vix femunciales. *Corolla* monopetala, pallide lutea, venis obfcuris : *Tubus* infundibuliformis ; *Limbus* fexpartitus : *laciniæ* ovato-oblongæ, obtufiufculæ, integerrimæ : *exteriores* prope apicem interdum rubicundæ ; *interiorum* una cæteris paulo latior.

18. I.

*chinenſis.* 18. I. foliis enſiformibus, floribus remotis, panicula dichotoma, floribus pedunculatis. *Sp. pl.* 52.

Chineſe Ixia.

*Nat.* of China.

*Cult.* 1759, by Mr. Philip Miller. *Mill. dict. edit.* 7.
*no.* 1.

*Fl.* June and July.      H. ♃.


## GLADIOLUS. *Gen. pl.* 57.

*Cor.* 6-partita, irregularis, inæqualis. *Stigmata* 3.

*commu-* 1. G. foliis enſiformibus, floribus diſtantibus. *Sp. pl.* 52.
*nis.*      *Thunb. Gladiolus, n.* 9.

α flore rubro.
Common red Corn-flag.

β flore incarnato.
Fleſh-colour'd Corn-flag.

γ flore albo.
White Corn-flag.

*Nat.* of the South of Europe.

*Cult.* 1596, by Mr. John Gerard. *Hort. Ger.*

*Fl.* June and July.      H. ♃.


*tubiflorus.* 2. G. foliis lineari-lanceolatis villoſis ſubplicatis ſcapo
longioribus, tubo longiſſimo, ſpathis hirſutis.

Gladiolus tubiflorus. *Linn. ſuppl.* 96. *Thunb. Gladiolus, n.* 23.

Long-tub'd Corn-flag.

*Nat.* of the Cape of Good Hope. Mr. *Francis Maſſon.*
*Introd.* 1774.

*Fl.* June.      G. H. ♃.

*Descr.* *Folia* ſpithamæa et ultra. *Scapus* teres, vil-
loſus, digitalis. *Flores* feſſiles, ſubſecundi. *Bracteæ*
tres ad baſin ſinguli germinis, hirſutæ, acuminatæ:
exterior ſeſquiuncialis; interiores dimidio breviores.
*Corollæ tubus* filiformi-cylindraceus, ſpathis duplo
longior,

longior, e violaceo albus. *Limbus* infundibuliformis, fexpartitus: *laciniæ* lanceolatæ, acuminatæ, tubo quater breviores, ftramineæ: *tres fuperiores* extra
medium reflexæ, bafi notatæ macula bifurca rubra,
et in medio lateralium macula angulata ruberrima;
*laciniæ tres inferiores* immaculatæ. *Stigmata* fubcu-
neiformia, fupra villofa.

3. G. foliis oblongo-lanceolatis villofis plicatis, tubo fpa-   *plicatus.*
   this longiore.
Gladiolus plicatus. *Sp. pl.* 53. *Thunb. Gladiolus, n.* 24.
Hairy Corn-flag.
*Nat.* of the Cape of Good Hope.
*Cult.* 1757, by Mr. Ph. Miller. *Mill. ic.* 103. *t.* 155.
  *f.* 1.
*Fl.* May and June.                   G. H. ♃.

4. G. foliis lineari-lanceolatis villofis plicatis, tubo fpa-   *ftrictus.*
   thæ æquali.
α corolla faturate cærulea, tubo et bafi laciniarum e
   purpureo nigris.
Upright blue Corn-flag.
β corolla pallide purpurafcente, tubo cæruleo.
Upright purple Corn-flag.
*Nat.* of the Cape of Good Hope. Mr. *Francis Maſſon.*
*Introd.* 1774.
*Fl.* June.                      G. H. ♃.

5. G. foliis lineari-cruciatis, corollis campanulatis. *Sp.*   *triftis.*
   *pl.* 53.
Square-ftalk'd Corn-flag.
*Nat.* of the Cape of Good Hope.
*Cult.* 1745, by Mr. Philip Miller. *Trew. ehret.* 10.
  *t.* 39.
*Fl.* May and June.              G. H. ♃.
                          6. G.

*carina-*
*tus.*

6. G. foliis linearibus utrinque carinatis glabris, tubo
spathis limbisque breviore, stigmatibus indivisis complicatis.

Spotted-stalk'd Corn-flag.

*Nat.* of the Cape of Good Hope. Mr. *Francis Masson.*
*Introd.* 1774.

*Fl.* April and May.                     G. H. ♃.

DESCR. *Caulis* sesquipedalis, teres. *Folia* caule longiora. *Flores* suaveolentes. *Tubus* subcylindricus,
semuncia paulo longior, albidus. *Limbus* subcam-
panulatus, subringens: *laciniæ tres superiores* obo-
vato-oblongæ, tubo duplo longiores, pallide viola-
ceæ; *tres inferiores* angustiores: intermedia longitudine superiorum, infra medium flavicans; laterales breviores, medio flavæ.

*blandus.*

7. G. foliis lineari-lanceolatis nervosis glabris, floribus
spicatis, lacinia suprema reflexa, stigmatibus subbi-
lobis.

Blush-colour'd Corn-flag.

*Nat.* of the Cape of Good Hope. Mr. *Francis
Masson.*
*Introd.* 1774.

*Fl.* June.                               G. H. ♃.

DESCR. *Corollæ tubus* angustus, compressus, parum
incurvus, sesquiuncialis, pallide rubens. *Limbi lacinia suprema* oblonga, acuta, medio concava,
ibique dilatata, supra medium reflexa, ex albo in-
carnata, longitudine tubi; *laciniæ duæ laterales* ob-
longo-lanceolatæ, patentes, apice reflexæ, incarnatæ,
suprema paulo breviores; *laciniæ tres inferiores* li-
neari-lanceolatæ, erecto-patentes, supra medium parum reflexæ, ex albido-incarnatæ, in medio macula
rubra transversa notatæ, superioribus paulo breviores: intermedia cæteris acutior. *Filamenta* alba,
laciniis

laciniis corollæ dimidio breviora. *Antheræ* oblongæ, lineares, erectæ, cæruleæ. *Stylus* albus, filamentis longior. *Stigmata* tria, patula, apice dilatata in laminas bilobas, margine villofas.

8. G. foliis linearibus glabris, floribus fpicatis diftantibus, lacinia fuprema recta, ftigmatibus fpathulatis indivifis.                                                    *anguftus.*
Gladiolus anguftus. *Sp. pl.* 53.
Narrow-leav'd Corn-flag.
*Nat* of the Cape of Good Hope.
*Cult.* 1757, by Mr. Philip Miller. *Mill. ic.* 95. *t.* 142. *f.* 2.
*Fl.* May and June.                              G. H. ♃.

9. G. foliis lanceolato-enfiformibus planis, fauce labii fuperioris trilaminato: laminis unguiformibus perpendicularibus, bracteis acuminatis.                           *flavus.*
Yellow Corn-flag.
*Nat.* of the Cape of Good Hope. Mr. *William Paterfon.*
*Introd.* 1780, by the Countefs of Strathmore.
*Fl.* February and March.                         G. H. ♃.
OBS. Differt a Gladiolo fecurigere bracteis acuminatis, corolla tota intenfe flava, et foliis paulo latioribus.

10. G. foliis lineari-enfiformibus planis, fauce labii fuperioris trilaminato: laminis unguiformibus perpendicularibus, bracteis obtufis.                              *fecuriger.*
Copper-colour'd Corn-flag.
*Nat.* of the Cape of Good Hope. Mr. *Francis Maffon.*
*Introd.* 1774.
*Fl.* May.                                        G. H. ♃.

F                                    DESCR.

DESCR. *Folia* glabra, fpithama longiora. *Scapus* te-
retiufculus, foliis paulo brevior, interdum ramofus.
*Bracteæ* breves, ovatæ, obtufæ, interdum apice inci-
fæ, membranaceæ. *Corollæ* pallide fulvæ. *Tubus*
infundibuliformis, uncia brevior. *Limbus* fexpartitus: *laciniæ* oblongo-ovatæ, obtufæ, interdum obfolete emarginatæ, tubo breviores: tres *fuperiores*
in fauce notatæ macula flava margine rubicundo
circumfcripta, ibique auctæ laminis tribus compreffis, obtufis, perpendicularibus, flavis, diametro
fefquilineari.

## ANTHOLYZA. *Gen. pl.* 58.

*Cor.* tubulofa, irregularis, recurvata. *Capf.* infera.

*ringens.*    1. A. corollæ labiis divaricatis, fauce compreffa. *Sp.*
*pl.* 54.
Narrow-leav'd Antholyza.
*Nat.* of the Cape of Good Hope.
*Cult.* 1759, by Mr. Philip Miller. *Mill. dict. edit.* 7.
*no.* 1.
*Fl.* May and June.                              G. H. ♃.

*plicata.*    2. A foliis plicatis, caule ramofo hirfuto, corolla ringente
ftaminibus breviore. *Linn. fuppl.* 96.
Plaited-leav'd Antholyza.
*Nat.* of the Cape of Good Hope. Mr. *Francis Maffon.*
*Introd.* 1774.
*Fl.* April.                                     G. H. ♃.

*Cunonia.*   3. A. corollis rectis: labii quinquepartiti lobis duobus
extimis latioribus adfcendentibus. *Sp. pl.* 54.
Scarlet-flower'd Antholyza.
*Nat.* of the Cape of Good Hope.
*Cult.* 1756, by Mr. Philip Miller. *Mill. ic.* 75. *t.* 113.
*Fl.* May and June.                              G. H. ♃.
                                                  4. A.

4. A. corollis incurvatis: labii quinquepartiti lobis duo- *æthiopica.*
    bus alternis patulis majoribus lanceolatis. *Sp.*
    *pl.* 54.
Broad-leav'd Antholyza.
*Nat.* of the Cape of Good Hope.
*Cult.* 1759, by Mr. Philip Miller.
*Fl.* May and June.          G. H. ♃.

5. A. corollis infundibuliformibus, foliis ensiformibus. *Meriana.*
    *Syst. veget.* 87.
Red-flower'd Antholyza.
*Nat.* of the Cape of Good Hope.
*Cult.* 1750, by Mr. Philip Miller. *Trew. ehret.* 11.
    *t.* 40.
*Fl.* May and June.          G. H. ♃.

6. A. corollis infundibuliformibus, foliis linearibus. *Syst.*   *Meria-*
    *veget.* 87.                                          *nella.*
Dwarf Antholyza.
*Nat.* of the Cape of Good Hope.
*Introd.* 1754, by Capt. Hutchinson. *Mill. ic.* 198.
    *t.* 297. *f.* 2.
*Fl.* May and June.          G. H. ♃.

## A R I S T E A.

*Petala* 6. *Stylus* declinatus. *Stigma* infundibulifor-
    me, hians. *Capſ.* infera, polyſperma.

1. ARISTEA.                                   *cyanea.*
Ixia africana. *Sp. pl.* 51.
Moræa africana. *Syst. veget.* 93.
Graſs-leav'd Ariſtea.

                            *Nat.*

*Nat.* of the Cape of Good Hope.
*Introd.* 1774, by Mr. Francis Maſſon.
*Fl.* April——June.     G. H. ♃.

I R I S.    *Gen. pl.* 59.

*Cor.* 6-partita : *Petalis* alternis reflexis.   *Stigmata* petaliformia.

\* *Barbatæ nectariis petalorum reflexorum.*

*Juſiana.*    1. I. corolla barbata, caule foliis longiore unifloro. *Sp. pl.* 55.
Chalcedonian Iris.
*Nat.* of the Levant.
*Cult.* 1596, by Mr. John Gerard. *Hort. Ger.*
*Fl.* March and April.     H. ♃.

*florenti-*    2. I. corollis barbatis, caule foliis altiore ſubbifloro, flo-
*na.*      ribus feſſilibus. *Sp. pl.* 55.
Florentine Iris.
*Nat.* of the South of Europe.
*Cult.* 1596, by Mr. John Gerard. *Hort. Ger.*
*Fl.* May and June.     H. ♃.

*germani-*    3. I. corollis barbatis, caule foliis longiore multifloro,
*ca.*      floribus inferioribus pedunculatis. *Sp. pl.* 55.
German Iris, or Flower-de-luce.
*Nat.* of Germany.
*Cult.* 1596, by Mr. John Gerard. *Hort. Ger.*
*Fl.* May and June.     H. ♃.

*lurida.*    4. I. corollis barbatis, caule foliis altiore multifloro, petalis exterioribus revolutis ; interioribus erecto-in-
flexis ſubundulatis ſubemarginatis.
Dingy Iris.
                 *Nat.*

*Nat.* of the South of Europe.

*Cult.* 1758, by Mr. Philip Miller.

*Fl.* April.                                    H. ♃.

DESCR. *Petala exteriora* retroflexa, atro-purpurea, infra medium lituris flavicantibus ornata; barba lutea. *Petala interiora* obovato-oblonga, exterioribus paulo breviora; lamina purpurafcente; ungue fqualide lutefcente. *Stigmata* fqualide lutefcentia, fuperne pallide purpurafcentia.

OBS. Forte varietas Ireos fambucinæ, fed omnino inodora.

5. I. corollis barbatis, paule foliis altiore multifloro, petalis deflexis planis; erectis emarginatis. *Sp. pl.* 55. *Jacqu. hort.* 1. *p.* 1. *t.* 2.            *fambucina.*

Elder-fcented Iris.

*Nat.* of the South of Europe.

*Cult.* 1748, by Mr. Philip Miller. *Mill. dict. edit.* 5. *no.* 28.

*Fl.* June.                                    H. ♃.

6. I. corollis barbatis, caule foliis altiore multifloro, petalis deflexis replicatis; erectis emarginatis. *Sp. pl.* 56.            *fqualens.*

Iris variegata. *Jacqu. auftr.* 1. *p.* 7. *t.* 5.

Brown-flower'd Iris.

*Nat.* of Germany.

*Cult.* 1768, by Mr. Philip Miller. *Mill. dict. edit.* 8.

*Fl.* June.                                    H. ♃.

7. I. corollis barbatis, caule fubfoliofo longitudine foliorum multifloro. *Sp. pl.* 56. *Curtis magaz.* 16.            *variegata.*

Variegated Iris.

*Nat.* of Hungary.

F 3                              *Cult.*

*Cult.* 1597.   *Ger. herb.* 51. *f.* 1.
*Fl.* May and June.                  H. ♃.

*biflora.*    8. I. corollis barbatis, caule foliis breviore trifloro. *Sp. pl.* 56.
Two-flower'd Iris.
*Nat.* of Portugal and Spain.
*Cult.* 1596, by Mr. John Gerard.   *Hort. Ger.*
*Fl.* April and May.                 H. ♃.

*criſtata.*    9. I. corollis barbatis: barba criſtata, caule ſubunifloro
longitudine foliorum, germinibus trigonis, petalis
ſubæqualibus.
Creſted Iris.
*Nat.* of North America.
*Introd.* 1756, by Peter Collinſon, Eſq. *Coll. mſcr.*
*Fl.* May.                      H. ♃.
DESCR. *Radix* repens. *Caulis* compreſſus, vix digitalis, inferne veſtitus foliis enſiformibus. *Petala
exteriora* oblonga, obtuſa, integra, cærulea, medio
lutea, criſtis tribus longitudinalibus undulatis loco
barbæ; *Petala interiora* parum anguſtiora, tota cærulea. *Filamenta* et *Antheræ* pallide flaveſcentes.
*Stigmata* dilute cærulea, petalis breviora.

*pumila.*    10. I. corollis barbatis, caule foliis breviore unifloro.
*Sp. pl.* 56. *Curt. mag.* 9. *Jacqu. auſtr.* 1. *p.* 5. *t.* 1.
Dwarf Iris.
*Nat.* of Auſtria.
*Cult.* 1596, by Mr. John Gerard.   *Hort. Ger.*
*Fl.* April.                   H. ♃.

*dichoto-*    11. I. corolla tenuiſſime barbata, caule tereti elongato
*ma.*           paniculato, ramis alternis divaricatis bi-ſeu quadrifloris. *Linn. ſuppl.* 97.
                                Fork'd

Fork'd Iris.
*Nat.* of Siberia.
*Introd.* 1784, by Mr. John Bell.
*Fl.* Auguft. H. ♃.

** *Imberbes, petalis deflexis lævibus.*

12. I. corollis imberbibus, petalis interioribus ftigmate mi- *Pfeud A-*
noribus, foliis enfiformibus. *Sp. pl.* 56. *Curtis lond.* *corus.*
Yellow Iris, or Flower-de-luce.
*Nat.* of Britain.
*Fl.* June. H. ♃.

13. I. corollis imberbibus, petalis interioribus patentiffi- *fœtidiffi-*
mis, caule uniangulato, foliis enfiformibus. *Sp.* *ma.*
*pl.* 57.
Stinking Iris, or Gladwyn.
*Nat.* of Britain.
*Fl.* June. H. ♃.

14. I. corollis imberbibus, germinibus trigonis, caule *fibirica.*
tereti, foliis linearibus. *Sp. pl.* 57. *Jacqu. auftr.*
1. *p.* 6. *t.* 3. *Curtis magaz.* 50.
Siberian Iris.
*Nat.* of Siberia, Auftria, and Switzerland.
*Cult.* 1596, by Mr. John Gerard. *Hort. Ger.*
*Fl.* May and June. H. ♃.

15. I. corollis imberbibus, germinibus fubtrigonis, caule *verficolor.*
tereti flexuofo, foliis enfiformibus. *Sp. pl.* 57. *Curt.*
*mag.* 21.
Various-colour'd Iris.
*Nat.* of North America.
*Cult.* 1732, by James Sherard, M. D. *Dill. elth.*
*tab.* 155.
*Fl.* May and June. H. ♃.

F 4 16. I.

*virgini-*  16. I. corollis imberbibus, germinibus trigonis, caule
*ca.*             ancipiti. *Sp. pl.* 58.
          Virginian Iris.
          *Nat.* of North America.
          *Cult.* 1758, by Mr. Philip Miller.
          *Fl.* June.                                                    H. ♃.

*martini-* 17. I. corollis imberbibus, germinibus trigonis, petalis
*cenſis.*          baſi foveolis glandulofis. *Sp. pl.* 58.
          Martinico Iris.
          *Nat.* of the Iſland of St. Lucia.
          *Introd.* 1782, by Mr. Alexander Anderſon.
          *Fl.* June.                                                    S. ♃.

*ſpuria.*  18. I. corollis imberbibus, germinibus ſexangularibus,
              caule tereti, foliis ſublinearibus. *Sp. pl.* 58. *Jacqu.*
              *auſtr.* 1, *p.* 6. *t.* 4.
          Spurious Iris.
          *Nat.* of Germany.
          *Cult.* 1759, by Mr. Philip Miller. *Mill. dict. edit.* 7.
          *no.* 14.
          *Fl.* July.                                                    H. ♃.

*ochroleu-* 19. I. corollis imberbibus, germinibus ſexangularibus,
*ca.*             caule ſubtereti, foliis enſiformibus ſtriatis. *Linn.*
              *mant.* 175.
          Pale yellow-flower'd Iris.
          *Nat.* of the Levant.
          *Cult.* 1759, by Mr. Philip Miller. *Mill. dict. edit.* 7.
          *no.* 9.
          *Fl.* July.                                                    H. ♃.

*halophila.* 20. I. corollis imberbibus, germinibus ſexangularibus,
              caule tereti, foliis enſiformibus : radicalibus longiſ-
              ſimis.

                                                                        Iris

Iris halophila. *Pallas iter* 2. *p.* 733. *vol.* 3. *p.* 713.
 tab. *B. fig.* 2.
Long-leav'd Iris.
*Nat.* of Siberia.
*Introd.* 1780, by Peter Simon Pallas, M.D.
*Fl.* July——September. H. ♃.

21. I. corollis imberbibus, germinibus fexangularibus, *graminea.*
 caule ancipiti, foliis linearibus. *Sp. pl.* 58. *Jacqu.*
 *auftr.* 1. *p.* 5. *t.* 2.
Grafs-leav'd Iris.
*Nat.* of Auftria.
*Cult.* 1597. *Ger. herb.* 52. *f.* 5.
*Fl.* June. H. ♃.

22. I. cotollis imberbibus, caule unifloro foliis breviore, *verna.*
 radice fibrofa. *Sp. pl.* 58.
Spring Iris.
*Nat.* of North America.
*Cult.* 1739, by Mr. Philip Miller. *Mill. dict. vol.* 2.
 *no.* 43.
*Fl.* April and May. H. ♃.

23. I. corollis imberbibus, foliis tetragonis. *Sp. pl.* 58. *tuberofa.*
Snake's-head Iris.
*Nat.* of the Levant.
*Cult.* 1597. *Ger. herb.* 94. *f.* 2.
*Fl.* March and April. H. ♃.

24. I. corollis imberbibus, floribus binis, foliis fubulato- *Xiphium.*
 canaliculatis caule brevioribus. *Sp. pl.* 58.
 α Iris bulbofa latifolia caule donata. *Bauh. pin.* 38.
 Great Bulbofe rooted Iris.
 β Iris bulbofa cæruleo violacea. *Bauh. pin.* 40.
 Small Bulbofe-rooted Iris.
 *Nat.*

*Nat.* of the South of Europe.
*Cult.* 1633. *Ger. emac.* 99. *f.* 1.
*Fl.* June. H. ♃.

*perſica.* 25. I. corolla imberbi, petalis interioribus breviſſimis
patentiſſimis. *Syſt. veg. ed.* 13. *p.* 79. *Curt. mag.* 1.
Perſian Iris.
*Nat.* of Perſia.
*Cult.* 1629. *Park. parad.* 172. *no.* 2.
*Fl.* March. H. ♃.

*Siſyrin-* 26. I. corollis imberbibus, foliis canaliculatis, bulbis ge-
*chium.* minis ſuperimpoſitis. *Sp. pl.* 59.
Crocus-rooted Iris.
*Nat.* of Spain and Portugal.
*Cult.* 1597. *Ger. herb.* 94. *f.* 1.
*Fl.* May. H. ♃.


M O R Æ A. *Gen. pl.* 60.

*Cor.* hexapetala : *Petala* 3-interiora patentia; reli-
qua Ireos.

*vegeta.* 1. M. foliis canaliculatis. *Spec. plant.* 59.
Iris plumaria. *Syſt. veget.* 89.
α Moræa ſpatha biflora, caule planifolio, floribus mino-
ribus. *Mill. ic.* 159. *t.* 238. *f.* 1.
Small-flower'd Graſs-leav'd Moræa.
β Moræa ſpatha uniflora, caule planifolio, floribus ma-
joribus. *Mill. ic.* 159. *t.* 238. *f.* 2.
Great-flower'd Graſs-leav'd Moræa.
*Nat.* of the Cape of Good Hope.
*Cult.* 1758, by Mr. Philip Miller. *Mill. ic. loc. cit.*
*Fl.* May. G. M. ♃.

2. M.

2. M. foliis gladiatis. *Linn. mant.* 28.  *iridioi-*
Sword-leav'd Moræa.  *des.*
*Nat.* of the Cape of Good Hope.
*Cult.* 1758, by Mr. Philip Miller. *Mill. ic.* 159.
*t.* 239. *f.* 1.
*Fl.* May and June.  G. H. ♃.

3. M. caule ancipiti uni-feu bifloro, foliis enfiformibus:  *lugens.*
infimis fubfalcatis, floribus terminalibus. *Linn.*
*fuppl.* 99.
Dark-flower'd Moræa.
*Nat.* of the Cape of Good Hope.
*Introd.* 1786, by Mr. Francis Maffon.
*Fl.*  G. H. ♄.

## WACHENDORFIA. *Gen. pl.* 61.

*Cor.* hexapetala, inæqualis, infera. *Capf.* trilocularis, fupera.

1. W. fcapo fimplici. *Sp. pl.* 59.  *thyrfiflo-*
Simple-ftalk'd Wachendorfia.  *ra.*
*Nat.* of the Cape of Good Hope.
*Cult.* 1759, by Mr. Philip Miller. *Mill. dict. edit.* 7.
*Addenda.*
*Fl.* May and June.  G. H. ♃.

2. W. fcapo polyftachyo. *Sp. pl.* 59.  *pan cu'a-*
Panicled Wachendorfia.  *ta.*
*Nat.* of the Cape of Good Hope.
*Introd.* 1767, by Mr. William Malcolm.
*Fl.* February.  G. H. ♃.

COMME-

## COMMELINA. *Gen. pl.* 62.

*Cor.* 6-petala. *Nectaria* 3, cruciata, filamentis propriis inferta.

* *Dipetalæ, ob duo petala majora.*

*communis.*    1. C. corollis inæqualibus, foliis ovato-lanceolatis acutis, caule repente glabro. *Sp. pl.* 60.
Common American Commelina.
*Nat.* of America.
*Cult.* 1732, by James Sherard, M.D. *Dill. elth.* 93. *t.* 78. *f.* 89.
*Fl.* June and July.        H. ⊙.

*africana.*    2. C. corollis inæqualibus, foliis lanceolatis glabris, caule decumbente. *Sp. pl.* 60.
African Commelina.
*Nat.* of the Cape of Good Hope.
*Cult.* 1759, by Mr. Philip Miller. *Mill. dict. edit.* 7. *no.* 3.
*Fl.* May——October.        S. ♃.

*erecta.*    3. C. corollis inæqualibus, foliis ovato-lanceolatis, caule erecto subhirsuto simpliciffimo. *Syst. veget.* 94.
Upright Virginian Commelina.
*Nat.* of Virginia.
*Cult.* 1732, by James Sherard, M.D. *Dill. elth.* 91. *t.* 77. *f.* 88.
*Fl.* August and September.        H. ♃.

** *Tripetalæ:* Zanoniæ Plumieri, *petalis* 3 *majoribus.*

*tuberosa.*    4. C. corollis æqualibus, foliis sessilibus ovato-lanceolatis subciliatis. *Sp. pl.* 61.
Tuberous-rooted Commelina.
*Nat.* of Mexico.

*Cult.*

*Cult.* 1732, by James Sherard, M.D. *Dill. elth.* 94.
*t.* 79. *f.* 90.
*Fl.* June and July. S. ♃.

5. C. corollis æqualibus, pedunculis incraffatis, foliis    *Zanonia.*
lanceolatis: vaginis tumidis margine hirfutis, brac-
teis geminis. *Syft. veget.* 94.
Gentian-leav'd Commelina.
*Nat.* of the Weft Indies.
*Cult.* 1759, by Mr. Philip Miller. *Mill. dict. edit.* 7.
*no.* 5.
*Fl.* July——December. S. ♃.

6. C. corollis æqualibus, foliis lanceolatis, floribus pa-    *fpirata.*
niculatis. *Linn. mant.* 176.
Spear-leav'd Commelina.
*Nat.* of the Eaft Indies.
*Introd.* 1783, by John Earl of Bute.
*Fl.* July and Auguft. S. ☉.

## C A L L I S I A. *Gen. pl.* 63.

*Cal.* 3-phyllus. *Petala* 3. *Antheræ* geminæ. *Capf.*
bilocularis.

1. CALLISIA. *Spec. plant.* 62.    *repens.*
Creeping Callifia.
*Nat.* of the Weft Indies.
*Introd.* 1776, by John Fothergill, M.D.
*Fl.* June and July. S. ♃.

## S C H O E N U S *Gen. pl.* 65.

*Glumæ* paleaceæ, univalves, congeftæ. *Cor.* o. *Sem.* 1,
fubrotundum, inter glumas.

1. S. culmo tereti, foliis margine dorfoque aculeatis. *Marifcus.*
*Sp. pl.* 62.
Prickly

Prickly Bog-rufh, or baftard Cyperus.
*Nat.* of England.
*Fl.* July and Auguft.                                   H. ♃.

*mucrona-*    2. S. culmo tereti nudo, fpiculis ovatis fafciculatis, invo-
*tus.*              lucro fubhexaphyllo, foliis canaliculatis.   *Syft.*
                    *veget.* 95.
                    Clufter'd Bog-rufh.
                    *Nat.* of the South of Europe, and the Levant.
                    *Introd.* 1781, by P. M. A. Brouffonet, M. D.
                    *Fl.*                                   H. ♃.

*nigricans.*  3. S. culmo tereti nudo, capitulo ovato, involucri di-
                    phylli valvula altera fubulata longa.   *Sp. pl.* 64.
                    Black Bog-rufh.
                    *Nat.* of Britain.
                    *Fl.* July.                              H. ♃.

*ferrugi-*    4. S. culmo tereti nudo, fpica duplici, involucri valvula
*neus.*             majore fpicam æquante.   *Sp. pl.* 64.
                    Rufty Bog-rufh.
                    *Nat.* of Britain.
                    *Fl.* July.                              H. ♃.

*compref-*    5. S. culmo fubtriquetro nudo, fpica difticha, involucro
*fus.*              monophyllo.   *Sp. pl.* 65.
                    Comprefs'd Bog-rufh.
                    *Nat.* of Britain.
                    *Fl.* July.                              H. ♃.

*albus.*      6. S. culmo fubtriquetro foliofo, floribus fafciculatis,
                    foliis fetaceis.   *Sp. pl.* 65.
                    White-flower'd Bog-rufh.
                    *Nat.* of Britain.
                    *Fl.* July.                              H. ♃.

                                                    CYPERUS.

## CYPERUS. *Gen. pl.* 66.

*Glumæ* paleaceæ, diſtiche imbricatæ. *Cor.* o. *Sem.* 1, nudum.

1. C. culmo compreſſo baſi viſcido, foliis aſperis apice triquetris.      *viſcoſus.*
Clammy Cyperus.
*Nat.* of Jamaica.
*Introd.* 1781, by Monſ. Thouin.
*Fl.* May——Auguſt.      S. ♃.

2. C. culmo obſolete triquetro proſtrato, ſpicis ſubquaternis ſeſſilibus.      *pannonicus.*
Cyperus. pannonicus, *Jacqu. auſtr.* 5. *p.* 29. *tab. app.* 6. *Linn. ſupp.* 103.
Dwarf Cyperus.
*Nat.* of Hungary.
*Introd.* 1781, by Abbé Pourret.
*Fl.* July and Auguſt.      H. ⊙.

3. C. culmo triquetro folioſo, umbella folioſa ſupradecompoſita, pedunculis nudis, ſpicis alternis. *Sp. pl.* 67.      *longus.*
Sweet Cyperus.
*Nat.* of England.
*Fl.* July.      H. ♃.

4. C. culmo triquetro nudo, umbella triphylla, pedunculis ſimplicibus inæqualibus, ſpicis confertis lanceolatis. *Sp. pl.* 68.      *flaveſcens.*
Yellow Cyperus.
*Nat.* of France and Germany.
*Introd.* 1776, by Monſ. Thouin.
*Fl.* July——September.      H. ♂.

5. C.

*fuſcus.*      5. C. culmo triquetro nudo, umbella trifida, pedunculis
                ſimplicibus inæqualibus, ſpicis confertis linearibus.
                *Sp. pl.* 69.
                Brown Cyperus.
                *Nat.* of France, Germany, and Switzerland.
                *Introd.* 1777, by Mr. Thomas Blackie.
                *Fl.* July——September.                          H. ☉.

*alterni-*     6. C. culmo triquetro nudo apice alternatim folioſo, pe-
*folius.*         dunculis lateralibus proliferis.   *Linn. mant.* 28.
                Alternate-leav'd Cyperus.
                *Nat.* of the Iſland of Madagaſcar.
                *Introd.* 1781, by Monſ. Thouin.
                *Fl.* February and March.                       S. ♃.

## S C I R P U S.   *Gen. pl.* 67.

*Glumæ* paleaceæ, undique imbricatæ.  *Cor.* o.  *Sem.* 1,
                imberbe.

### * *Spica unica.*

*paluſtris.*   1. S. culmo tereti nudo, ſpica ſubovata terminali.  *Sp.*
                *pl.* 70.
                Marſh Club-ruſh.
                *Nat.* of Britain.
                *Fl.* July.                                     H. ♃.

*cæſpito-*     2. S. culmo ſtriato nudo, ſpica bivalvi terminali longi-
*ſus.*            tudine calycis, radicibus ſquamula interſtinctis.  *Sp.*
                *pl.* 71.
                Dwarf Club-ruſh.
                *Nat.* of Britain.
                *Fl.* July.                                     H. ♃.

*acicula-*     3. S. culmo tereti nudo ſetiformi, ſpica ovata terminali
*ris.*            bivalvi, ſeminibus nudis.  *Sp. pl.* 71.
                                                                Needle

Needle Upright Club-ruſh.
*Nat.* of Britain.
*Fl.* July.                    H. ♃.

4. S. culmis teretibus nudis alternis, caule folioſo flacci-  *fluitans.*
    do. *Sp. pl.* 71.
Floating Club-ruſh.
*Nat.* of Britain.
*Fl.* July and Auguſt.                    H. ♃.

  * * *Culmo tereti polyſtachio.*
5. S. culmo tereti nudo, ſpicis ovatis pluribus peduncu-  *lacuſtris.*
    latis terminalibus. *Sp. pl.* 72.
Tall Club-ruſh, or Bull-ruſh.
*Nat.* of Britain.
*Fl.* July and Auguſt.                    H. ♃.

6. S. culmo tereti nudo, ſpicis ſubglobofis glomeratis pe-  *Holoſ-*
    dunculatis, involucro diphyllo inæquali mucronato.  *chœnus.*
    *Sp. pl.* 72.
Round-headed Club-ruſh.
*Nat.* of England.
*Fl.* July.                    H. ♃.

7. S. culmo tereti nudo, capitulo laterali conglobato,  *romanus.*
    bractea reflexa. *Sp. pl.* 72. *Jacqu. auſtr.* 5. *p.* 23.
    *t.* 448.
Roman Club-ruſh.
*Nat.* of the South of Europe.
*Introd.* 1779, by Joſeph Nicholas de Jacquin, M.D.
*Fl.* July.                    H. ♃.

8. S. culmo nudo ſetaceo, ſpica terminàli ſeſſili. *Syſt.*  *ſetaceus.*
    *veget.* 9).
                    G                    Leaſt

Leaft Club-rufh.
*Nat.* of England.
*Fl.* July and Auguft.　　　　　　　　　　　　H. ☉.

*** *Culmo triquetro, panicula nuda.*

*mucrona-* 　9. S. culmo triangulo nudo acuminato, fpicis conglo-
*tus.* 　　　　meratis feffilibus lateralibus. *Syft. veget.* 100.
Pointed Club-rufh.
*Nat.* of England.
*Fl.* July and Auguft,　　　　　　　　　　　　H. ♃.

**** *Culmo triquetro, panicula foliacea.*

*mariti-* 　10. S. culmo triquetro, panicula conglobata foliacea, fpi-
*mus.* 　　　　cularum fquamis trifidis : intermedia fubulata. *Sp.*
　　　　　*pl.* 74. *Curtis lond.*
Round-rooted Sea Club-rufh.
*Nat.* of Britain.
*Fl.* July——September.　　　　　　　　　　　H. ♃.

*luzulæ.* 　11. S. culmo triquetro nudo, umbella foliofa prolifera,
　　　　　fpiculis fubrotundis. *Sp. pl.* 75.
Clufter'd Club-rufh.
*Nat.* of the Eaft Indies.
*Introd.* 1776, by Lady Ann Monfon.
*Fl.* Auguft and September.　　　　　　　　　S. ♃.

*fylvati-* 　12. S. culmo triquetro foliofo, umbella foliacea, peduncu-
*cus.* 　　　　lis nudis fupradecompofitis, fpicis confertis. *Sp.*
　　　　　*pl.* 75.
Wood Club-rufh, or Millet Cyperus Grafs.
*Nat.* of Britain.
*Fl.* July and Auguft.　　　　　　　　　　　H. ♃.

KYLLIN-

## KYLLINGIA. *Linn. fuppl.* 11.

*Amentum* ovatum f. oblongum, imbricatum. *Flores*
calyce corollaque bivalvi.

1. K. capitulis terminalibus fubternis glomeratis feffili-  *triceps.*
    bus. *Linn. fuppl.* 104.
    Schœnus niveus. *Syft. veget. ed.* 13. *p.* 81.
    Three headed Kyllingia.
    *Nat.* of both Indies.
    *Introd.* 1776, by Monf. Thouin.
    *Fl.* September——November.     S. ♃.

## ERIOPHORUM. *Gen. pl.* 68.

*Glumæ* paleaceæ, undique imbricatæ. *Cor.* o. *Sem.* 1,
*Lana* longiffima cinctum.

1. E. culmis vaginatis teretibus, fpica fcariofa. *Sp.*  *vagina-*
    *pl.* 76. *Curtis lond.*     *tum.*
    Mountain Cotton-grafs, or Hare's-tail Rufh.
    *Nat.* of Britain.
    *Fl.* March and April.     H. ♃.

2. E. culmis teretibus, foliis planis, fpicis pedunculatis.  *polyfta-*
    *Sp. pl.* 76. *Curtis lond.*     *chion.*
    Common Cotton-grafs.
    *Nat.* of Britain.
    *Fl.* June and July.     H. ♃.

## NARDUS. *Gen. pl.* 69.

*Cal.* o. *Cor.* 2-valvis.

1. N. fpica fetacea recta fecunda. *Syft. veget.* 102.  *ftricta.*
    Mat-grafs.
    *Nat.* of Britain.
    *Fl.* June and July.     H. ♃.

G 2      LYGEUM.

## LYGEUM. *Gen. pl.* 70.

*Spatha* 1-phylla. *Corollæ* binæ fupra idem germen.
*Nux* bilocularis.

*Spartum.*    1. LYGEUM. *Sp. pl.* 78.
Rufh-leav'd Lygeum.
*Nat.* of Spain.
*Introd.* 1776, by Meffrs. Kennedy and Lee.
*Fl.* May and June.            H. ♃.

*DIGYNIA.*

# D I G Y N I A.

## S A C C H A R U M. *Gen. pl.* 73.

*Cal.* 2-valvis, lanugine involucratus. *Cor.* 2-valvis.

1. S. floribus paniculatis, foliis planis. *Syfl. veget.* 103.    *officina-*
Common Sugar-cane.                                       *rum.*
*Nat.* of both Indies.
*Cult.* before 1597, by Mr. John Gerard. *Ger. herb.* 35.
*Fl.*                                           S. ♃.

## P E R O T I S.

*Cal.* nüllus. *Cor.* bivalvis : valvulæ æquales ariftatæ.

1. PEROTIS.                                      *latifolia.*
Saccharum fpicatum. *Sp. pl.* 79.
Spiked Perotis.
*Nat.* of the Eaft Indies.
*Introd.* 1777, by Daniel Charles Solander, L.L.D.
*Fl.* Auguft and September.                 S. ☉.

## P H A L A R I S. *Gen. pl.* 74.

*Cal.* 2-valvis, 1-florus, carinatus, æqualis, marginibus
interioribus rectis.

1. P. panicula mutica fubovata fpiciformi, glumis calyci-    *canarien-*
nis navicularibus integris, corolla quadrivalvi : val-    *fis.*
vulis exterioribus lanceolatis glabris; interioribus
villofis.
Phalaris canarienfis. *Sp. pl.* 79.
Manured Canary-grafs.
*Nat.* of Britain.
*Fl.* June——Auguft.                        H. ☉.

*aquatica.*   2. P. panicula mutica cylindracea fpiciformi, glumis calycinis navicularibus fubdenticulatis, corolla trivalvi:
valvulis interioribus villofis; exteriore minuta fubulata.

Phalaris aquatica.   *Sp. pl.* 79.
Water Canary-grafs.
*Nat.* of Egypt.
*Introd.* 1778, by Monf. Thouin.
*Fl.* June and July.                                    H. ☉.

*phleoides.*   3. P. panicula mutica cylindrica fpiciformi, glumis calycinis carinatis integerrimis fcabriufculis, corolla
bivalvi glabriufcula.

α glumis calycinis nudiufculis.
Phalaris phleoides.   *Sp. pl.* 80.
Phleum pratenfe β.   *Hudf. angl.* 26.
Naked-cupp'd Canary-grafs.

arenaria.   β glumis calycinis ciliatis.
Phleum arenarium.   *Sp. pl.* 88.
Phalaris arenaria.   *Hudf. angl.* 23.
Sea Canary-grafs.
*Nat.* of England.
*Fl.* July and Auguft.                                 H. ☉

*utricula-
ta.*   4. P. panicula ovata fpiciformi, glumis calycinis navicularibus: dorfo dilatato, arifta receptaculi glumis
longiore.

Phalaris utriculata.   *Sp. pl.* 80.
Bladder'd Canary-grafs.
*Nat.* of Italy.
*Introd.* 1777, by Monf. Thouin.
*Fl.* June——Auguft.                                    H. ☉.

*paradoxa.*   5. P. panicula mutica oblonga fpiciformi, glumis calycinis

nis navicularibus unidentatis, corolla bivalvi gla-
bra, flofculis infimis præmorfis.
Phalaris paradoxa. *Sp. pl.* 1665.
Briftly-fpiked Canary-grafs.
*Nat.* of the Levant.
*Introd.* 1771, by Monf. Richard.
*Fl.* June and July.                          H. ☉.


6. P. panicula mutica cylindrica fpiciformi, glumis caly-    *panicula-*
cinis carinatis fuperne gibbis, corolla bivalvi glabra.    *ta.*
Phleum paniculatum. *Hudf. angl.* 26.
Panicled Canary-grafs.
*Nat.* of England.
*Fl.* July.                          H. ☉.


P A S P A L U M.  *Gen. pl.* 75.
*Cal.* 2-valvis, orbiculatus. *Cor.* ejufdem magnitudi-
nis. *Stigmata* penicilliformia.

1. P. fpicis alternis, rachi membranacea, floribus alter-    *fcrobicu-*
nis, calycibus multinerviis extus fcrobiculatis. *Linn.*    *latum.*
*mant.* 29.
Dimpled Pafpalum.
*Nat.* of the Eaft Indies.
*Introd.* 1778, by Sir Jofeph Banks, Bart.
*Fl.* July——September.                          S. ♃

2. P. fpicis paniculatis verticillato-aggregatis. *Sp. pl.* 81.    *panicula-*
Panicled Pafpalum.    *tum.*
*Nat.* of Jamaica.
*Introd.* 1782, by Monf. Thouin.
*Fl.* July and Auguft.                          S. ☉.

3. P. fpicis duabus: altera fubfeffili, floribus acuminatis.    *diftichum.*
*Sp. pl.* 82.

G 4                          Two

Two-fpiked Pafpalum.
*Nat.* of Jamaica.
*Introd.* 1776, by John Fothergill, M. D.
*Fl.* July.                                        S. ♂.


PANICUM.   *Gen. pl.* 76.
*Cal.* 3-valvis : valvula tertia minima.

\* *Spicata.*

*fericeum.*  1. P. fpica tereti, involucris fetaceis villofis unifloris lon-
               gitudine flofculorum, foliis planis.
             Silky Panic-grafs.
             *Nat.* of the Weft Indies.
             *Introd.* 1780, by John Earl of Bute.
             *Fl.* June——September.                S. ☉.


*verticil-*  2. P. fpica verticillata, racemulis quaternis, involucellis
*latum.*        unifloris bifetis, culmis diffufis.  *Sp. pl.* 82. *Curtis
               lond.*
             Rough Panic-grafs.
             *Nat.* of England.
             *Fl.* July.                             H. ☉.


*glaucum.*   3. P. fpica tereti, involucellis bifloris fafciculato-pilofis,
               feminibus undulato rugofis.  *Syft. veget.* 105.
             Glaucous Panic-grafs.
             *Nat.* of both the Indies, and North America.
             *Introd.* 1771, by Monf. Richard.
             *Fl.* June and July.                    H. ☉.


*viride.*    4. P. fpica tereti, involucellis bifloris fafciculato-pilofis,
               feminibus nervofis.  *Sp. pl.* 83.  *Curtis lond.*
             Green Panic-grafs.
             *Nat.* of England.
             *Fl.* July and Auguft.                  H. ☉.
                                                     5. P.

5. P. fpica compofita: fpiculis glomeratis fetis immix- *italicum.*
tis, pedunculis hirfutis. *Sp. pl.* 83.
Italian Panic-grafs.
*Nat.* of both Indies.
*Cult.* 1739, by Mr. Philip Miller. *Rand. chelf. no.* 3.
*Fl.* July and Auguft.                                S. ☉.

6. P. fpicis alternis fecundis: fpiculis fubdivifis, glumis *Crus cor-*
fubariftatis hifpidis, rachi trigona. *Sp. pl.* 84.    *vi.*
Crows-foot Panic-grafs.
*Nat.* of the Eaft Indies.
*Introd.* 1781, by Monf. Thouin.
*Fl.* July and Auguft.                                S. ☉.

7. P. fpicis alternis conjugatifque: fpiculis fubdivifis, *Crus*
glumis ariftatis hifpidis, rachi quinquangulari. *Sp.* *galli.*
*pl.* 83. *Curtis lond.*
Thick-fpiked Cock's-foot Panic-grafs.
*Nat.* of England.
*Fl.* July.                                           H. ☉.

8. P. fpicis alternis fecundis muticis ovatis fcabris, rachi *colonum.*
teretiufcula. *Sp. pl.* 84.
Purple Panic-grafs.
*Nat.* of the Eaft Indies.
*Cult.* 1699, by the Dutchefs of Beaufort. *Br. Muf.*
*Sloan. mff.* 525 and 3349.
*Fl.* July and Auguft.                                S. ☉.

9. P. fpicis digitatis bafi interiore nodofis, flofculis ge- *fangui-*
minis muticis, vaginis foliorum punctatis. *Sp.* *nale.*
*pl.* 84. *Curtis lond.*
Slender-fpiked Cock's-foot **Panic**-grafs.
*Nat.* of England.
*Fl.* Auguft.                                         H. ☉.
                                                     10. P.

*Dactylon.*  10. P. fpicis digitatis patentibus bafi interiore villofis,
floribus folitariis, farmentis repentibus. *Sp. pl.* 85.
Creeping Panic-grafs.
*Nat.* of England.
*Fl.* July.                                              H. ♃.

*filiforme.*  11. P. fpicis fubdigitatis approximatis erectis linearibus:
rachi flexuofa: dentibus bifloris: altero feffili.
*Syft. veget.* 106.
Clofe-fpiked Panic-grafs.
*Nat.* of North America.
*Introd.* 1781, by Monf. Thouin.
*Fl.* July——September.                                   H. ☉.

* * *Paniculata.*

*colora-*  12. P. panicula patente, ftaminibus piftillifque coloratis,
*tum.*          culmo ramofo. *Linn. mant.* 30. *Jacqu. ic.*
Color'd Panic-grafs.
*Nat.* of Egypt.
*Introd.* 1771, by Monf. Richard.
*Fl.* July and Auguft.                                   S. ☉.

*repens.*  13. P. panicula virgata, foliis divaricatis. *Sp. pl.* 87.
Slender Panic-grafs.
*Nat.* ——————. Cultivated in the Eaft Indies.
*Claude Ruffel,* Efq.
*Introd.* 1777, by Patrick Ruffel, M. D.
*Fl.* July——September.                                   S. ☉.

*miliace-*  14. P. panicula laxa flaccida, foliorum vaginis hirtis, glu-
*um.*          mis mucronatis nervofis. *Syft. veget.* 106.
Millet Panic-grafs.
*Nat.* of the Eaft Indies.
*Cult.* 1597. *Ger. herb.* 73.
*Fl.* July.                                              S. ☉.
                                                        15. P.

15. P. panicula capillari erecta patente, foliorum vagi-     *capillare.*
     nis hirtis. *Syst. veget.* 106.
     Hair-panicled Panic-grafs.
     *Nat.* of Virginia and Jamaica.
     *Cult.* 1758, by Mr. Philip Miller.
     *Fl.* June——Auguft.              S. ☉.

16. P. panicula racemis lateralibus fimplicibus, foliis    *latifoli-*
     ovato-lanceolatis collo pilofis. *Sp. pl.* 86.      *um.*
     Broad-leav'd Panic-grafs.
     *Nat.* of North America.
     *Introd.* 1765, by Mr. John Cree.
     *Fl.* Auguft and September.          H. ♃.

17. P. paniculatum ramofiffimum, foliis ovato-oblongis    *arbore-*
     acuminatis. *Sp. pl.* 87.                   *fcens.*
     Tree Panic-grafs.
     *Nat.* of the Eaft Indies.
     *Introd.* 1776, by Monf. Thouin.
     *Fl.* March and April.             S. ♄.

18. P. panicula virgata, glumis acuminatis lævibus: ex-    *virga-*
     tima dehifcente. *Sp. pl.* 87.              *tum.*
     Long-panicled Panic-grafs.
     *Nat.* of North America.
     *Introd.* 1781, by Mr. William Curtis.
     *Fl.* Auguft and September.          H. ♃.

## P H L E U M.    *Gen. pl.* 77.

*Cal.* 2-valvis, feffilis, uniflorus, truncatus, furcatus:
     *Valvulæ* æquales: marginibus rectis. *Cor.* 2-val-
     vis, inclufa.

1. PHLEUM.                            *pratenfe.*

                    *α* Phleum

vulgare.　*a* Phleum ſpica cylindrica longiſſima, culmo reƈto.　*Sp.*
　　　　*pl.* 87.
　　　　Meadow Cat's-tail Graſs.

nodoſum.　β Phleum ſpica cylindrica, culmo adſcendente, foliis
　　　　obliquis, radice bulboſa.　*Syſt. nat.* 92.
　　　　Bulboſe Cat's-tail Graſs.
　　　　*Nat.* of Britain.
　　　　*Fl.* July.　　　　　　　　　　　　　　H. ♃.

## ALOPECURUS.　*Gen. pl.* 78.

*Cal.* 2-valvis.　*Cor.* 1-valvis.

*indicus.*　1. A. ſpica tereti, involucellis ſetaceis faſciculatis bifloris, pedunculis villoſis.　*Syſt. veget.* 108.
　　　　Panicum alopecuroides.　*Sp. pl.* 82.
　　　　Giant Fox-tail Graſs.
　　　　*Nat.* of Jamaica.
　　　　*Cult.* 1748.　*Mill. diƈt. edit.* 5.　Panicum 6.
　　　　*Fl.* July——Oƈtober.　　　　　　　　S. ☉.

*bulboſus.*　2. A. culmo ereƈto, ſpica cylindrica, radice bulboſa.
　　　　*Syſt. veget.* 108.
　　　　Bulbous Fox-tail Graſs.
　　　　*Nat.* of England.
　　　　*Fl.* July.　　　　　　　　　　　　　H. ♃.

*pratenſis.*　3. A. culmo ſpicato ereƈto, glumis villoſis, corollis muticis,　*Syſt. veget.* 108.　*Curtis lond.*
　　　　Meadow Fox-tail Graſs.
　　　　*Nat.* of Britain.
　　　　*Fl.* May.　　　　　　　　　　　　　H. ♃.

*agreſtis.*　4. A. culmo ſpicato ereƈto, glumis lævibus.　*Syſt. veget.* 108.

　　　　　　　　　　　　　　　　　　Alopecurus

Alopecurus myofuroides. *Curtis lond.*
Field Fox-tail Grafs.
*Nat.* of Britain.
*Fl.* July and Auguft.                                    H. ♃.

5. A. culmo fpicato infracto, corollis muticis. *Syft. veget.*   *genicula-*
   108. *Curtis lond.*                                  *tus.*
Flote Fox-tail Grafs.
*Nat.* of Britain.
*Fl.* May——Auguft.                                    H. ♃.

M I L I U M. *Gen. pl.* 79.

*Cal.* 2-valvis, uniflorus : valvulis fubæqualibus. *Cor.*
  breviffima. *Stigmata* penicilliformia.

1. M. panicula fubfpicata, floribus ariftatis. *Sp. pl.* 91.  *lendige-*
Yellow-fpiked Millet-grafs.                             *rum.*
*Nat.* of England.
*Fl.* June and July.                                    H. ☉.

2. M. racemis digitatis, calycum valvula exteriore cilia-  *cimici-*
  ta. *Linn. mant.* 184.                                 *num.*
Spotted Millet-grafs.
*Nat.* of India.
*Introd.* 1778, by Sir Jofeph Banks, Bart.
*Fl.* July——September.                                    S. ☉.

3. M. floribus paniculatis difperfis muticis. *Sp. pl.* 90.  *effufum.*
  *Curtis lond.*
Common Millet-grafs.
*Nat.* of Britain.
*Fl.* June and July.                                    H. ♃.

4. M. floribus paniculatis ariftatis. *Sp. pl.* 90.     *paradox-*
Black-feeded Millet-grafs.                           *um.*
                                       *Nat.*

*Nat.* of the South of France.
*Introd.* 1771, by Monſ. Richard.
*Fl.* July.                                          H. ☉.

## AGROSTIS. *Gen. pl.* 80.

*Cal.* 2-valvis, uniflorus, corolla paulo minor. *Stigmata* longitudinaliter hiſpida.

\* *Ariſtatæ.*

ſpica ven-
ti.

1. A. petalo exteriore ariſta recta ſtricta longiſſima,
panicula patula. *Syſt. veget.* 110.
Silky Bent-graſs.
*Nat.* of England.
*Fl.* July.                                          H. ☉.

panicea.

2. A. panicula ſubſpicata, ramis ramuliſque faſciculatis,
valvulis calycinis alteraque corollina ariſtatis : ariſta corollina breviſſima.
Alopecurus ariſtatus.  *Hudſ. angl.* 28.

major.   α paniculæ ramis divulſis.
Alopecurus monſpelienſis.  *Sp. pl.* 89.
Great bearded Bent-graſs.

minor.   β paniculæ ramis adpreſſis.
Alopecurus paniceus.  *Sp. pl.* 90.
Small bearded Bent-graſs.
*Nat.* of England.
*Fl.* July.                                          H. ☉.

miliacea.

3. A. petalo exteriore ariſta terminali recta ſtricta medio-
cri.  *Sp. pl.* 91.
Millet Bent-graſs.
*Nat.* of Spain.
*Introd.* 1778, by Monſ. Thouin.
*Fl.* July.                                          H. ♃.

4. A.

4. A. calycibus elongatis, petalorum ariſta dorſali re-   *canina.*
curva, culmis proſtratis ſubramoſis. *Syſt. veget.* 110.
Brown Bent-graſs.
*Nat.* of Britain.
*Fl.* July and Auguſt.                                   H. ♃.

**\* \* *Muticæ.***

5. A. paniculæ ramulis patentibus muticis, culmo re-   *ſtolonife-*
pente, calycibus æqualibus. *Syſt. veget.* 111.         *ra.*
Creeping Bent-graſs.
*Nat.* of Britain.
*Fl.* July and Auguſt.                                   H. ♃.

6. A. panicula capillari patente, calycibus ſubulatis æqua-   *capilla-*
libus hiſpidiuſculis coloratis, floſculis muticis. *Sp.*       *ris.*
*pl.* 93.
Fine Bent-graſs.
*Nat.* of Britain.
*Fl.* Auguſt.                                             H. ♃.

7. A. panicula laxa, calycibus muticis æqualibus, culmo   *alba.*
repente. *Syſt. veget.* 111.
White Marſh Bent-graſs.
*Nat.* of Britain.
*Fl.* July.                                               H. ♃.

8. A. panicula oblonga congeſta, calycibus corolliſque   *mexica-*
acuminatis ſubæqualibus muticis. *Linn. mant.* 31.       *na.*
Mexican Bent-graſs.
*Nat.* of South America.
*Introd.* 1780, by Mr. Gilbert Alexander.
*Fl.* June——September.                                   S. ♃.

9. A. panicula contracta mutica, racemis lateralibus   *indica.*
erectis alternis. *Sp. pl.* 94.

                                                Indian

Indian Bent-graſs.
*Nat.* of India.
*Introd.* 1773, by John Earl of Bute.
*Fl.* July and Auguſt.                                    S. ☉.


*** *Cruciatæ.*

*lenta.*   10. A. ſpicis ſubternis umbellatis, floſculis muticis oblon-
gis acutis: valvulis calycinis ſubæqualibus, foliis
vaginiſque glabris.
Forked Bent-graſs.
*Nat.* of the Eaſt Indies.  *John Gerard Koenig*, M.D.
*Introd.* 1778, by Sir Joſeph Banks, Bart.
*Fl.* July and Auguſt.                                    S. ☉.


*compla-*   11. A. ſpicis umbellatis glabris, valvulis exterioribus ca-
*nata.*        lycinis ariſtatis, foliis complanatis vaginiſque glabris.
Flat-ſtalk'd Bent-graſs.
*Nat.* of Jamaica.   Mr. *Gilbert Alexander.*
*Introd.* 1779.
*Fl.* July and Auguſt.                                    S. ♃. .


A I R A.   *Gen. pl.* 81.

*Cal.* 2-valvis, 2-florus.  *Floſculi* abſque interjeċto
rudimento.

* *Muticæ.*

*aquatica.*   1. A. panicula patente, floribus muticis lævibús calyce
longioribus, foliis planis.  *Sp. pl.* 95.  *Curtis lond.*
Water Hair-graſs.
*Nat.* of Britain.
*Fl.* June and July.                                    H. ♃.


** *Ariſtatæ.*

*cæſpitoſa.*   2. A. foliis planis, panicula patente, petalis baſi villoſis
ariſtatiſque: ariſta reċta brevi.  *Sp. pl.* 96.

Turſy

Turfy Hair-grafs.
*Nat.* of Britain.
*Fl.* Auguft. H. ♃.

3. **A.** foliis fetaceis, culmis fubnudis, panicula divaricata, *flexuofa.*
   pedunculis flexuofis. *Sp. pl.* 96.
   Heath Hair-grafs.
   *Nat.* of Britain.
   *Fl.* Auguft. H. ♃.

4. **A.** foliis fetaceis : fummo fpathaceo paniculam inferne *canefcens.*
   obvolvente. *Sp. pl.* 97.
   Gray Hair-grafs.
   *Nat.* of England.
   *Fl.* July. H. ♃.

5. **A.** foliis fetaceis : vaginis angulatis, floribus panicu- *præcox.*
   lato-fpicatis, flofculis bafi ariftatis. *Sp. pl.* 97.
   *Curtis lond.*
   Early Hair-grafs.
   *Nat.* of Britain.
   *Fl.* June. H. ☉.

6. **A.** foliis fetaceis, panicula divaricata, floribus ariftatis *caryo-*
   diftantibus. *Sp. pl.* 97. *phyllea.*
   Silvery-leav'd Hair-grafs.
   *Nat.* of Britain.
   *Fl.* July. H. ☉.

## M E L I C A. *Gen. pl.* 82.

*Cal.* 2-valvis, 2-florus. *Rudimentum* floris inter
       flofculos.

1. **M.** flofculi inferioris petalo exteriore ciliato. *Sp. ciliata.*
   *pl.* 97.

H                      Ciliated

Ciliated Melic-grafs.
*Nat.* of the North of Europe.
*Introd.* 1771, by Monf. Richard.
*Fl.* July. H. ♃.

*nutans.* 2. M. petalis imberbibus, panicula nutante fimplici. *Sp. pl.* 98.
Mountain Melic-grafs.
*Nat.* of Britain.
*Fl.* June and July. H. ♃.

*uniflora.* 3. M. panicula rara, calycibus bifloris: flofculo altero hermaphrodito; altero neutro. *Retzii obf.* 1. *p.* 10. *n.* 9. *Curtis lond.*
Melica nutans. *Hudf. angl.* 37.
Single-flowered Wood Melic-grafs.
*Nat.* of Britain.
*Fl.* May and June. H. ♃.

*cærulea.* 4. M. panicula coarctata, floribus cylindricis. *Syft. veget.* 113. *Curtis lond.*
Aira cærulea. *Sp. pl.* 95.
Purple Melic-grafs.
*Nat.* of Britain.
*Fl.* Auguft. H. ♃.

*altiffima.* 5. M. petalis imberbibus, panicula ramofiffima. *Sp. pl.* 98.
Tall Melic-grafs.
*Nat.* of Siberia.
*Introd.* 1770, by Monf. Richard.
*Fl.* Auguft. H. ♃.

P O A.

## P O A. *Gen. pl.* 83.

*Cal.* 2-valvis, multiflorus. *Spicula* ovata : valvulis margine fcariofis acutiufculis.

1. P. panicula diffufa, fpiculis fexfloris linearibus. *Sp.*    *aquatica.* *pl.* 98. *Curtis lond.*
Water Meadow-grafs.
*Nat.* of England.
*Fl.* July.      H. 4.

2. P. panicula diffufa ramofiffima, fpiculis fexfloris cor-    *alpina.* datis. *Sp. pl.* 99.
Alpine Meadow-grafs.
*Nat.* of Britain.
*Fl.* June and July.      H. 4.

3. P. panicula fubdiffufa, fpiculis trifloris bafi pubefcenti-    *trivialis.* bus, culmo erecto tereti. *Sp. pl.* 99. *Curtis lond.*
Common Meadow-grafs.
*Nat.* of Britain.
*Fl.* June——Auguft.      H. 4.

4. P. panicula diffufa, fpiculis quadrifloris pubefcentibus,    *anguftifo-* culmo erecto tereti. *Sp. pl.* 99.                   *lia.*
Narrow-leav'd Meadow-grafs.
*Nat.* of Britain.
*Fl.* July.      H. ☉.

5. P. panicula erecta, fpiculis trifloris glabris, corollis    *Gerardi.* acuminatis calyce duplo longioribus. *Ger. prov.* 91.
*t.* 2. *f.* 1.
Poa Gerardi. *Allion. pedem.* 2. *p.* 245.
Panicled Meadow-grafs.
*Nat.* of the Alps of France, Italy, and Switzerland.
*Introd.* 1775, by the Doctors Pitcairn and Fothergill.
*Fl.* April and May.      H. ☉.

*pratenfis.*  6. P. panicula diffufa, fpiculis quinquefloris glabris,
culmo erecto tereti. *Sp. pl.* 99. *Curtis lond.*
Great Meadow-grafs.
*Nat.* of Britain.
*Fl.* June.                                        H. ♃.

*annua.*  7. P. panicula diffufa angulis rectis, fpiculis obtufis,
culmo obliquo compreffo. *Sp. pl.* 99. *Curtis lond.*
Annual Meadow-grafs.
*Nat.* of Britain.
*Fl.* April——September.                            H. ☉.

*maritima.*  8. P. panicula difticho-fecunda ovata inferne patula, ra-
mis fubternatis, fpiculis glabris, foliis planis, cul-
mis adfcendentibus.
Poa maritima. *Hudf. angl.* 42.
Sea Meadow-grafs.
*Nat.* of Britain.
*Fl.* June and July.                               H. ♃.

*Eragrof-*  9. P. panicula patente, pedicellis flexuofis, fpiculis fer-
*tis.*        ratis decemfloris, glumis trinerviis. *Syft. veget.* 114.
Spreading Meadow-grafs.
*Nat.* of Italy.
*Introd.* 1776, by Monf. Thouin.
*Fl.* July.                                        H. ☉.

*abyffinica.*  10. P. panicula capillari laxa erecta, fpiculis quadrifloris
lævibus lineari-lanceolatis, foliis glabris fubconvo-
lutis.
Poa abyffinica. *Jacqu. ic. mifcell.* 2. *p.* 364. *Syft.*
*veget.* 114.
Poa Tef. *Zuccagni monogr.* cum fig.
Tef. *Ludolf. æthiop.* l. 1. *c.* 9. *n.* 2. *comment. p.* 137.
                                             Smooth

Smooth upright Meadow grafs.
*Nat.* ————. Cultivated in Abyffinia. *James Bruce*, Efq.
*Introd.* 1775.
*Fl.* Auguft and September.                     G. H. ☉.

11. P. panicula laxa patentiffima capillari, foliis pilofis,    *capillaris.*
    culmo ramofiffimo. *Sp. pl.* 100.
Hair-panicled Meadow-grafs.
*Nat.* of Virginia and Canada.
*Introd.* 1781, by Monf. Thouin.
*Fl.* October and November.                     H. ☉.

12. P. panicula oblonga capillari fubverticillata, floribus    *tenella.*
    fexfloris minutiffimis nutantibus. *Syf. veget.* 114.
Small Meadow-grafs.
*Nat.* of the Eaft Indies.
*Introd.* 1781, by Monf. Thouin.
*Fl.* July and Auguft.                     S. ☉.

13. P. panicula lanceolata fubramofa fecunda: ramulis    *rigida.*
    alternis fecundis. *Syf. veget.* 114. *Curtis lond.*
Hard Meadow-grafs.
*Nat.* of England.
*Fl.* July.                     H. ☉.

14. P. panicula fecunda coarctata, culmo obliquo com-    *compreffa.*
    preffo. *Sp. pl.* 101.
Creeping Meadow-grafs.
*Nat.* of Britain.
*Fl.* June.                     H. ☉.

15. P. panicula attenuata, fpiculis fub-bifloris mucrona-    *nemorali*
    tis fcabris, culmo incurvo. *Sp. pl.* 102.

H 3                     Wood

Wood Meadow-grafs.
*Nat.* of Britain.
*Fl.* June.                                          H. ♃.

*bulbofa.*   16. P. panicula fecunda patentiufcula, fpiculis quadri-
floris.  *Sp. pl.* 102.
Bulbous Meadow-grafs.
*Nat.* of England.
*Fl.* July.                                          H. ♃.

*criftata.*   17. P. panicula fpicata, calycibus fubpilofis fubquadriflo-
ris pedunculo longioribus, petalis ariftatis.  *Syft.*
*veget.* 115.
Aira criftata.  *Sp. pl.* 94.
Crefted Meadow-grafs.
*Nat.* of Britain.
*Fl.* July and Auguft.                               H. ♃.

*ciliaris.*   18. P. panicula contracta, glumarum valvulis interiori-
bus pilofo-ciliatis.  *Syft. veget.* 115.
Ciliated Meadow-grafs.
*Nat.* of Jamaica.
*Introd.* 1776, by Monf. Thouin.
*Fl.* July and Auguft.                               S. ☉.

B R I Z A.   *Gen. pl.* 84.

*Cal.* 2 valvis, multiflorus.  *Spicula* difticha: valvulis cordatis, obtufis : interiore minuta.

*minor.*   1. B. fpiculis triangulis, calyce flofculis (7) longiore.
*Sp. pl.* 102.
Small Quaking-grafs.
*Nat.* of England.
*Fl.* June——Auguft.                                  H. ♃.
                                                     2. B.

2. B. fpiculis ovatis, calyce flofculis (7) æquali. *Sp.* *virens.*
pl. 103.
Spanifh Quaking-grafs.
*Nat.* of Spain.
*Introd.* 1787, by Mr. Zier.
*Fl.* July.                                H. ☉.

3. B. fpiculis ovatis, calyce flofculis (7) breviore. *Sp.* *media.*
pl. 103.
Middle Quaking-grafs.
*Nat.* of Britain.
*Fl.* June and July.                       H. ♃.

4. B. fpiculis cordatis, flofculis feptendecim. *Sp. pl.* 103. *maxima.*
Great Quaking-grafs.
*Nat.* of the South of Europe.
*Cult.* 1633. *Ger. emac.* 87. *f.* 3.
*Fl.* June and July.                       H. ☉.

5. B. fpiculis lanceolatis, flofculis viginti. *Sp. pl.* 103. *Eragroſ-*
Branch'd Quaking-grafs.                     *tis.*
*Nat.* of the South of Europe.
*Introd.* 1776, by Monf. Thouin.
*Fl.* July and Auguft.                      H. ☉.


D A C T Y L I S. *Gen. pl.* 86.

*Cal.* 2-valvis, compreffus : altera valvula majore carinata.

1. D. fpicis fparfis fecundis numerofis, floribus arĉte im-   *cynoſu-*
bricatis, culmo erecto.                                       *roides.*
Dactylis cynofuroides. *Sp. pl.* 104. (exclufis fynonymis Loefiingii et Raji)
American Cock's-foot-grafs.

H 4                          *Nat.*

*Nat.* of North America.
*Introd.* 1781, by Mr. William Curtis.
*Fl.* Auguft and September.                    H. ♃.

*ſtriƈta.*    2. D. ſpicis terminalibus ſubgeminis, floribus remotis
            adpreſſis, culmis foliiſque ſtriƈtis.
            Daƈtylis cynoſuroides. *Loefl. it.* 115. *Hudſ. angl.* 43.
            Sea Cock's-foot-graſs.
            *Nat.* of England.
            *Fl.* Auguft and September.                H. ♃.

*patens.*    3. D. ſpicis ſparſis ſecundis paucis, floribus arƈte imbri-
            catis, caule decumbente, foliis patentiſſimis.
            Spreading Cock's-foot-graſs.
            *Nat.* of North America.
            *Introd.* 1781, by Mr. William Curtis.
            *Fl.* July and Auguft.                    H. ♃.

*glomera-    4. D. panicula ſecunda glomerata. *Sp. pl.* 105.
ta.*          Rough Cock's-foot-graſs.
            *Nat.* of Britain.
            *Fl.* June and Auguft.                    H. ♃.


                C Y N O S U R U S.  *Gen. pl.* 87.
            *Cal.* 2-valvis, multiflorus, *Recept.* proprium unila-
                    terale, foliaceum.

*criſtatus.*  1. C. braƈteis pinnatifidis. *Sp. pl.* 105.
            Creſted Dog's-tail-graſs.
            *Nat.* of Britain.
            *Fl.* Auguft.                            H. ♃.

*echinatus.*  2. C. braƈteis pinnato-paleaceis ariſtatis. *Sp. pl.* 105.
                                                    Rough

Rough Dog's-tail-grafs.
*Nat.* of England.
*Fl.* Auguft.                                    H. ⊙.

3. C. fpica compofita : fpiculis fparfis : fructiferis erec-   *erucæfor-*
   tis, calycibus uni-biflorifque : glumis obtufis navi-   *mis.*
   cularibus : carina obtufa, corollis acuminatis.
Phalaris erucæformis. *Sp. pl.* 80.
Linear-fpiked Dog's-tail-grafs.
*Nat.* of Ruffia, and Hudfon's Bay.
*Introd.* 1773, by Monf. Richard.
*Fl.* July.                                       H. ⊙.

4. C. fpica fecunda, calycis gluma interiore fpiculis fub-   *Lima.*
   jecta. *Sp. pl.* 105.
Imbricated Dog's-tail-grafs.
*Nat.* of Spain.
*Introd.* 1776, by Monf. Thouin.
*Fl.* July and Auguft.                            H. ⊙.

5. C. fpiculis alternis fecundis feffilibus rigidis obtufis   *durus.*
   appreffis. *Sp. pl.* 105.
Rigid Dog's-tail-grafs.
*Nat.* of the South of Europe.
*Introd.* 1776, by Monf. Thouin.
*Fl.* July.                                       H. ⊙.

6. C. bracteis integris. *Sp. pl.* 106. *Jacqu. ic. mifcell.*   *cæruleus.*
   2. *p.* 66.
Blue Dog's-tail-grafs.
*Nat.* of Britain.
*Fl.* July.                                       H. ♃.

7. C. fpicis digitatis incurvatis, culmo erecto compreffo,   *coraca-*
   foliis fuboppofitis. *Sp. pl.* 106.                  *nus.*

Thick-

Thick-fpiked Dog's-tail-grafs.
*Nat.* of India.
*Cult.* 1714, by the Dutchefs of Beaufort. *Br. Muf.*
H. S. 138. *fol.* 27.
*Fl.* July——September.                                    S. ⊙.

*ægyptius.*   8. C. fpicis digitatis quaternis obtufis patentiffimis mu-
cronatis, calycibus mucronatis, caule repente, foliis
oppofitis. *Syft. veg.* 117.
Creeping Dog's-tail-grafs.
*Nat.* of Africa, Afia, and America.
*Introd.* 1770, by Monf. Richard.
*Fl.* July——September.                                    H. ⊙.

*indicus.*   9. C. fpicis digitatis linearibus, culmis compreffis decli-
natis bafi nodofis, foliis alternis. *Sp. pl.* 106.
Indian Dog's-tail-grafs.
*Nat.* of both Indies.
*Cult.* 1714, by the Dutchefs of Beaufort. *Br. Muf.*
H. S. 137. *fol.* 21.
*Fl.* Auguft.                                             S. ⊙.

*virgatus.*  10. C. panicula ramis fimplicibus, floribus feffilibus fub-
fexfloris : ultimo fterili ; infimis fubariftatis. *Sp.*
*pl.* 106.
Fine-fpiked Dog's-tail-grafs.
*Nat.* of Jamaica.
*Cult.* 1727, by Mr. Philip Miller. *R. S. no.* 273.
*Fl.* July——September.                                    S. ♃.

*aureus.*    11. C. paniculæ fpiculis fterilibus pendulis ternatis, flori-
bus ariftatis. *Sp. pl.* 107.
Golden-fpiked Dog's-tail-grafs.
*Nat.* of the South of Europe, and the Levant.
*Introd.* 1770, by Monf. Richard.
*Fl.* July.                                               H. ⊙.

FESTUCA.

## FESTUCA. *Gen. pl.* 88.

*Cal.* 2-valvis. *Spicula* oblonga, teretiuscula: glumis acuminatis.

**\* *Panicula secunda.***

1. F. panicula secunda : spiculis erectis, calycis valvula altera integra, altera acuminata. *Syst. veget.* 118. *bromoi-des.*
Barren Fescue-grafs.
*Nat.* of Britain.
*Fl.* May and June.                         H. ☉.

2. F. panicula secunda coarctata ariftata, culmo tetra-gono nudiufculo, foliis fetaceis. *Sp. pl.* 108. *ovina.*
α Gramen foliolis junceis brevibus majus, radice nigra.
*Bauh. pin.* 5.
Sheep's Fescue-grafs.
β Feftuca fpiculis viviparis. *Fl. fuec. ed.* 1. *n.* 94. *vivipara.*
Viviparous Fescue-grafs.
*Nat.* of Britain.
*Fl.* June and July.                         H. ♃.

3. F. panicula secunda scabra, fpiculis fexfloris ariftatis : *rubra.*
flofculo ultimo mutico, culmo femitereti. *Sp. pl.* 109.
Red Fescue-grafs.
*Nat.* of Britain.
*Fl.* June.                                     H. ♃.

4. F. panicula secunda oblonga, fpiculis fexfloris oblon-gis lævibus, foliis fetaceis. *Sp. pl.* 108. *duriufcu-la.*
Hard Fescue-grafs.
*Nat.* of Britain.
*Fl.* May and June.                         H. ♃.

5. F.

*elatior.* 5. F. panicula ſecunda erecta, ſpiculis ſubariſtatis: ex-
terioribus teretibus. *Sp. pl.* 111.
Tall Feſcue-graſs.
*Nat.* of Britain.
*Fl.* July.                                                H. ♃.

*myurus.* 6. F. panicula ſpicata, calycibus minutiſſimis muticis,
floribus ſcabris: ariſtis longis. *Syſt. veget.* 118.
Wall Feſcue-graſs,
*Nat.* of England.
*Fl.* June and July.                                       H. ☉.

*uniglu-* 7. F. panicula ſubſimplici coarctata ſecunda erecta ariſta-
*mis.*       ta, calyce univalvi, floſculis diſtantibus.
Lolium bromoides. *Hudſ. angl.* 55.
Sea Feſcue-graſs.
*Nat.* of England.
*Fl.* June and July.                                       H. ☉.

** *Panicula æquali.*

*decum-* 8. F. panicula erecta, ſpiculis ſubovatis muticis, calyce
*bens.*     floſculis majore, culmo decumbente. *Sp. pl.* 110.
Decumbent Feſcue-graſs.
*Nat.* of Britain.
*Fl.* July and Auguſt.                                     H. ♃.

*fluitans.* 9. F. panicula ramoſa erecta, ſpiculis ſubſeſſilibus tere-
tibus muticis. *Sp. pl.* 111. *Curtis lond.*
Flote Feſcue-graſs.
*Nat.* of Britain.
*Fl.* June and July.                                       H. ♃.

*calycina.* 10. F. panicula coarctata, ſpiculis linearibus, calyce floſculis longiore, foliis baſi barbatis. *Sp. pl.* 110.
Bearded-leav'd Feſcue-graſs.

                                                           *Nat.*

*Nat.* of Spain.
*Introd.* 1781, by Monf. Thouin.
*Fl.* June and July. H. ⊙.

## B R O M U S. *Gen. pl.* 89.

*Cal.* 2-valvis. *Spicula* oblonga, teres, difticha : arifta
infra apicem.

1. B. panicula patente, fpiculis ovatis : ariftis rectis, fe- *fecalinus.*
minibus diftinctis. *Syft. veget.* 119.
Field Brome-grafs.
*Nat.* of England.
*Fl.* May and June. H. ⊙.

2. B. panicula erectiufcula, fpicis ovatis pubefcentibus : *mollis.*
ariftis rectis, foliis molliffime villofis. *Syft. veget.*
119. *Curtis lond.*
Soft Brome-grafs.
*Nat.* of Britain.
*Fl.* June. H. ⊙.

3. B. panicula nutante, fpicis ovatis : ariftis divaricatis. *fquarro-*
*Sp. pl.* 112. *fus.*
Corn Brome-grafs.
*Nat.* of England.
*Fl.* July. H. ♃.

4. B. panicula ramofa nutante fcabriufcula, fpiculis linea- *afper.*
ribus fubteretibus decemfloris pilofis ariftatis, culmo foliifque hirtis.
Bromus afper. *Linn. fuppl.* 111. *Syft. veget. ed.* 14.
*p.* 119.
Bromus hirfutus. *Curtis lond.*
Bromus nemoralis. *Hulf. angl. ed.* 2. *p.* 51.

Bromus

Bromus ramofus. *Hudf. angl. ed.* 1. *p.* 40. *Syft. veget. ed.* 13. *p.* 102. (non 103.)

Bromus montanus. *Pollich palat.* 116. *Retzii fcandin.* 124. *Retzii obferv.* 2. *p.* 7. *n.* 4.

Wood Brome-grafs.

*Nat.* of England.

*Fl.* July and Auguft.                    H. ☉.

*fterilis.*    5. B. panicula patula, fpiculis oblongis diftichis, glumis fubulato-ariftatis. *Sp. pl.* 113. *Curtis lond.*

Barren Brome-grafs.

*Nat.* of Britain.

*Fl.* June and July.                    H. ☉.

*arvenfis.*    6. B. panicula nutante, fpiculis ovato-oblongis. *Sp. pl.* 113.

Corn Brome-grafs.

*Nat.* of Britain.

*Fl.* July.                    H. ♃.

*teEtorum.*    7. B. panicula nutante, fpiculis linearibus. *Sp. pl.* 114.

Nodding-panicled Brome-grafs.

*Nat.* of Europe.

*Introd.* 1776, by Monf. Thouin.

*Fl.* July and Auguft.                    H. ♂.

*giganteus.*    8. B. panicula nutante, fpiculis quadrifloris': ariftis brevioribus. *Sp. pl.* 114. *Curtis lond.*

Tall Brome-grafs.

*Nat.* of Britain.

*Fl.* Auguft.                    H. ♃.

*rubens.*    9. B. panicula fafciculata, fpiculis fubfeffilibus villofis: ariftis erectis. *Sp. pl.* 114.

Spanifh Brome-grafs.

                                                    *Nat.*

*Nat.* of Spain.
*Introd.* 1776, by Monf. Thouin.
*Fl.* June.                                              H. ♃.

10. B. panicula rariore patulo-erecta, fpiculis linearibus :    *madriten-*
        intermediis geminis, pedicellis fuperne incraffatis.    *fis.*
        *Syft. veget.* 120.
     Bromus muralis.    *Hudf. angl. ed.* 2. *p.* 50.
     Bromus ciliatus.    *Hudf. angl. ed.* 1. *p.* 40.
     Wall Brome-grafs.
     *Nat.* of England.
     *Fl.* June and July.                                H. ☉.

11. B. fpiculis alternis fubfeffilibus teretibus, culmo indi-    *pinnatus.*
        vifo.  *Sp. pl.* 115.
     Spiked Brome-grafs.
     *Nat.* of England.
     *Fl.* May and June.                                H. ♃.

12. B. fpicis duabus erectis alternis.   *Sp. pl.* 115.    *diftachy-*
     Two-fpiked Brome-grafs.                               *os.*
     *Nat.* of the South of Europe, and the Levant.
     *Introd.* 1772, by Monf. Richard.
     *Fl.* June.                                        H. ☉.


STIPA.   *Gen. pl.* 90.

*Cal.* 2-valvis, uniflorus.    *Cor.* valvula exteriore arifta
        terminali : bafi articulata.

1. S. ariftis lanatis.   *Sp. pl.* 115.                *pennata.*
    Soft Feather-grafs.
    *Nat.* of England.
    *Fl.* July and Auguft.                             H. ♃.

                                              2. S.

*juncea.* 2. S. ariftis nudis rectis, calycibus femine longioribus,
foliis intus lævibus. *Syft. veget.* 121.
Rufh-leav'd Feather-grafs.
*Nat.* of France and Switzerland.
*Introd.* 1772, by Monf. Richard.
*Fl.* July.                                    H. ♂.

## AVENA. *Gen. pl.* 91.

*Cal.* 2-valvis, multiflorus : arifta dorfali contorta.

*fibirica.* 1. A. paniculata, calycibus unifloris, feminibus hirfutis,
ariftis calyce triplo longioribus. *Sp. pl.* 117.
Siberian Oat-grafs.
*Nat.* of Siberia.
*Introd.* 1777, by Meffrs. Kennedy and Lee.
*Fl.* July and Auguft.                        H. ♃.

*elatior.* 2. A. paniculata, calycibus bifloris, flofculo hermaphro-
dito fubmutico, mafculo ariftato. *Syft. veget.* 121.
*Curtis lond.*
Tall Oat-grafs.
*Nat.* of Britain.
*Fl.* July and Auguft.                        H. ♃.

*penfylva-nica.* 3. A. panicula attenuata, calycibus bifloris, feminibus vil-
lofis, ariftis calyce duplo longioribus. *Sp. pl.* 117.
Penfylvanian Oat-grafs.
*Nat.* of Penfylvania.
*Introd.* 1785, by William Pitcairn, M.D.
*Fl.* July.                                     H. ☉.

*læflingi-ana.* 4. A. panicula contracta, flofculis binis hirfutis: altero
pedunculato, apice biariftato, arifta intermedia
longiori. *Syft. veget.* 121.
Spanifh Oat-grafs.
                                       *Nat.*

*Nat.* of Spain.
*Introd.* 1770, by Monf. Richard.
*Fl.* June and July. H. ☉.

5. A. paniculata, calycibus difpermis, feminibus lævibus, *fativa.*
altero ariftato. *Syft. veget.* 122.
α Avena nigra. *Bauh. pin.* 23.
Cultivated Black Oat.
β Avena alba. *Bauh. pin.* 23.
Cultivated White Oat.
*Nat.*
*Fl.* July. H. ☉.

6. A. paniculata, calycibus trifloris, receptaculo calycem *nuda.*
excedente, petalis dorfo ariftatis; tertio flofculo mu-
tico. *Syft. veget.* 122.
Naked Oat.
*Nat.*
*Fl.* July. H. ☉.

7. A. paniculata, calycibus trifloris, flofculis omnibus *fatua.*
ariftatis bafique pilofis. *Syft. veget.* 122.
Wild Oat-grafs.
*Nat.* of Britain.
*Fl.* Auguft. H. ♃.

8. A. fubfpicata, calycibus fubtrifloris bafi pilofis, foliis *pubéfcens.*
planis pubefcentibus. *Syft. veget.* 122.
Soft Oat-grafs.
*Nat.* of Britain.
*Fl.* June and July. H. ♃.

9. A. paniculata, calycibus quinquefloris: exterioribus *fterilis.*
flofculis ariftifque bafi pilofis; interioribus muticis.
*Sp. pl.* 118. *Jacqu. ic. collect.* 1. *p.* 90.
Bearded Oat-grafs.
I *Nat.*

Nat. of Barbary.
Introd. 1777, by Monf. Thouin.
Fl. July.                             H. ☉.

*flavescens.*  10. A. panicula laxa, calycibus trifloris brevibus, flosculis
omnibus ariftatis. *Sp. pl.* 118. *Curtis lond.*
Yellow Oat-grafs.
Nat. of Britain.
Fl. July.                             H. ♃.

*fragilis.*  11. A. fpicata calycibus quadrifloris, flofculo longioribus.
*Syft. veget.* 122.
Brittle Oat-grafs.
Nat. of Spain and Portugal.
Introd. 1770, by Monf. Richard.
Fl. July.                             H. ☉.

*pratenfis.*  12. A. fpicata, calycibus quinquefloris. *Sp. pl.* 119.
Meadow Oat-grafs.
Nat. of Britain.
Fl. July.                             H. ♃.

### LAGURUS. *Gen. pl.* 92.

*ovatus.*

Cal. 2-valvis: arifta villofa. Cor. petalo exteriore
ariftis 2 terminalibus; tertia dorfali retorta.

1. L. fpica ovata ariftata. *Sp. pl.* 119.
Oval-fpiked Lagurus.
Nat. of the South of Europe.
Cult. 1640. *Park. theat.* 1166. *n.* 1.
Fl. July——September.               H. ☉.

ARUNDO.

## ARUNDO. *Gen. pl.* 93.

*Cal.* 2-valvis. *Flofculi* congefti, lana cincti.

1. A. calycibus multifloris, fpicis ternis feffilibus. *Sp.*     *Bambos.*
    *pl.* 120.
Bamboo Reed-grafs, or Cane.
*Nat.* of India.
*Cult.* 1730, by Mr. Philip Miller. *R. S. no.* 409.
*Fl.*                                  S. ♄.

2. A. calycibus quinquefloris, panicula diffufa, culmo     *Donax.*
    fruticofo. *Syft. veget.* 123.
Manur'd Reed-grafs.
β Arundo indica laconica verficolor. *Mill. dict.*     verfico-
Strip'd Reed-grafs.                               lor.
*Nat.* of the South of Europe.
*Cult.* 1648, in Oxford Garden. *Hort. oxon. edit.* 1.
    *p.* 6.
*Fl.* July and Auguft.                    H. ♃.

3. A. calycibus quinquefloris, panicula laxa. *Sp. pl.* 120.     *phragmi-*
Common Reed-grafs.                         *tis.*
*Nat.* of Britain.
*Fl.* July——September.                 H. ♃.

4. A. calycibus unifloris, panicula erecta, foliis fubtus     *epigejos.*
    glabris. *Sp. pl.* 120.
Small Reed-grafs.
*Nat.* of England.
*Fl.* July.                            H. ♃.

5. A. calycibus unifloris lævibus, corollis lanuginofis,     *Calama-*
    culmo ramofo. *Syft. veget.* 123.               *groftis.*
Wood Reed-grafs.
*Nat.* of Britain.
*Fl.* June and July.                  H. ♃.

*arenaria.*  6. A. calycibus unifloris, foliis involutis mucronato-
pungentibus. *Sp. pl.* 121.
Sea Reed-grafs.
*Nat.* of Britain.
*Fl.* June and July.                              H. ♃.

*colorata.*  7. A. calycibus carinatis unifloris, corollis glabris bafi
penicillis duobus lanuginofis, foliis planis.
Phalaris arundinacea. *Sp. pl.* 80.
Reed Canary-grafs.
*Nat.* of Britain.
*Fl.* July.                                       H. ♃.

## L O L I U M.  *Gen. pl.* 95.

*Cal.* 1-phyllus, fixus, multiflorus.

*perenne.*  1. L. fpica mutica, fpiculis compreffis multifloris. *Sp.*
*pl.* 122.
Perennial Darnel, or Rye-grafs.
*Nat.* of Britain.
*Fl.* June.                                       H. ♃.

*temulen-*  2. L. fpica ariftata, fpiculis compreffis multifloris. *Syft.*
*tum.*        *veget.* 124.
Annual Darnel-grafs.
*Nat.* of Britain.
*Fl.* July and Auguft.                            H. ☉.

## R O T T B O E L L I A.  *Linn. fuppl.* 13.

*Rachis* articulata, teretiufcula, in pluribus filiformis.
*Cal.* ovato-lanceolatus, planus, fimplex vel biparti-
tus. *Flofculi* alterni in rachi flexuofa.

*incurva-*  1. R. fpica tereti fubulata, gluma calycina fubulata ad-
*ta.*        preffa bipartita. *Linn. fuppl.* 114.

Ægilops

Ægilops incurvata. *Sp. pl.* 1490.
Sea Hard-grafs.
*Nat.* of England.
*Fl.* July and Auguft.     H. ☉.

E L Y M U S. *Gen. pl.* 96.
*Cal.* lateralis, 2-valvis, aggregatus, multiflorus.

1. E. fpica erecta arcta, calycibus tomentofis flofculo   *arena-*
    longioribus. *Sp. pl.* 122.                 *rius.*
    Sea Lyme-grafs.
    *Nat.* of Britain.
    *Fl.* May and June.     H. ♃.

2. E. fpica pendula arcta, fpiculis binatis calyce longio-   *fibiricus.*
    ribus. *Sp. pl.* 123.
    Siberian Lyme-grafs.
    *Nat.* of Siberia.
    *Cult.* 1758, by Mr. Philip Miller.
    *Fl.* June and July.     H. ♃.

3. E. fpica nutante patula, fpiculis inferioribus ternatis,   *canaden-*
    fuperioribus binatis. *Sp. pl.* 123.            *fis.*
    Canadian Lyme-grafs.
    *Nat.* of Canada and Virginia.
    *Introd.* before 1699, by the Rev. John Banifter.
    *Morif. hift.* 3. *p.* 180. *no.* 10.
    *Fl.* July and Auguft.     H. ♃.

4. E. fpica erecta, fpiculis trifloris, involucro ftriato.   *virgini-*
    *Syft. veget.* 125.                         *cus.*
    Virginian Lyme-grafs.
    *Nat.* of Virginia.
    *Introd.* 1781, by Mr. William Curtis.
    *Fl.*     H. ♃.

    5. E.

*Caput medufæ.*  5. E. ſpiculis bifloris, involucris ſetaceis patentiſſimis. *Sp. pl.* 123.
Portugal Lyme-graſs.
*Nat*₁ of Portugal.
*Introd.* 1784, by Monſ. Thouin.
*Fl.* July.                                    H. ☉.

*Hyſtrix.*  6. E. ſpica erecta, ſpiculis involucro deſtitutis patenti-
bus. *Syſt. veget.* 125.
Rough Lyme-graſs.
*Nat.* of the Levant.
*Introd.* 1770, by Monſ. Richard.
*Fl.* July and Auguſt.                        H. ☉⸴

TRIANDRIA DIGYNIA. Elymus.

5. E. ſpiculis bifloris, involucris ſetaceis patentiſſimis. *Sp. pl.* 123.
*cereai*  Portugal Lyme-graſs.
hyber  *Nat*₁ of Portugal.
num.  *Introd.* 1784, by Monſ. Thouin.
vernu  *Fl.* July.                            H. ☉.

6. E. ſpica erecta, ſpiculis involucro deſtitutis patenti-
bus. *Syſt. veget.* 125.                      . ☉.
Rough Lyme-graſs.
*Nat.* of the Levant.
*Introd.* 1770, by Monſ. Richard.
*Fl.* July and Auguſt.                        H. ☉⸴

*vulgare.*  1. H. floſculis omnibus hermaphroditis ariſtatis: ordini-
bus duobus erectioribus. *Sp. pl.* 125.
Spring Barley.
*Nat.*
*Fl.* July.                                   H. ☉.

                                              2. H.

2. H. flofculis omnibus hermaphroditis ariftatis, femini-  *hexafti-*
 bus fexfariam æqualiter pofitis. *Sp. pl.* 125. *chon.*
 Winter Barley.
 *Nat.*
 *Fl.* July. H. ⊙.

3. H. flofculis lateralibus mafculis muticis, feminibus  *diftichon.*
 angularibus imbricatis. *Sp. pl.* 125.
 Common Barley.
 *Nat.*
 *Fl.* July. H. ⊙.

4. H. flofculis, lateralibus mafculis muticis, feminibus  *zeocriton.*
 angularibus patentibus corticatis. *Sp. pl.* 125.
 Battledore Barley.
 *Nat.*
 *Fl.* July. H. ⊙.

5. H. flofculis omnibus fertilibus ternis ariftatis, invo-  *bulbofum.*
 lucris fetaceis bafi ciliatis. *Sp. pl.* 125.
 Bulbous Barley.
 *Nat.* of Italy and the Levant.
 *Introd.* 1770, by Monf. Richard.
 *Fl.* July. H. ♃.

6. H. flofculis lateralibus mafculis ariftatis, involucris  *murinum.*
 intermediis ciliátis. *Sp. pl.* 126. *Curtis lond.*
 Wall Barley.
 *Nat.* of Britain.
 *Fl.* April——Auguft. H. ⊙.

7. H. flofculis omnibus ariftatis: lateralibus mafculis,  *pratenfe.*
 involucellis omnibus fcabridis.

I 4 Hordeum

Hordeum pratenſe.  *Hudſ. angl.* 56.
Hordeum murinum β.  *Sp. pl.* 126.
Meadow Barley.
*Nat.* of England.
*Fl.* June and July.                                            H. ♃.

*jubatum.*  8. H. ariſtis involucriſque ſetaceis longiſſimis. *Sp. pl.* 126.
Long-bearded Barley.
*Nat.* of Canada and Hudſon's-bay.
*Introd.* 1782, by the Hudſon's-bay Company.
*Fl.* July and Auguſt.                                          H. ♂.

## TRITICUM. *Gen. pl.* 99.

*Cal.* 2-valvis, ſolitarius, ſubtriflorus.  *Flos* obtuſiuſcu-
lus, acutus.

*æſtivum.*  1. T. calycibus quadrifloris ventricoſis glabris imbricatis
ariſtatis.  *Sp. pl.* 126.
Summer Wheat.
*Nat.*
*Fl.* July.                                                     H. ☉.

*hyber-*    2. T. calycibus quadrifloris ventricoſis lævibus imbri-
*num.*          catis ſubmuticis.  *Sp. pl.* 126.
Winter or Lammas Wheat.
*Nat.*
*Fl.* July.                                                     H. ♂.

*turgi-*    3. T. calycibus quadrifloris ventricoſis villoſis imbri-
*dum.*          catis obtuſis.  *Syſt. veget.* 126.
Turgid or Cone Wheat.
*Nat.*
*Fl.* July.                                                     H. ♂.

4. T.

4. T. calycibus bifloris nudis, flofculis longiffime ariftatis, racheos dentibus barbatis. *Sp. pl.* 127. *polonicum.*
Polifh Wheat.
*Nat.*
*Cult.* 1692. *Pluk. phyt. t.* 231. *f.* 6.
*Fl.* July. H. ☉.

5. T. calycibus truncatis quadrifloris, flofculis ariftatis hermaphroditis: intermedio neutro. *Sp. pl.* 127. *Spelta.*
Spelt Wheat.
*Nat.*
*Fl.* July and Auguft. H. ♂.

6. T. calycibus fubtrifloris: primo ariftato; intermedio fterili. *Syft. veget.* 126. *monococ-cum.*
One grained Wheat.
*Nat.*
*Cult.* 1658, in the Oxford garden. *Hort. oxon. ed.* 2.
*p.* 32.
*Fl.* July and Auguft. H. ☉.

7. T. fpica ovata compreffa bifaria, glumis calycinis co-rollinifque glabris, ariftis flofculo brevioribus. *proftra-tum.*
Triticum proftratum. *Linn. fuppl.* 114.
Agropyron triticeum. *Gærtner nov. comm. petrop.*
XIV. *pars* 1. *pag.* 540. *tab.* 19. *fig.* 4.
Trailing Wheat-grafs.
*Nat.* of Siberia.
*Introd.* 1780, by Monf. Thouin.
*Fl.* June. H. ☉.

8. T. calycibus truncatis quinquefloris, foliis involutis. *junceum.*
*Sp. pl.* 128.
Sea Wheat-grafs.
*Nat.* of Britain.
*Fl.* July. H. ♃.

9. T.

*repens.*     9. T. calycibus quadrifloris fubulatis acuminatis, foliis
              planis. *Syft. veget.* 127.
              Dog's-grafs or Couch-grafs.
              *Nat.* of Britain.
              *Fl.* July and Auguft.                    H. 4.

*tenellum.*   10. T. calycibus fubquadrifloris: flofculis muticis acu-
              tis, foliis fetaceis. *Sp. pl.* 127.
              Dwarf Wheat-grafs.
              *Nat.* of Spain.
              *Introd.* 1781, by Monf. Thouin.
              *Fl.* July.                                H. ☉.

*unilate-*    11. T. calycibus unilateralibus alternis muticis. *Linn.*
*rale.*       *mant.* 35.
              Poa loliacea. *Hudf. angl.* 43. (fed fynonyma caute
              diftinguenda).
              Spiked Sea Wheat-grafs.
              *Nat.* of Britain.
              *Fl.* June and July.                      H. ☉.

*unioloides.* 12. T. fpiculis lineari-lanceolatis carinatis diftichis.
              Cynofurus ficulus. *Jacqu. obf.* 2. *p.* 22.· *t.* 43.
              Gramen paniculis elegantiffimis, denfis, ficulum.
              *Tourn. inft.* 522.
              Gramen filiceum paniculis integris. *Bocc. fic.* 62.
              *t.* 33.
              Linear-fpiked Wheat-grafs.
              *Nat.* of Italy.
              *Cult.* 1758, by Mr. Philip Miller.
              *Fl.* July——September.                    H. ☉.

*TRIGYNIA.*

# *TRIGYNIA.*

## MONTIA. *Gen. pl.* 101.

*Cal.* 2-phyllus. *Cor.* 1-petala, irregularis. *Capf.*
    1-locularis, 3-valvis.

1. MONTIA. *Sp. pl.* 129. *Curtis lond.*          *fontana.*
Water Chick-weed.
*Nat.* of Britain.
*Fl.* April and May.          H. ☉.

## HOLOSTEUM. *Gen. pl.* 104.

*Cal.* 5-phyllus. *Petala* 5. *Capf.* 1-locularis, fubcy-
    lindracea, apice dehifcens.

1. H. floribus umbellatis. *Sp. pl.* 130.      *umbella-*
Umbel'd Holofteum.                   *tum.*
*Nat.* of England.
*Fl.* July and Auguft.          H. ☉.

## KOENIGIA. *Linn. mant.* 3.

*Cal.* 3-phyllus. *Cor.* nulla. *Sem.* 1, ovatum, nudum.

1. KOENIGIA. *Linn. mant.* 35.        *iflandica.*
Iceland Koenigia.
*Nat.* of Iceland.
*Introd.* 1773, by Sir Jofeph Banks, Bart.
*Fl.* April.          H. ☉.

## POLYCARPON. *Gen. pl.* 105.

*Cal.* 5-phyllus. *Petala* 5, minima, ovata. *Capf.*
    1-locularis, 3-valvis.

1. POLYCARPON. *Sp. pl.* 131.        *tetraphyl-*
                   Four-  *lum.*

Four-leav'd Polycarpon.
*Nat.* of England.
*Fl.* July. H. ⊙.

## MOLLUGO. *Gen. pl.* 106.

*Cal.* 5-phyllus. *Cor.* o. *Capf.* 3-locularis, 3-valvis.

*verticil-lata.*

1. M. foliis verticillatis cuneiformibus acutis, caule fub-
divifo decumbente, pedunculis unifloris. *Sp. pl.* 131.
Whorl'd Mollugo.
*Nat.* of Virginia.
*Cult.* 1759, by Mr. Philip Miller. *Mill. dict. edit.* 7.
*no.* 1.
*Fl.* June——Auguft. H. ⊙.

## MINUARTIA. *Gen. pl.* 107.

*Cal.* 5-phyllus. *Cor.* o. *Capf.* 1-locularis, 3-valvis.
*Sem.* nonnulla.

*dichoto-ma.*

1. M. floribus confertis dichotomis. *Sp. pl.* 132.
Fork'd Minuartia.
*Nat.* of Spain.
*Introd.* 1771, by Monf. Richard.
*Fl.* June and July. H. ⊙.

*Claſſis IV.*

# TETRANDRIA

## *MONOGYNIA.*

P R O T E A.  *Gen. pl.* 110, 111.

*Cor.* 4-petala (petalis ſubinde vario modo cohæren-
tibus).  *Antheræ* inſertæ petalis infra apicem.
*Sem.* 1, ſuperum, nudum.

    * Pinnatæ: *foliis pinnatis, filiformibus.*

1. P. foliis bipinnatis filiformibus hirtis, pedunculis ca-   *Serraria.*
pitulis longioribus, ſquamis calycinis ovato-lanceo-
latis hirtis. *Thunb. Protea, n.* 6. *Syſt. veget.* 136.
Cut-leav'd Protea.
*Nat.* of the Cape of Good Hope.
*Introd.* 1786, by Mr. Francis Maſſon.
*Fl.*                       G. H. ♄.

2. P. foliis bipinnatis filiformibus, capitulis ſpicatis diſ-   *ſpicata.*
tinĉtis. *Thunb. Protea, n.* 11. *Syſt. veget.* 137.
Spiked Protea.
*Nat.* of the Cape of Good Hope.
*Introd.* 1786, by Mr. Francis Maſſon.
*Fl.*                       G. H. ♄.

    ** Dentatæ: *foliis dentatis, calloſis.*

3. P. foliis quinquedentatis glabris, caule erecto, capitulo   *Conocar-*
terminali. *Thunb. Prot. n.* 14. *Syſt. veget.* 137.   *pa.*
Tooth'd-leav'd Protea.
                                 *Nat.*

*Nat.* of the Cape of Good Hope.
*Introd.* 1774, by Mr. Francis Maſſon.
*Fl.*                                         G. H. ♄.

\*\*\* Aceroſæ: *foliis filiformibus, ſubulatis.*

*pinifolia.*   4. P. foliis filiformibus, floribus racemoſis ecalyculatis
glabris.   *Thunb. Protea, n.* 20.   *Syſt. veget.* 138.
Pine-leav'd Protea.
*Nat.* of the Cape of Good Hope.
*Introd.* 1780, by Mr. William Forſyth.
*Fl.*                                         G. H. ♄.

*racemoſa.*   5. P. foliis filiformibus, floribus racemoſis calyculatis to-
mentoſis.   *Thunb. Protea, n.* 21.   *Syſt. veget.* 138.
Downy-flower'd Protea.
*Nat.* of the Cape of Good Hope.
*Introd.* 1787, by Mr. Francis Maſſon.
*Fl.*                                         G. H. ♄.

\*\*\*\* Lanceolatæ: *foliis ellipticis et lanceolatis.*

*aulacea.*   6. P. foliis ellipticis, floribus racemoſis ecalyculatis,
*Thunb. Protea, n.* 33. *tab.* 2.   *Syſt. veget.* 139.
Widow-wail-leav'd Protea.
*Nat.* of the Cape of Good Hope. Mr. *Francis Maſſon.*
*Introd.* 1774.
*Fl.*                                         G. H. ♄.

*umbella-*   7. P. foliis ellipticis, capitulis terminalibus, bracteis mul-
*ta.*            tifidis.   *Thunb. Protea, n.* 34.   *Syſt. veget.* 139.
Umbel'd Protea.
*Nat.* of the Cape of Good Hope.
*Introd.* 1787, by Mr. Francis Maſſon.
*Fl.*                                         G. H. ♄.

8. P.

8. P. foliis ellipticis glabris, capitulo terminali tomentofo.    *linearis.*
   *Thunb. Protea, n. 35. tab. 4. Syft. veget.* 139.
   Linear-leav'd Protea.
   *Nat.* of the Cape of Good Hope. Mr. *Francis Maffon.*
   *Introd.* 1774.
   *Fl.*                       G. H. ♄.

9. P. foliis lineari-cuneiformibus fericeis, capitulo ter-    *cinerea.*
   minali fericeo.
   Gray Protea.
   *Nat.* of the Cape of Good Hope. Mr. *Francis Maffon.*
   *Introd.* 1774.
   *Fl.* July and Auguft.          G. H. ♄.

10. P. foliis lanceolatis acutis capituloque terminali ro-    *Scolymus.*
   tundo glabris. *Thunb. Protea, n.* 36. *Syft. veg.* 139.
   Small fmooth-leav'd Protea.
   *Nat.* of the Cape of Good Hope.
   *Introd.* 1780, by the Countefs of Strathmore.
   *Fl.* May and June.           G. H. ♄.

11. P. foliis lanceolato-ellipticis capituloque terminali    *mellifera.*
   oblongo glabris. *Thunb. Protea, n.* 37. *Syft.*
   *veget.* 139.
   Honey-bearing Red-bark'd Protea.
   *Nat.* of the Cape of Good Hope.
   *Introd.* 1774, by Mr. Francis Maffon.
   *Fl.* December——February.      G. H. ♄.

12. P. foliis lanceolato-cuneiformibus incanis, capitulo    *plumofa.*
   terminali oblongo, petalis inferne glabris, fuperne
   pilofis : pilis longiffimis.
   Feather-flowered Protea.
   *Nat.* of the Cape of Good Hope. Mr. *Francis Maffon.*
   *Introd.* 1774.
   *Fl.* June——Auguft.         G. H. ♄.
                             13. P.

*parviflo-ra.*   13. P. foliis ellipticis obtufis callofis obliquis, capitulis
terminalibus ramulorum glabris.  *Thunb. Protea,*
n. 40. tab. 4.  *Syft. veget.* 140.
Small-flower'd Protea.
*Nat.* of the Cape of Good Hope.
*Introd.* 1774, by Mr. Francis Maſſon.
*Fl.* Auguſt.                          G. H. ♃.

*pallens.*   14. P. foliis ellipticis glabris acutis callofis, capitulo
terminali involucrato, involucro longo acuto pal-
lido.  *Thunb. Protea,* n. 41.  *Syft. veget.* 140.
Pale Protea.
*Nat.* of the Cape of Good Hope.
*Introd.* 1774, by Mr. Francis Maſſon.
*Fl.* June.                            G. H. ♃.

*conifera.*   15. P. foliis ellipticis glabris acutis callofis, capitulo ter-
minali involucrato, involucro longo acuto con-
colore.  *Thunb. Protea,* n. 42.  *Syft. veget.* 140.
Cone-bearing Protea.
*Nat.* of the Cape of Good Hope.
*Cult.* 1752, by Mr. Philip Miller.  *Mill. dict. edit.* 6.
no. 3.
*Fl.* April.                           G. H. ♃.

*Leviſa-nus.*   16. P. foliis ellipticis glabris obtufis, capitulo terminali
involucrato, involucro brevi obtuſo.  *Thunb. Prot.*
n. 43.  *Syft. veget.* 140.
Branching Protea.
*Nat.* of the Cape of Good Hope.
*Introd.* 1774, by Mr. Francis Maſſon.
*Fl.* April.                           G. H. ♃.

*ſtrobilina.*   17. P. foliis elliptico-oblongis retuſo-callofis capituloque
terminali glabris.  *Thunb. Protea,* n. 44.  *Syft.*
*veget.* 140.

Obtuſe-

Obtufe-leav'd Protea.
*Nat.* of the Cape of Good Hope.
*Introd.* 1774, by Mr. Francis Maffon.
*Fl.* March. G. H. ♄.

18. P. foliis lanceolatis fericeis, caule fruticofo, capitulis *faligna.*
oblongis involucratis. *Thunb. Protea, n.* 47.
*Syft. veget.* 140.
Willow-leav'd Protea.
*Nat.* of the Cape of Good Hope.
*Introd.* 1774, by Mr. Francis Maffon.
*Fl.* G. H. ♄.

19. P. foliis lanceolatis argenteo-tomentofis ciliatis, *argentea.*
caule arboreo, capitulis globofis. *Thunb. Protea,*
*n.* 48. *Syft. veget.* 141.
Silvery Protea.
*Nat.* of the Cape of Good Hope.
*Cult.* 1693. *Philofoph. tranf. no.* 198. *p.* 664.
*Fl.* Auguft. G. H. ♄.

\* \* \* \* \* Oblongæ: *foliis oblongis.*
20. P. foliis oblongis glabris, capitulo oblongo, fquamis *fpeciofa.*
calycinis apice barbatis. *Thunb. Protea, n.* 53.
*Syft. veget.* 141.
Oblique-leav'd Protea.
*Nat.* of the Cape of Good Hope.
*Introd.* 1786, by Mr. Francis Maffon.
*Fl.* G. H. ♄.

21. P. foliis ovatis glabris callofis, capitulo ovato, corol- *totta.*
lis cylindricis hirtis. *Thunb. Protea, n.* 54. *Syft.*
*veget.* 141.
Upright fmooth Protea.
*Nat.* of the Cape of Good Hope.
*Introd.* 1774, by Mr. Francis Maffon.
*Fl.* G. H. ♄.

K 22. P.

*hirta.*    22. P. foliis ovatis glabris, floribus lateralibus. *Thunb.*
   *Protea, n.* 55. *Syſt. veget.* 141.
  Hairy Protea.
  *Nat.* of the Cape of Good Hope.
  *Introd.* 1774, by Mr. Francis Maſſon.
  *Fl.*           G. H. ♄.

*pubera.*    23. P. foliis ovatis capituliſque terminalibus tomentoſis.
   *Thunb. Protea, n.* 56. *Syſt. veget.* 141.
  Downy-leav'd Protea.
  *Nat.* of the Cape of Good Hope.
  *Introd.* 1774, by Mr. Francis Maſſon.
  *Fl.*           G. H. ♄.

   ****** Rotundatæ: *foliis ſubrotundis.*

*cyna-*    24. P. foliis ſubrotundis petiolatis glabris. *Thunb. Protea,*
*roides.*   *n.* 59. *Syſt. veget.* 142.
  Round-leav'd Protea.
  *Nat.* of the Cape of Good Hope.
  *Introd.* 1774, by Mr. Francis Maſſon.
  *Fl.*           G. H. ♄.

## GLOBULARIA. *Gen. pl.* 112.

*Cal.* communis imbricatus; *proprius* infundibuliformis,
inferus. *Recept.* paleaceum.

*longifolia.*    1. G. caule fruticoſo, foliis omnibus lineari-lanceolatis
  integerrimis, capitulis axillaribus.
  Alypum ſive Herba terribilis procerior, cortice ci-
   nereo ſcabro, folio acuminato longiore. *Sloan.*
   *jam.* 124. *hiſt.* 1. *p.* 19. *t.* 5. *f.* 3.
  Long-leav'd Globularia.
  *Nat.* of Madeira.
  *Introd.* 1775, by Sir Joſeph Banks, Bart.
  *Fl.* July and Auguſt.     G. H. ♄.

            2. G.

2. G. caule fruticofo, foliis lanceolatis tridentatis in-  *Alypum.*
   tegrifque, capitulis terminalibus.
   Globularia Alypum.  *Sp. pl.* 139.
   Three-tooth'd-leav'd Globularia.
   *Nat.* of the South of Europe.
   *Cult.* 1739.  *Mill. dict. vol.* 2. *n.* 4.
   *Fl.* Auguft——November.            G. H. ♄.

3. G. caule herbaceo, foliis radicalibus tridentatis ; cau-  *vulgaris.*
   linis lanceolatis.  *Sp. pl.* 139.
   Common Globularia, or Blue Daify.
   *Nat.* of Europe.
   *Cult.* 1739.  *Mill. dict. vol.* 2. *n.* 1.
   *Fl.* May and June.                H. ♃.

4. G. foliis radicalibus crenato-aculeatis ; caulinis inte-  *fpinofa.*
   gerrimis mucronatis.  *Sp. pl.* 139.
   Prickly-leav'd Globularia.
   *Nat.* of Spain.
   *Cult.* before 1752, by Mr. Philip Miller.  *Mill. dict.*
   *edit.* 6. *n.* 6.
   *Fl.* May.                         G. H. ♃.

5. G. caule fubnudo, foliis cuneiformibus tricufpidatis :  *cordifolia.*
   intermedio minimo.  *Sp. pl.* 139.  *Jacqu. auftr.* 3.
   *p.* 26. *t.* 245.
   Wedge-leav'd Globularia.
   *Nat.* of Hungary, Auftria, and Switzerland.
   *Cult.* 1759, by Mr. Philip Miller.  *Mill. dict. edit.* 7.
   *no.* 5.
   *Fl.* June and July.               H. ♃.

6. G. caule nudo, foliis integerrimis lanceolatis.  *Sp.*    *nudicau-*
   *pl.* 140.  *Jacqu. auftr.* 3. *p.* 17. *t.* 230.             *lis.*
   Naked-ftalk'd Globularia.

*Nat.*

*Nat.* of Auftria.
*Cult.* 1739, by Mr. Philip Miller.  *Mill. dict. vol.* 2.
no. 2.
*Fl.* July.                                           H. ♃.

### CEPHALANTHUS. *Gen. pl.* 113.

*Cal.* communis o ; *proprius* fuperus, infundibuliformis. *Recept.* globofum, nudum.  *Sem.* 1, lanu-
ginofum.

occidenta-  1. C. foliis oppofitis ternatifque.  *Sp. pl.* 138.
lis.        American Button-wood.
            *Nat.* of North America.
            *Introd.* 1735, by Peter Collinfon, Efq. *Collinf. mfs.*
            *Fl.* Auguft.                              H. ♄.

### DIPSACUS. *Gen. pl.* 114.

*Cal.* communis polyphyllus ; *proprius* fuperus. *Recept.*
paleaceum.

fullonum.  1. D. foliis feffilibus ferratis.  *Sp. pl.* 140.
           α Dipfacus fylveftris. *Curtis lond.* *Jacqu. auftr.* 5. *p.* 1.
             t. 402.
             Wild Teafel.
fativus.   β Dipfacus fativus.  *Bauh. pin.* 385.
             Manured Teafel.
             *Nat.* of Britain.
             *Fl.* July.                              H. ♂.

lacinia-   2. D. foliis connatis finuatis.  *Sp. pl.* 141. *Jacqu.*
tus.           auftr. 5. *p.* 2. *t.* 403.
             Cut-leav'd Teafel.
             *Nat.* of Germany.
             *Cult.* 1683, by Mr. James Sutherland.  *Sutherl. hort.*
             edin. 104. no. 5.
             *Fl.* July and Auguft.                   H. ♂.
                                                      3. D.

3. D. foliis petiolatis appendiculatis. *Sp. pl.* 141. *Curtis pilofus. lond. Jacqu. auſtr.* 3. *p.* 27. *t.* 248.
Small Teaſel.
*Nat.* of Britain.
*Fl.* Auguſt. H. ♂.

## SCABIOSA. *Gen. pl.* 115.

*Cal.* communis polyphyllus; *proprius* duplex, ſuperus.
*Recept.* paleaceum ſ. nudum.

### * *Corollulis quadrifidis.*

1. S. corollulis quadrifidis æqualibus, calycibus imbri- *alpina.*
catis, floribus cernuis, foliis pinnatis : foliolis lanceo-
latis ſerratis. *Syſt. veget.* 143.
Alpine Scabious.
*Nat.* of the Alps of Switzerland and Italy.
*Cult.* 1570, by Mr. Hugh Morgan. *Lobel. adv.* 233.
*Fl.* June and July. H. ♃.

2. S. corollulis quadrifidis ſubradiantibus, calycibus im- *rigida.*
bricatis obtuſis, foliis lanceolatis ſerratis auriculatis.
*Syſt. veget.* 143.
Rough-leav'd Scabious.
*Nat.* of the Cape of Good Hope.
*Cult.* 1731, by Mr. Ph. Miller. *Mill. dict. edit.* 1. *n.* 11.
*Fl.* July. G. H. ♄.

3. S. corollulis quadrifidis æqualibus, calycibus paleiſque *tranſyl-*
ariſtatis, foliis radicalibus lyratis; caulinis pinnatifi- *vanica.*
dis. *Syſt. veget.* 143. *Jacqu. hort.* 2. *p.* 50. *t.* 111.
Tranſylvanian Scabious.
*Nat.* of Tranſylvania.
*Cult.* 1699, by Mr. Jacob Bobart. *Moriſ. hiſt.* 3.
*p.* 46. *no.* 13.
*Fl.* July. H. ☉.

4. S.

*fyriaca.*   4. S. corollulis quadrifidis æqualibus, calycibus imbricatis
paleifque ariftatis, caule dichotomo, foliis lanceolatis.
*Syft. veget.* 144.
Syrian Scabious.
*Nat.* of Syria.
*Cult.* 1699, by Mr. Jacob Bobart.   *Morif. hift.* 3.
*p.* 46. *no.* 14 *f.* 6. *t.* 14. *f.* 14.
*Fl.* July.                                   H. ☉.

*attenua-*   5. S. corollulis quadrifidis æqualibus, calycibus imbrica-
*ta.*        tis : fquamis oblongis obtufis, foliis linearibus gla-
bris integris bafique pinnatifidis.
Scabiofa attenuata.   *Linn. fuppl.* 118.
Narrow-leav'd Scabious.
*Nat.* of the Cape of Good Hope. Mr. *Francis Maffon.*
*Introd.* 1774.
*Fl.* July——September.                        G. H. ♄.

*leucan-*    6. S. corollulis quadrifidis fubæqualibus, fquamis calycinis
*tha.*        ovatis imbricatis, foliis pinnatifidis. *Syft. veget.* 144.
Snowy Scabious.
*Nat.* of the South of France.
*Cult.* 1739.   *Mill. dict. vol.* 2. *n.* 3.
*Fl.* September and October.                  H. ♃.

*Succifa.*   7. S. corollulis quadrifidis æqualibus, caule fimplici, ra-
mis approximatis, foliis lanceolato-ovatis.   *Syft.*
*veget.* 144.   *Curtis lond.*
Devil's-bit Scabious.
*Nat.* of Britain.
*Fl.* Auguft——October.                        H. ♃.

*integrifo-*  8. S. corollulis quadrifidis radiantibus, foliis indivifis :
*lia.*        radicalibus ovatis ferratis ; rameis lanceolatis, caule
herbaceo.   *Syft. veget.* 144.
Red-flower'd Annual Scabious.
                                              *Nat.*

*Nat.* of France and Switzerland.
*Cult.* 1748, by Mr. Philip Miller. *Mill. dict. edit.* 5.
*Fl.* June——Auguft.                                    H. ⊙.

9. S. corollulis quadrifidis radiantibus, foliis lanceolatis   *tatarica.*
   pinnatifidis : lobis fubimbricatis, caule hifpido.
   *Syft. veget.* 144.
   Giant Scabious.
   *Nat.* of Ruffia.
   *Introd.* 1779, by Chev. Thunberg.
   *Fl.* June and July.                                 H. ♂.

10. S. corollulis quadrifidis radiantibus, foliis pinnatifi-   *arvenfis.*
    dis incifis, caule hifpido. *Syft. veget.* 144. *Curtis*
    *lond.*
    Field Scabious.
    *Nat.* of Britain.
    *Fl.* July——October.                                H. ♃.

11. S. corollulis quadrifidis radiantibus, foliis omnibus   *fylvatica.*
    indivifis ovato-oblongis ferratis, caule hifpido.
    *Syft. veget.* 144. *Jacqu. auftr.* 4. *p.* 32. *t.* 362.
    Broad-leav'd Scabious.
    *Nat.* of Auftria and Switzerland.
    *Cult.* 1748. *Mill. dict. ed.* 5.
    *Fl.* July.                                         H. ♃.

### ** *Corollulis quinquefidis.*

12. S. corollulis quinquefidis, calycibus breviffimis, foliis   *gramun-*
    caulinis bipinnatis filiformibus. *Syft. veget.* 145.   *tia.*
    Cut-leav'd Scabious.
    *Nat.* of the South of France.
    *Cult.* 1739, by Mr. Philip Miller. *Rand. chel.*
    *no.* 10.
    *Fl.* July and Auguft.                              H. ♃.

*columba-*
*ria.*

13. S. corollulis quinquefidis radiantibus, foliis radicali-
bus ovatis crenatis ; caulinis pinnatis fetaceis.
*Sp. pl.* 143.
Fine-leav'd Scabious.
*Nat.* of Britain.
*Fl.* July and Auguft.                                     H. ♃.

*ſicula.*

14. S. corollis quinquefidis æqualibus calyce brevioribus,
foliis lyrato-pinnatifidis.   *Linn. mant.* 196.
Scabiofa divaricata.   *Jacqu. hort.* 1. *p.* 5. *t.* 15.
Sicilian Scabious.
*Nat.* of Sicily.
*Introd.* 1783, by Abbé Pourret.
*Fl.* Auguft.                                             H. ☉.

*mariti-*
*ma.*

15. S. corollulis quinquefidis radiantibus calyce brevio-
ribus, foliis pinnatis : fummis linearibus integer-
rimis.   *Sp. pl.* 144.
Sea Scabious.
*Nat.* of Italy and France.
*Cult.* 1683, by Mr. James Sutherland.   *Sutherl.*
*hort. edin.* 310. *no.* 2.
*Fl.* July.                                               H. ☉.

*ſtellata.*

16. S. corollulis quinquefidis radiantibus, foliis diffeƈtis,
receptaculis florum fubrotundis.   *Syſt. veget.* 145.
Starry Scabious.
*Nat.* of Spain.
*Cult.* 1596, by Mr. John Gerard.   *Hort. Ger.*
*Fl.* July and Auguft.                                    H. ☉.

*prolifera.*

17. S. corollulis quinquefidis radiantibus, floribus fubfeffilibus, caule prolifero, foliis indivifis.   *Syſt. veg.* 145.
Prolific Scabious.

                                                          *Nat.*

*Nat.* of Egypt.
*Cult.* 1768. *Mill. dict. edit.* 8.
*Fl.* July and Auguft. H. ☉.

18. S. corollulis quinquefidis radiantibus, foliis diffectis,    *atropur-*
   receptaculis florum fubulatis. *Syft. veget.* 145.    *purea.*
Sweet Scabious.
*Nat.*
*Cult.* 1629. *Park. parad.* 324. *n.* 3.
*Fl.* July——September. H. ♂.

19. S. corollulis quinquefidis radiantibus, foliis pinna-    *argentea.*
tifidis : laciniis linearibus, pedunculis longiffimis,
caule tereti. *Syft. veget.* 145.
Silvery Scabious.
*Nat.* of the Levant.
*Cult.* 1713, by Mr. Thomas Fairchild. *Philofoph.*
   *tranf. n.* 337. *p.* 58. *no.* 88.
*Fl.* June——October. H. ♃.

20. S. corollulis quinquefidis æqualibus, foliis fimplici-    *africana.*
bus incifis, caule fruticofo. *Syft. veget.* 146.
African Scabious.
*Nat.* of Africa.
*Introd.* 1690, by Mr. Bentick. *Br. Muf. Sloan.*
   *mfs.* 3370.
*Fl.* July——October. G. H. ♄.

21. S. corollulis quinquefidis, foliis lanceolatis fubinte-    *cretica.*
gerrimis, caule fruticofo. *Syft. veget.* 146.
Cretan Scabious.
*Nat.* of Candia and Sicily.
*Cult.* 1596, by Mr. John Gerard. *Hort. Ger.*
*Fl.* June——October. G. H. ♄.

22. S.

*gramini-*
*folia.*

22. S. corollulis quinquefidis radiantibus, foliis lineari-
lanceolatis integerrimis, caule herbaceo. *Syft.*
*veget.* 146.
Grafs-leav'd Scabious.
*Nat.* of the Alps of Switzerland and Italy.
*Cult.* 1683, by Mr. James Sutherland. *Sutherl.*
*hort. edin.* 310. *no.* 3.
*Fl.* July.                                      G. H. ♃.

*palæftina.*

23. S. corollulis quinquefidis radiantibus : laciniis omni-
bus trifidis, foliis indivifis fubferratis : fummis bafi
pinnatifidis. *Linn. mant.* 37. *Jacqu. hort.* 1.
*p.* 42. *t.* 96.
Paleftine Scabious.
*Nat.* of Paleftine.
*Introd.* 1771, by Monf. Richard.
*Fl.* July and Auguft.                           H. ☉.

*ochroleu-*
*ca.*

24. S. corollulis quinquefidis radiantibus, foliis bipinna-
tis linearibus. *Syft. veget.* 146. *Jacqu. auftr.* 5.
*p.* 19. *t.* 439.
Pale-white Scabious.
*Nat.* of Germany.
*Cult.* 1739, by Mr. Philip Miller. *Rand. chel. no.* 19.
*Fl.* July and Auguft.                           H. ♂.

*pappofa.*

25. S. corollulis quinquefidis inæqualibus, caule her-
baceo erecto, foliis pinnatifidis, feminibus ariftatis
plumofoque-pappofis. *Syft. veget.* 146.
Downy-headed Scabious.
*Nat.* of the South of Europe.
*Cult.* 1739, by Mr. Philip Miller. *Rand. chel. no.* 23.
*Fl.* July.                                       H. ☉.

KNAUTIA.

## KNAUTIA. *Gen. pl.* 116.

*Cal.* communis oblongus, fimplex, 5-10-florus; proprius fimplex, fuperus. *Corollulæ* irregulares. *Recept.* nudum.

1. K. foliis incifis, corollulis quinis calyce longioribus. *orientalis.*
  *Syft. veget.* 47.
  Oriental Knautia.
  *Nat.* of the Levant.
  *Cult.* 1713. *Philofoph. tranf. n.* 337. *p.* 58. *no.* 87.
  *Fl.* June——September. H. ☉.

## SPERMACOCE. *Gen. pl.* 119.

*Cor.* 1-petala, infundibulif. *Semina* 2, bidentata.

1. S. glabra, foliis linearibus, ftaminibus inclufis, flori- *tenuior.*
  bus verticillatis. *Syft. veget.* 148.
  Slender Button-weed.
  *Nat.* of America.
  *Cult.* 1732, by James Sherard, M. D. *Dill. elth.* 370.
  *t.* 277. *f.* 359.
  *Fl.* June——Auguft. H. ☉.

2. S. glabra, foliis lanceolatis, verticillis globofis. *Sp.* *verticil-*
  *pl.* 148.                                                    *lata.*
  Whorl'd-flower'd Button-weed.
  *Nat.* of Africa.
  *Cult.* 1732, by James Sherard, M. D. *Dill. elth.* 369.
  *t.* 277. *f.* 358.
  *Fl.* June——Auguft. S. ♄.

3. S. hifpida, foliis obovatis obliquatis. *Syft. veget.* 148. *hifpida.*
  Briftly procumbent Button-weed.

*Nat.*

*Nat.* of the Eaft Indies.
*Introd.* 1781, by Sir Jofeph Banks, Bart.
*Fl.* Auguft and September.                                    S. ☉.


## SHERARDIA.  *Gen. pl.* 120.

*Cor.* 1-petala, infundibulif.  *Semina* 2, tridentata.

*arvenfis.*   1. S. foliis omnibus verticillatis, floribus terminalibus.
              *Sp. pl.* 149.  *Curtis lond.*
              Corn Sherardia, or Field Madder.
              *Nat.* of Britain.
              *Fl.* May and June.                              H. ☉.


## ASPERULA.  *Gen. pl.* 121.

*Cor.* 1-petala, infundibulif.  *Semina* 2, globofa.

*odorata.*    1. A. foliis octonis lanceolatis, florum fafciculis pedun-
              culatis.  *Sp. pl.* 150.  *Curtis lond.*
              Sweet-fcented Woodroof.
              *Nat.* of Britain.
              *Fl.* April——June.                              H. ♃.

*arvenfis.*   2. A. foliis fenis, floribus feffilibus aggregatis terminali-
              bus.  *Sp. pl.* 150.
              Field Woodroof.
              *Nat.* of France and Germany.
              *Introd.* 1772, by Monf. Richard.
              *Fl.* July.                                      H. ☉.

*taurina.*    3. A. foliis quaternis ovato-lanceolatis, floribus fafcicu-
              latis terminalibus.  *Sp. pl.* 150.
              Broad-leav'd Woodroof.
              *Nat.* of the Alps of Switzerland and Italy.
              *Cult.* 1739, by Mr. Philip Miller.  *Rand. chel.* Ru-
              beola 3.
              *Fl.* April——June.                              H. ♃.

                                                              4. A.

4. A. foliis quaternis oblongis lateribus revolutis ob-    *craffifolia.*
tufiufculis pubefcentibus.  *Linn. mant.* 37.
Thick-leav'd Woodroof.
*Nat.* of the Levant.
*Introd.* 1775, by Monf. Thouin.
*Fl.* June.                                    H. ♃.

5. A. foliis linearibus: inferioribus fenis; intermediis    *tinctoria.*
quaternis, caule flaccido, floribus plerifque trifidis.
*Sp. pl.* 150.
Narrow-leav'd Woodroof.
*Nat.* of France, Sweden, and Siberia.
*Cult.* 1764, by Mr. James Gordon.
*Fl.* June and July.                           H. ♃.

6. A. foliis quaternis linearibus: fuperioribus oppofitis,    *cynanchi-*
caule erecto, floribus quadrifidis.  *Syft. veget.* 149.    *ca.*
Small Woodroof, or Squinancy-wort.
*Nat.* of England.
*Fl.* July.                                    H. ♃.

7. A. foliis quaternis ellipticis enerviis læviufculis, pe-    *lævigata.*
dunculis divaricatis trichotomis, feminibus fcabris.
*Syft. veget.* 149.
Shining Woodroof.
*Nat.* of the South of Europe.
*Introd.* 1775, by Monf. Thouin.
*Fl.* June.                                    H. ♃.

HOUSTONIA. *Gen. pl.* 124.

*Cor.* 1-petala, infundibuliformis. *Capfula* 2-locula-
ris, 2-fperma, fupera.

1. H. foliis radicalibus ovatis, caule compofito, pedun-    *cærulea.*
culis primis bifloris. *Sp. pl.* 152.

                                              Blue-

Blue-flower'd Houstonia.
*Nat.* of North America.
*Introd.* 1785, by Mr. Archibald Menzies.
*Fl.* most part of the Summer.                    H. ♃.


## G A L I U M.   *Gen. pl.* 125.

*Cor.* 1-petala, plana.   *Sem.* 2, subrotunda.

* *Fructu glabro.*

*rubioides.*   1. G. foliis quaternis lanceolato-ovatis æqualibus subtus
                scabris, caule erecto, fructibus glabris. *Sp. pl.* 152.
                Madder-leav'd Ladies Bedstraw.
                *Nat.* of the South of Europe.
                *Introd.* 1775, by the Doctors Pitcairn and Fothergill.
                *Fl.* July.                              H. ♃.

*palustre.*    2. G. foliis quaternis obovatis inæqualibus, caulibus
                diffusis.  *Sp. pl.* 153.
                White Ladies Bedstraw.
                *Nat.* of Britain.
                *Fl.* July.                              H. ♃.

*monta-*      3. G. foliis subquaternis linearibus lævibus, caule debili
*num.*             scabro, seminibus glabris.  *Sp. pl.* 155.
                Mountain Ladies Bedstraw.
                *Nat.* of England.
                *Fl.* July.                              H. ♃.

*uligino-*     4. G. foliis senis lanceolatis retrorsum serrato-aculeatis
*sum.*             mucronatis rigidis, corollis fructu majoribus.  *Sp.*
                *pl.* 153.
                Marsh Ladies Bedstraw.
                *Nat.* of Britain.
                *Fl.* July and August.                  H. ♃.

                                                        5. G.

5. G. foliis fenis lanceolatis carinatis fcabris retrorfum *fpurium.*
   aculeatis, geniculis fimplicibus, fructibus glabris.
   *Sp. pl.* 154.
   Corn Ladies Bedftraw.
   *Nat.* of England.
   *Fl.* June.                                        H. ☉.

6. G. foliis octonis hifpidis linearibus acuminatis fubim- *pufillum.*
   bricatis, pedunculis dichotomis.  *Sp. pl.* 154.
   Leaft Ladies Bedftraw.
   *Nat.* of England.
   *Fl.* July and Auguft.                             H. ♃.

7. G. foliis octonis linearibus fulcatis, ramis floriferis *verum.*
   brevibus.  *Sp. pl.* 155.
   Yellow Ladies Bedftraw, or Cheefe-rening.
   *Nat.* of Britain.
   *Fl.* July and Auguft.                             H. ♃.

8. G. foliis octonis ovato-linearibus fubferratis patentif- *Mollugo.*
   fimis mucronatis, caule flaccido, ramis patentibus.
   *Sp. pl.* 155.
   Great Ladies Bedftraw.
   *Nat.* of Britain.
   *Fl.* June and July.                               H. ♃.

9. G. foliis octonis lanceolatis lævibus margine fcabris, *fylvati-*
   pedunculis capillaribus, caule erecto glabro tereti.  *cum.*
   Galium fylvaticum.  *Sp. pl.* 155.
   Wood Ladies Bedftraw.
   *Nat.* of the South of Europe.
   *Cult.* 1713.  *Philofoph. tranf. n.* 337. *p.* 42. *no.* 32.
   *Fl.* July.                                        H. ♃.

                                                10. G.

*linifoli-*
*um.*

10. G. foliis fubfeptenis lineari-lanceolatis lævibus, pedunculis capillaribus, caule erecto tetragono.

Galium linifolium.    *Mill. dict.*

Rubia lævis linifolia montis virginis.    *Barrel. ic.* 583.

Flax-leav'd Ladies Bedftraw.

*Nat.* of the South of Europe.

*Cult.* 1759, by Mr. Philip Miller: *Mill. dict. edit.* 7.
*no.* 8.

*Fl.* June and July.    H. ♃.

*rigidum.*

11. G. foliis verticillatis linearibus fupra fcabris, paniculis divaricatis, caule erecto tereti pilofo-fcabriuf-
culo.

Rigid Ladies Bedftraw.

*Nat.*

*Introd.* 1778.

*Fl.* June.    H. ♃.

*arifta-*
*tum.*

12. G. foliis octonis lanceolatis lævibus, panicula capillari, petalis ariftatis, feminibus glabris.    *Syft.*
*veget.* 150.

Galium lævigatum.    *Sp. pl.* 1667.

Bearded Ladies Bedftraw.

*Nat.* of Italy.

*Introd.* 1778, by Mr. Thomas Blackie.

*Fl.* July.    H. ♃.

*glaucum.*

13. G. foliis verticillatis linearibus, pedunculis dichotomis, caule lævi.    *Syft. veget.* 151. *Jacqu. auftr.* 1.
*p.* 51. *t.* 81.

Glaucous Ladies Bedftraw.

*Nat.* of the South of Europe.

*Cult.* 1713.    *Philofoph. tranf. n.* 337. *p.* 191. *no.* 48.

*Fl.* June——September.    H. ♃.

* *Fructu*

* *Fructu hispido.*

14. G. foliis quaternis lanceolatis trinerviis glabris, caule *boreale.* erecto, feminibus hispidis. *Sp. pl.* 156.
Crofs-leav'd Ladies Bedftraw.
*Nat.* of Britain.
*Fl.* June——Auguft. H. ♃.

15. G. foliis quaternis fubovalibus pilofis enerviis, femi- *pilofum.* nibus pilofis.
Hairy Ladies Bedftraw.
*Nat.* of North America.
*Introd.* 1778, by John Fothergill, M. D.
*Fl.* June and July. H. ♃.

16. G. foliis octonis lanceolatis carinatis fcabris retror- *Aparine.* fum aculeatis, geniculis villofis, fructu hifpido.
*Sp. pl.* 157. *Curtis lond.*
Common Ladies Bedftraw, or Cleavers.
*Nat.* of Britain.
*Fl.* May——Auguft. H. ⊙.

17. G. foliis verticillatis linearibus, pedunculis bifidis, *parifienfe.* fructibus hifpidis. *Sp. pl.* 157.
Small Ladies Bedftraw.
*Nat.* of England.
*Fl.* July and Auguft. H. ⊙.

CRUCIANELLA. *Gen. pl.* 126.

*Cor.* 1-petala, infundibulif. tubo filiformi ; limbo unguiculato. *Cal.* 2-phyllus. *Sem.* 2, linearia.

1. C. erecta, foliis fenis linearibus, floribus fpicatis. *Sp.* *angufti-* *pl.* 157. *folia.*
Narrow-leav'd Crucianella.

L *Nat.*

*Nat.* of the South of France.
*Cult.* 1659, in Oxford Garden. *Hort. oxon. edit.* 2.
*p.* 161.
*Fl.* June and July. H. ☉.

*latifolia.* 2. C. procumbens, foliis quáternis lanceolatis, floribus
spicatis. *Sp. pl.* 158.
Broad-leav'd Crucianella.
*Nat.* of the South of France.
*Cult.* 1633. *Ger. emac.* 1119. *f.* 4.
*Fl.* June and July. H. ☉.

*maritima.* 3. C. procumbens suffruticosa, foliis quaternis mucrona-
tis, floribus oppositis quinquefidis. *Syst. veget.* 151.
Sea Crucianella.
*Nat.* of the South of France.
*Cult.* 1640. *Park. theat.* 275. *no.* 6.
*Fl.* June and July. G. H. ♄.

R U B I A. *Gen. pl.* 127.

*Cor.* 1-petala, campanulata. *Baccæ* 2, monospermæ.

*tinctorum.* 1. R. foliis annuis, caule aculeato. *Syst. veget.* 152.
Dyer's Madder.
*Nat.* of the South of Europe.
*Cult.* 1597. *Ger. herb.* 961. *f.* 1.
*Fl.* June. H. ♃.

*peregri-* 2. R. foliis perennantibus linearibus supra lævibus. *Syst.*
*na.* *veget.* 152.
Wild Madder.
*Nat.* of England.
*Fl.* July. H. ♃.

3. R.

3. R. foliis perennantibus fenis ellipticis lucidis, caule *lucida.*
    lævi. *Syſt. veget.* 152.
Shining-leav'd Madder.
*Nat.* of Majorca.
*Introd.* 1762, by Mr. James Gordon.
*Fl.* July.                     G. H. ♄.

4. R. foliis perennantibus ellipticis margine carinaque *fruticoſa.*
    aculeatis, caule frutefcente afpero.
Rubia fruticofa. *Jacqu. ic. colleÄ.* 1. *p.* 71.
Prickly-leav'd Madder.
*Nat.* of the Canary Iflands. Mr. *Francis Maſſon.*
*Introd.* 1779.
*Fl.* September.                G. H. ♄.

5. R. foliis perennantibus linearibus fupra fcabris. *Syſt.* *anguſtifo-*
    *veget.* 152.                              *lia.*
Narrow-leav'd Madder.
*Nat.* of Minorca.
*Introd.* 1772, by Monf. Richard.
*Fl.* July and Auguſt.           G. H. ♃.

## CATESBÆA. *Gen. pl.* 130.

*Cor.* 1-petala, infundib. longiffima, fupera. *Stamina*
    intra faucem. *Bacca* polyfperma.

1. CATESBÆA. *Sp. pl.* 159.              *ſpinoſa.*
Lilly Thorn.
*Nat.* of the Ifland of Providence.
*Introd.* 1726, by Mr. Mark Catefby. *Mill. diÄ. edit.* 8.
*Fl.* moft part of the Summer.        S. ♄.

## IXORA. *Gen. pl.* 131.

*Cor.* 1-petala, infundib. longa, fupera. *Stamina* fupra
faucem. *Bacca* 4-fperma.

*coccinea.* 1. I. foliis ovalibus femiamplexicaulibus, floribus fafci-
culatis. *Sp. pl.* 159.
Scarlet Ixora.
*Nat.* of the Eaft Indies.
*Introd.* 1690, by Mr. Bentick. *Br. Muf. Sloan. mff.*
3370.
*Fl.* July and Auguft.　　　　　　　　　　　　S. ♄.

## MITCHELLA. *Gen. pl.* 134.

*Cor.* 1-petalæ, fuperæ, binæ eidem germini. *Stigm.* 4.
*Bacca* bifida, 4-fperma.

*repens.* 1. MITCHELLA. *Sp. pl.* 161.
Creeping Mitchella.
*Nat.* of North America.
*Introd.* about 1761, by Mr. John Bartram.
*Fl.* June.　　　　　　　　　　　　　　　　H. ♄.

## CALLICARPA. *Gen. pl.* 135.

*Cal.* 4-fidus. *Cor.* 4-fida. *Bacca* 4-fperma.

*america-* 1. C. foliis ferratis fubtus tomentofis. *Syft. veget.* 153.
*na.* American Callicarpa.
*Nat.* of North America.
*Introd.* 1724, by Mr. Mark Catefby. *Mill. dict. edit.* 8.
Johnfonia.
*Fl.*　　　　　　　　　　　　　　　　　　S. ♄.

WITHERINGIA. *L'Heritier fert. angl.*

*Cor.* fubcampanulata: tubo quadrigibbo. *Cal.* minimus, obfolète 4-dentatus. *Peric.* 2-loculare.

1. WITHERINGIA. *L'Herit. fert. angl. tab.* 1.      *folanacea.*
Yellow-flower'd Witheringia.
*Nat.* of South America.
*Cult.* before 1742, by Robert James Lord Petre.
*Fl.* moft part of the year.                S. ♃.
DESCR. *Caulis* herbaceus, vix pedalis, teres, e lateribus petiolorum decurrentibus angulatus, fordide ruber, villofiufculus. *Folia* alterna, gemina, ovato-oblonga, acuta, integerrima, pilofiufcula, palmaria. *Petioli* vix unciales, fupra duobus canaliculis exarati, rubicundi. *Umbellæ* multifloræ, axillares, feffiles. *Pedunculi* teretes, glabri, femunciales. *Corolla* dilute lutea: *Tubus* fuburceolatus, gibbis quatuor obtufe tetragonus, diametro fefquilineari; laciniæ limbi trilineares. *Filamenta* albida, extus glabra, intus hirfuta.


BLÆRIA. *Gen. pl.* 139.

*Cal.* 4-partitus. *Cor.* 4-fida. *Stamina* receptaculo inferta. *Capf.* 4-locularis, polyfperma.

1. B. antheris muticis exfertis, calycibus tetraphyllis,      *ericoides.*
bracteis ternis longitudine calycis, foliis quaternis oblongo-acerofis pilofis imbricatis.
Blæria ericoides. *Sp. pl.* 162.
Heath-leav'd Blæria.
*Nat.* of the Cape of Good Hope.
*Introd.* 1774, by Mr. Francis Maffon.
*Fl.* Auguft——November.                G. H. ♄.

*muscosa.*   2. B. antheris muticis fubexfertis, calycibus monophyllis
pilofis, corollis campanulatis fuperne pilofis, floribus axillaribus, ftigmatıbus peltatis.

Mofs-leav'd Blæria.

*Nat.* of the Cape of Good Hope.   Mr. *Francis Maſſon.*

*Introd.* 1774.

*Fl.* June——Auguſt.                                    G. H. ♄.


BUDDLEA.   *Gen. pl.* 140.

*Cal.* 4-fidus.   *Cor.* 4-fida.   *Stamina* ex incifuris.
*Capſ.* 2-fulca, 2-locularis, polyfperma.

*globosa.*   1. B. foliis lanceolatis, capitulis folitariis.

Budleja globofa.   *Hope in act. harlem. vol.* 20. *part.* 2.
*pag.* 417. *tab.* 11.

Palquin.   *Feuillee it.* 3. *p.* 51. *t.* 38.

Round-headed Buddlea.

*Nat.* of Chili.

*Introd.* 1774, by Meſſrs. Kennedy and Lee.

*Fl.* May and June.                                    H. ♄.


*salvifolia.*   2. B. foliis lanceolato-ovatis cordatis rugofis.

Lantana falvifolia.   *Sp. pl.* 875.

Sage-leav'd Buddlea.

*Nat.* of the Cape of Good Hope.

*Cult.* 1760, by Mr. Philip Miller.

*Fl.* Auguſt and September.                            G. H. ♄.


PLANTAGO.   *Gen. pl.* 142.

*Cal.* 4-fidus.   *Cor.* 4-fida : limbo reflexo.   *Stamina*
longiſſima.   *Capſ.* 2-locularis, circumfciſſa.

\* *Scapo nudo.*

*major.*   1. P. foliis ovatis glabris, fcapo tereti, fpica floſculis imbricatis.   *Sp. pl.* 163.   *Curtis lond.*

α Plantago

α Plantago latifolia vulgaris. *Park. theat.* 493.
Great Plantain.

β Plantago major panicula fparfa. *Bauh. hiſt.* 3. *p.* 503.
Befom Plantain.

γ Plantago latifolia rofea, floribus quaſi in fpica difpoſitis.
*Bauh. pin.* 189.
Rofe Plantain.
*Nat.* of Britain.
*Fl.* May——July.                    H. ♃.


2. P. foliis ovatis fubdenticulatis pubefcentibus novem-        *maxima.*
nerviis, fpica cylindrica imbricata, fcapo tereti.
Plantago maxima. *Jacqu. ic. coll.* 1. *p.* 82.
Plantago. *Gmel. ſib.* 4. *p.* 71. *t.* 35.
Broad-leav'd Plantain.
*Nat.* of Siberia.
*Cult.* 1763, by Mr. James Gordon.
*Fl.* July and Auguſt.                    H. ♃.


3. P. foliis ovatis glàbris, fcapo angulato, fpica flofculis      *aſiatica.*
diſtinctis. *Sp. pl.* 163.
Afiatic Plantain.
*Nat.* of Siberia.
*Introd.* 1787, by Monf. Thouin.
*Fl.* July.                    H. ☉.


4. P. foliis ovato-lanceolatis pubefcentibus, fpica cylin-       *media.*
drica, fcapo tereti. *Sp. pl.* 163. *Curtis lond.*
Hoary Plantain.
*Nat.* of Britain.
*Fl.* May——July.                    H. ♃.

*virginica.* 5. P. foliis lanceolato-ovatis pubefcentibus fubdenticu-
latis, fpicis floribus remotis, fcapo tereti. *Syft.
veget.* 155.
Virginian Plantain.
*Nat.* of North America.
*Introd.* 1775, by Monf. Thouin.
*Fl.* June——September.                           H. ☉.

*altiffima.* 6. P. foliis lanceolatis quinquenerviis dentatis glabris,
fcapo fubangulato, fpica oblongo-cylindrica. *Sp.
pl.* 164.
Tall Plantain.
*Nat.* of Italy.
*Introd.* 1774, by Jofeph Nicholas de Jacquin, M.D.
*Fl.* June and July.                            H. ♃.

*lanceola-
ta.* 7. P. foliis lanceolatis, fpica fubovata nuda, fcapo angu-
lato. *Sp. pl.* 164. *Curtis lond.*
Rib-wort Plantain.
*Nat.* of Britain.
*Fl.* May——July.                                H. ♃.

*Lagopus.* 8. P. foliis lanceolatis fubdenticulatis, fpica ovata hirfuta,
fcapo tereti. *Syft. veget.* 155.
Round-headed Plantain.
*Nat.* of Spain and Portugal.
*Cult.* 1683, by Mr. James Sutherland. *Sutherl. hort.
edin.* 273. *no.* 5.
*Fl.* June and July.                            H. ♃.

*lufitanica.* 9. P. foliis lato-lanceolatis trinerviis fubdentatis fubpilofis, fcapo angulato, fpica oblonga hirfuta. *Sp.
pl.* 1667.
Portugal Plantain.

                                                *Nat.*

*Nat.* of Spain.
*Introd.* 1781, by P. M. A. Brouffonet, M. D.
*Fl.* July and Auguft.                                    H. ♃.

10. P. foliis lanceolatis obliquis villofis, fpica cylindrica    *albicans.*
   erecta, fcapo tereti. *Syft. veget.* 156.
   Woolly Plantain.
   *Nat.* of France and Spain.
   *Introd.* 1770, by Monf. Richard.
   *Fl.* June——September.                              H. ♃.

11. P. foliis linearibus planis, fcapo tereti hirfuto, fpica    *alpina.*
   oblonga erecta. *Syft. veget.* 156. *Jacqu. hort.* 2.
   *p.* 58. *t.* 125.
   Alpine Plantain.
   *Nat.* of Auftria and Switzerland.
   *Introd.* 1774, by William Pitcairn, M. D.
   *Fl.* June and July.                                H. ♃.

12. P. foliis femicylindraceis integerrimis bafi lanatis,    *maritima.*
   fcapo tereti. *Sp. pl.* 165.
   Sea Plantain.
   *Nat.* of Britain.
   *Fl.* July.                                         H. ♃.

13. P. foliis fubulatis triquetris ftriatis fcabris, fcapo    *fubulata.*
   tereti. *Sp. pl.* 166.
   Awl-leav'd Plantain.
   *Nat.* of the South of Europe.
   *Introd.* 1773, by John Earl of Bute.
   *Fl.* July.                                         H. ♃.

14. P. foliis linearibus dentatis, fcapo tereti. *Sp. pl.* 166.    *Corono-*
   Buckfhorn Plantain.                                          *pus.*
   *Nat.* of Britain.
   *Fl.* June——Auguft.                                H. ☉.
                                                      15. P.

*læflingii.*   15. P. foliis linearibus fubdentatis, fcapo tereti, fpica
ovata : bracteis carinatis membranaceis. *Sp.*
*pl.* 166. *Jacqu. hort.* 2. *p.* 58. *t.* 126.
Narrow-leav'd Plantain.
*Nat.* of England.
*Fl.* July and Auguft.                                    H. ☉.


** * Caule ramofo.*

*Pfyllium.*   16. P. caule ramofo herbaceo, foliis fubdentatis recurva-
tis, capitulis aphyllis. *Sp. pl.* 167.
Clammy Plantain.
*Nat.* of the South of Europe, and the Canary Iflands.
*Cult.* 1562. *Turn. herb. part.* 2. *fol.* 105. *verfo.*
*Fl.* July.                                               H. ☉.


*fquarro-*   17. P. herbacea, caulibus ramofis diffufis decumbentibus,
*fa.*              foliis linearibus integerrimis, capitulis fquarrofis.
*Murray in commentat. gotting.* 1781. *p.* 38. *tab.* 3.
*Syft. veget.* 156.
Plantago ægyptiaca. *Jacqu. ic. collect.* 1. *p.* 45.
Leafy-fpiked Plantain.
*Introd.* 1787, by Mr. Zier.
*Fl.* Auguft and September.                              H. ☉.


*indica.*   18. P. caule ramofo herbaceo, foliis integerrimis reflexis
ciliatis, capitulis foliofis. *Syft. veget.* 156.
Indian Plantain.
*Nat.* of Egypt and India.
*Cult.* 1683, by Mr. James Sutherland. *Sutherl. hort.*
*edin.* 280. *no.* 3.
*Fl.* July and Auguft.                                   H. ☉.


*Cynops.*   19. P. caule ramofo fruticofo, foliis filiformibus integerrimis ftrictis, capitulis fubfoliatis. *Sp. pl.* 167.
Shrubby

Shrubby Plantain.

*Nat.* of the South of Europe.

*Cult.* 1596, by Mr. John Gerard.  *Hort. Ger.*

*Fl.* May——Auguſt.                          H. ♄.

## SCOPARIA. *Gen. pl.* 143.

*Cal.* 4-partitus.  *Cor.* 4-partita, rotata.  *Capſ.* 1-
       locularis, 2-valvis, polyſperma.

1. S. foliis ternis, floribus pedunculatis.  *Sp. pl.* 168.      *dulcis.*
   Sweet Scoparia.
   *Nat.* of Jamaica.
   *Cult.* 1730, by Mr. Philip Miller.  *R. S. no.* 450.
   *Fl.* June——September.                          S. ☉.

## CENTUNCULUS. *Gen. pl.* 145.

*Cal.* 4-fidus.  *Cor.* 4-fida, patens.  *Stamina* brevia.
       *Capſ.* 1-locularis, circumſciſſa.

1. CENTUNCULUS.  *Sp. pl.* 169.  *Curtis lond.*      *minimus.*
   Small Centunculus, or Baſtard Pimpernel.
   *Nat.* of Britain.
   *Fl.* June——July.                          H. ☉.

## SANGUISORBA. *Gen. pl.* 146.

*Cal.* 2-phyllus.  *Germen* inter calycem corollamque.

1. S. ſpicis ovatis.  *Sp. pl.* 169.                *officinalis.*
α Pimpinella ſanguiſorba major.  *Bauh. pin.* 160.
   Common Burnet-Saxifrage.
β Pimpinella major rigida præalta auriculata Sabauda.
      *Bocc. muſ. p.* 19. *t.* 9.
   Ear-leav'd Burnet-Saxifrage.
   *Nat.* α of Britain, and β of Italy.
   *Fl.* June——Auguſt.                          H. ♃.
                                               2. S.

*media.*  2. S. ſpicis cylindricis. *Sp. pl.* 169.
Short-ſpiked Burnet-Saxifrage.
*Nat.* of Canada.
*Introd.* 1785, by Mr. John Bell.
*Fl.* July——September. H. ♃.

*canaden-*  3. S. ſpicis longiſſimis. *Sp. pl.* 169.
*ſis.*  Canadian Burnet-Saxifrage.
*Nat.* of North America.
*Cult.* 1640. *Park. theat.* 583. *f.* 4.
*Fl.* June——September. H. ♃.

C I S S U S. *Gen. pl.* 147.

*Bacca* 1-ſperma, cinĉta calyce *Corollaque* quadripartita.

*vitiginea.*  1. C. foliis cordatis ſubquinquelobis tomentoſis. *Sp.*
*pl.* 170.
Vine-leav'd Ciſſus.
*Nat.* of India.
*Introd.* about 1772.
*Fl.* S. ♄.

*Sicyoi-*  2. C. foliis ſubcordatis nudis ſetaceo-ſerratis, ramulis te-
*des.*  retibus. *Syſt. veget.* 158.
Heart-leav'd Ciſſus.
*Nat.* of Jamaica.
*Cult.* before 1768, by Mr. Philip Miller. *Mill. diĉt.*
*edit.* 8.
*Fl.* S. ♄.

*acida.*  3. C. foliis ternatis obovatis glabris carnoſis inciſis.
*Syſt. veget.* 158.
Three-leav'd Ciſſus.
*Nat.* of Jamaica.

*Cult.*

*Cult.* 1692, in the Royal Garden at Hampton-court.
   *Pluk. phyt. t.* 152. *f.* 2.
*Fl.*                                                          S. ♄.


## EPIMEDIUM. *Gen. pl.* 148.

*Nectaria* 4, cyathiformia, petalis incumbentia.   *Cor.*
   4-petala.  *Cal.* caducus.  *Siliqua.*


1. EPIMEDIUM.  *Sp. pl.* 171.                                *alpinum.*
Barren-wort.
*Nat.* of the Alps of Italy.
*Cult.* 1590, by Mr. John Gerard.  *Ger. herb.* 389.
*Fl.* April and May.                                  H. ♃.


## CORNUS. *Gen. pl.* 149.

*Involucrum* 4-phyllum fæpius.   *Petala* fupera, 4.
   *Drupa* nucleo 2-loculari.

*** *Involucratæ.***

1. C. herbacea, ramis binis.  *Sp. pl.* 171.             *fuecica.*
Herbaceous Dogwood.
*Nat.* of Britain.
*Fl.* June.                                            H. ♃.


2. C. herbacea, ramis nullis.  *Sp. pl.* 172.  *L'Herit.*  *canaden-*
   *corn. n.* 2. *tab.* 1.                                    *fis.*
Canadian Dogwood.
*Nat.* of Canada.
*Introd.* 1774, by John Fothergill, M. D.
*Fl.* Auguft.                                          H. ♃.


3. C. arborea, involucro maximo : foliolis obcordatis.  *florida.*
   *Sp. pl.* 171.
Great-flower'd Dogwood.

                                               *Nat.*

*Nat.* of North America.

*Cult.* 1739, by Mr. Philip Miller. *Mill. dict. vol.* 2. *no.* 3.

*Fl.* April and May.    H. ♄.

*mascula.*    4. C. arborea, umbellis involucrum æquantibus. *Sp. pl.* 171.

Cornelian Cherry or Dogwood.

*Nat.* of Austria.

*Cult.* 1596, by Mr. John Gerard. *Hort. Ger.*

*Fl.* February——April.    H. ♄.

\* \* *Nudæ.*

*sangui-*    5. C. ramis rectis, foliis ovatis concoloribus, cymis
*nea.*          depreffis. *L'Herit. corn. n.* 5.

Cornus fanguinea. *Sp. pl.* 171.

Common Dogwood.

*Nat.* of Britain.

*Fl.* June and July.    H. ♄.

*sericea.*    6. C. ramis patulis, foliis ovatis fubtus ferrugineo feri-
                ceis, cymis depreffis. *L'Herit. corn. n.* 6. *tab.* 2.

Cornus fericea. *Linn. mant.* 199.

Cornus Amomum. *Mill. dict. du Roi hort. barbecc.*
    164. *obf. bot.* 7.

Cornus femina, floribus candidiffimis umbellatim dif-
    pofitis, baccis cæruleo-viridibus. *Gron. virg.* 20.

Cornus americana fylveftris, domefticæ fimilis, bacca
    cærulei coloris. *Pluk. phyt. t.* 169. *f.* 3.

Blue-berried Dogwood.

*Nat.* of North America.

*Cult.* 1759, by Mr. Philip Miller. *Mill. dict. edict.* 7.
    *n.* 5.

*Fl.* Auguft.    H. ♄.

7. C.

7. C. ramis recurvatis, foliis lato-ovatis fubtus canis,  *alba.*
   cymis depreffis. *L'Herit. corn. n.* 7.
   Cornus alba. *Linn. mant.* 40.
   Cornus tartarica. *Mill. dict.*
   White-berried Dogwood.
   *Nat.* of North America and Siberia.
   *Cult.* 1759, by Mr. Philip Miller. *Mill. dict. ed.* 7.
   *n.* 7.
   *Fl.* June——September.        H. ♄.

8. C. ramis ftrictis, foliis ovatis concoloribus nudiufculis, *ftricta.*
   cymis paniculatis. *L'Herit. corn. n.* 9. *tab.* 4.
   Upright Dogwood.
   *Nat.* of North America.
   *Cult.* 1758, by Mr. Philip Miller.
   *Fl.* June and July.        H. ♄.

9. C. ramis erectis, foliis ovatis fubtus canis, cymis pani- *panicula-*
   culatis. *L'Herit. corn. n.* 10. *tab.* 5.    *ta.*
   New Holland Dogwood.
   *Nat.* of North America.
   *Cult.* 1758, by Mr. Philip Miller.
   *Fl.* June and July.        H. ♄.

10. C. foliis alternis. *Linn. fuppl.* 125. *L'Herit. corn.* *alternifo-*
   *n.* 11. *tab.* 6.    *lia.*
α ramis rubris.        *corallina.*
   Red-twig'd Alternate-leav'd Dogwood.
β ramis viridibus.        *virefcens.*
   Green-twig'd Alternate-leav'd Dogwood.
   *Nat.* of North America.
   *Cult.* 1760, by Mr. James Gordon.
   *Fl.* September.        H. ♄.

SAMARA.

### SAMARA. *Linn. mant.* 144.

*Cal.* 4-phyllus. *Cor.* 4-petala. *Stam.* petalis inferta.
*Drupa* 1-fperma.

*pentan-*
*dra.*
1. S. foliis pentandris, foliis ellipticis.
Pentandrous Samara.
*Nat.* of the Cape of Good Hope.
*Introd.* about 1770.
*Fl.* November——February.       G. H. ♄.

### CHLORANTHUS. *Swartz.*

*Cal.* o. *Petalum* trilobum lateri germinis infidens.
*Antheræ* petalo accretæ. *Bacca* 1-fperma.

*inconfpi-*
*cuus.*
1. CHLORANTHUS. *Swartz in philofoph. tranfaƈ. vol.*
    77. *pag.* 359. *tab.* 14. *L'Herit. fert. angl. tab.* 2.
Nigrina fpicata. *Thunb. japon.* 65.
Tea-leav'd Chloranthus, or Chu-lan.
*Nat.* of China.
*Introd.* 1781, by James Lind, M. D.
*Fl.* moft part of the year.       S. ♄.

### MONETIA. *L'Heritier ftirp. nov.*

*Cal.* 4-fidus. *Pet.* 4. *Bacca?* 2-locularis. *Sem.* folitaria.

*barleri-*
*oides.*
1. MONETIA. *L'Heritier ftirp. nov. p.* 1. *t.* 1.
Azima tetracantha. *de Lamarck encyclop.* 1. *p.* 343.
Four-fpin'd Monetia.
*Nat.* of the Eaft Indies.
*Cult.* 1758, by Mr. Philip Miller.
*Fl.* July.       S. ♄.

### FAGARA.

## FAGARA. *Gen. pl.* 150.

*Cal.* 4-fidus. *Cor.* 4-petala. *Capf.* bivalvis, monofperma.

1. F. foliolis emarginatis. *Sp. pl.* 172.      *Pterota.*
Lentifcus-leav'd Fagara.
*Nat.* of Jamaica.
*Cult.* 1768, by Mr. Philip Miller. *Mill. diĉt. edit.* 8.
*Fl.* Auguft and September.        S. ♄.

2. F. foliolis crenatis. *Sp. pl.* 172.      *Piperita.*
Afh-leav'd Fagara.
*Nat.* of Japan.
*Intrôd.* 1773, by Sir James Cockburn, Bart.
*Fl.* September.        G. H. ♄.

3. F. articulis pinnarum fubtus aculeatis. *Sp. pl.* 172.   *Tragodes.*
Prickly-leav'd Fagara.
*Nat.* of the Weft Indies.
*Cult.* 1759, by Philip Miller. *Mill. diĉt. edit.* 7.
   Schinus 2.
*Fl.*        S. ♄.

## ÆGIPHILA. *Lin. mant.* 144.

*Cal.* 4-dentatus. *Cor.* 4-fida. *Stylus* femibifidus.
*Bacca* 4-fperma.

1. ÆGIPHILA. *Linn. mant.* 198.      *martini-*
Martinico Ægiphila.             *cenfis.*
*Nat.* of the Weft Indies.
*Introd.* 178·, by Mr. Francis Maffon.
*Fl.* November.        S. ♄.

M      CURTISIA.

## C U R T I S I A.

*Cal.* 4-part. *Pet.* 4. *Drupa* fupera, fubrotunda, fucculenta : *Nucleo* 4-5-loculari.

*faginea.*  1. CURTISIA.
Sideroxylon foliis acuminatis dentatis, fructu mono-
pyreno flavo. *Burm. afr.* 235. *t.* 82.
Beech-leav'd Curtifia, or Haffagay-tree.
*Nat.* of the Cape of Good Hope. Mr. *Francis Maffon.*
*Introd.* 1775.
*Fl.*                                               G. H. ♄.

## P T E L E A.    *Gen. pl.* 152.

*Cor.* 4-petala.  *Cal.* 4-partitus, inferus.  *Fructus*
membrana fubrotunda, centro monofpermos.

*trifoliata.*  1. P. foliis ternatis.  *Sp. pl.* 173.
Shrubby Trefoil.
*Nat.* of Carolina and Virginia.
*Cult.* 1724, by James Sherard, M. D.  *Dill. eltham.*
147.
*Fl.* June and July.                                H. ♄.

## L U D W I G I A.    *Gen. pl.* 153.

*Cor.* tetrapetala.  *Cal.* 4-partitus, fuperus.  *Capf.* 4-
gona, 4-locularis, infera, polyfperma.

*alternifo-*  1. L. foliis alternis lanceolatis, caule erecto.  *Syft. veget.*
*lia.*          161.
Alternate-leav'd Ludwigia.
*Nat.* of Virginia.
*Introd.* before 1752, by Thomas Dale, M. D.  *Mill.*
*dict. edit.* 6. *no* 1.
*Fl.* June and July.                                H. ⊙.

OLDEN-

OLDENLANDIA. *Gen. pl.* 154.

*Cor.* tetrapetala. *Cal.* 4-partitus, fuperus. *Capf.* 2-locularis, infera, polyfperma.

1. O. pedunculis multifloris, foliis lineari-lanceolatis.    *corymbofa.*
    *Sp. pl.* 174.
    Hyffop-leav'd Oldenlandia.
    *Nat.* of Jamaica.
    *Introd.* before 1739, by Mr. Robert Millar. *Mill. dict.*
        *vol.* 2.
    *Fl.* June——October.                                         S. ☉.

AMMANNIA. *Gen. pl.* 155.

*Cor.* 4-petala, calyci inferta, vel nulla. *Cal.* 1-phyllus, plicatus, 8-dentatus, inferus. *Capf.* 4-locularis.

1. A. foliis femiamplexicaulibus, caule tetragono, ramis    *latifolia.*
    erectis. *Syft. veget.* 162.
    Broad-leav d Ammannia.
    *Nat.* of the Weft Indies.
    *Introd.* about 1731, by W. Houftoun, M. D. *Mill.*
        *dict. ed.* 8. *n.* 1.
    *Fl.* July and Auguft.                                       S. ☉.

2. A. foliis femiamplexicaulibus, caule tetragono, ramis    *ramofior.*
    patentiffimis. *Syft. veget.* 162.
    Branching Ammannia.
    *Nat.* of Virginia.
    *Cult.* 1759. *Mill. dict. ed.* 7. *n.* 2.
    *Fl.* July.                                                  H. ☉.

3. A. foliis lanceolatis bafi attenuatis, caule ramofo,    *debilis.*
    floribus fafciculatis axillaribus, capfulis bilocularibus.
    Clufter-flower'd Ammannia.
    *Nat.* of the Eaft Indies.

*Introd.* 1778, by Sir Jofeph Banks, Bart.
*Fl.* July and Auguft.                                    S. ☉.
Descr. *Calyx* angulatus. *Petala* pallide purpurea.
  *Filamenta* fundo calycis inferta, eoque breviora. *An-*
  *theræ* ovatæ, flavæ. *Capfula* ovata, bilocularis.


## ISNARDIA. *Gen. pl.* 156.

*Cor.* nulla. *Cal.* 4-fidus. *Capf.* 4-locularis, cincta *calyce.*

*paluftris.*   1. Isnardia. *Sp. pl.* 175.
          Marfh Ifnardia.
          *Nat.* of Europe, North America, and the Weft Indies.
          *Introd.* 1776, by John Fothergill, M.D.
          *Fl.* July.                                    H. ☉.


## TRAPA. *Gen. pl.* 157.

*Cor.* 4-petala. *Cal.* 4-partitus. *Nux* fpinis 4 oppofitis cincta, quæ calycis folia fuere.

*natans.*   1. Trapa. *Sp. pl.* 175.
          Floating Water-caltrops.
          *Nat.* of Europe.
          *Introd.* 1781, by Daniel Charles Solander, LL.D.
          *Fl.* June——Auguft.                          H. ☉.


## ELÆAGNUS. *Gen. pl.* 159.

*Cor.* nulla. *Cal.* 4-fidus, campanulatus, fuperus.
  *Drupa* infra *calycem* campanulatum.

*anguftifo-*  1. E. inermis, foliis lanceolatis. *Syft. veget.* 163.
*lia.*        Narrow-leav'd Oleafter.
          *Nat.* of the South of Europe, and the Levant.
                                                      *Cult.*

*Cult.* 1633, by Mr. John Parkinſon. *Ger. emac.* 1491.
*f.* 2.
*Fl.* July.                                                   H. ♄.

2. E. inermis, foliis oblongis ovatis opacis. *Syſt. veget.* *orientalis.*
   163.
Oriental Oleaſter.
*Nat.* of the Levant.
*Introd.* 1783, by Mr. John Græffer.
*Fl.*                                                         G. H. ♄.

### STRUTHIOLA. *Linn. mant.* 4.

*Cor.* nulla. *Cal.* tubuloſus: ore glandulis 8. *Bacca*
           exſucca, monoſperma.

1. S. glabra. *Linn. mant.* 41.                               *erecta.*
Smooth Struthiola.
*Nat.* of the Cape of Good Hope.
*Cult.* 1758, by Mr. Philip Miller.
*Fl.* June——Auguſt.                                          G. H. ♄.

### RIVINA. *Gen. pl.* 162.

*Cor.* 4-petala, perſiſtens. *Cal.* nullus. *Bacca* 1-
           ſperma: *Semine* lentiformi.

1. R. floribus tetrandris, foliis ovatis pubeſcentibus.      *humilis.*
Rivina humilis. *Sp. pl.* 177.
Downy Rivina.
*Nat.* of the Weſt Indies.
*Cult.* before 1699, in Chelſea Garden. *Moriſ. hiſt.* 3.
   *p.* 522. *no.* 23.
*Fl.* moſt part of the year.                                 S. ♄.

2. R. floribus tetrandris, foliis ovatis lævibus.            *lævis.*
Rivina lævis. *Linn. mant.* 41, 512.
                   M 3                    Smooth

Smooth Rivina.
*Nat.* of the Weſt Indies.
*Cult.* 1733, by Mr. Philip Miller. *R. S. no.* 586.
*Fl.* moſt part of the year.                    S. ♄.

*octandra.*  3. R. floribus octandris, foliis ellipticis glabris.
Rivina octandra. *Sp. pl.* 177.
Climbing Rivina.
*Nat.* of the Weſt Indies.
*Cult.* before 1752, by Mr. Philip Miller. *Mill. dict.*
*edit.* 6. *n.* 2.
*Fl.*                                           S. ♄.

## CAMPHOROSMA. *Gen. pl.* 164.

*Cal.* urceolatus: dentibus 2 oppoſitis, alterniſque mi-
nimis. *Cor.* nulla. *Capſ.* 1-ſperma.

*monſpe-*   1. C. foliis hirſutis linearibus. *Sp. pl.* 178.
*liaca.*      Hairy Camphoroſma.
*Nat.* of the South of Europe.
*Cult.* 1739. *Mill. dict. vol.* 2. Camphorata 1.
*Fl.* Auguſt and September.                    G. H. ♄.

## ALCHEMILLA. *Gen. pl.* 165.

*Cal.* 8-fidus. *Cor.* 0. *Sem.* 1.

*vulgaris.*  1. A. foliis lobatis. *Sp. pl.* 178.
α Alchemilla vulgaris. *Bauh. pin.* 319.
Common Ladies Mantle.
*hybrida.*   β Alchemilla alpina pubeſcens minor. *Tournef. inſt.*
508.
Pubeſcent Ladies Mantle.
*Nat.* of Britain.
*Fl.* June —— Auguſt.                           H. ♃.

2. A.

2. A. foliis digitatis ferratis. *Sp. pl.* 179.    *alpina.*
Alpine Ladies Mantle.·
*Nat.* of Britain.
*Fl.* July.    H. ♃.

3. A. foliis quinatis multifidis glabris, *Sp. pl.* 179.    *penta-*
Five-leav'd Ladies Mantle.    *phyllea.*
*Nat.* of the Alps of Switzerland.
*Cult.* 1748, by Mr. Philip Miiler.    *Mill. dict. edit.* 5.
*no.* 5.
*Fl.* June and July.    H. ♃.


# D I G Y N I A.

## A P H A N E S.    *Gen. pl.* 166.

*Cal.* 4-fidus.    *Cor.* o.    *Sem.* 2, nuda.

1. APHANES.    *Sp. pl.* 179.    *arvenſis.*
Parſley Piert.
*Nat.* of Britain.
*Fl.* May.    H. ☉.


## H A M A M E L I S.    *Gen. pl.* 169.

*Involucr.* 3-phyllum.    *Cal.* proprius 4-phyllus.    *Pe-*
*tala* 4.    *Nux* 2-cornis, 2-locularis.

1. HAMAMELIS.    *Sp. pl.* 180.    *virginica.*
Witch Hazel.
*Nat.* of North America.
*Introd.* 1736, by Peter Collinſon, Eſq. *Coll. mſs.*
*Fl.* November——May.    H. ♄.

M 4    CUSCU-

## C U S C U T A. *Gen. pl.* 170.

*Cal.* 4-fidus. *Cor.* 1-petala. *Capf.* 2-locularis.

*europæa.* 1. C. floribus feffilibus. *Sp. pl.* 180.
Dodder.
*Nat.* of England.
*Fl.* July. H. ☉

## H Y P E C O U M. *Gen. pl.* 171.

*Cal.* 2-phyllus. *Petala* 4 : exterioribus duobus latioribus, 3-fidis. *Fructus* filiqua.

*procum-* 1. H. filiquis arcuatis compreffis articulatis. *Sp. pl.* 181.
*bens.* Procumbent Hypecoum.
*Nat.* of the South of Europe.
*Cult.* before 1597, by Mr. John Gerard. *Ger. herb.*
909. *n.* 3.
*Fl.* June and July. H. ☉

*pendulum.* 2. H. filiquis cernuis teretibus cylindricis. *Sp. pl.* 181.
Pendulous Hypecoum.
*Nat.* of the South of France.
*Cult.* 1640. *Park. theat.* 372. *f.* 2.
*Fl.* June and July. H. ☉

# *T E T R A G Y N I A.*

## I L E X. *Gen. pl.* 172.

*Cal.* 4-dentatus. *Cor.* rotata. *Stylus* o. *Bacca*
4-fperma.

*Aquifoli-* 1. I. foliis ovatis acutis fpinofis nitidis undulatis, floribus
*um.* axillaribus fubumbellatis.

Ilex

Ilex Aquifolium. *Sp. pl.* 181.

α Ilex aculeata baccifera. *Bauh. pin.* 425.    vulgaris.
Common Holly.

β foliis dentatis fpinofis integrifque.    hetero-
Various-leav'd Holly.    phylla.

γ foliis craffioribus æqualiter ferratis.    craffifo-
Thick-leav'd Holly.    lia.

δ foliis anguftioribus recurvatis.    recurva.
Slender Holly.

ε foliis fupra echinatis.    ferox.
Hedge-hog Holly.
*Nat.* of Britain.
*Fl.* May and June.    H. ♄.


2. I. foliis ovatis acutis fpinofis glabris planis, floribus ad    *opaca.*
bafin ramulorum annotinorum fparfis.
Carolina Holly.
*Nat.* of Carolina.
*Cult.* 1744, by Archibald Duke of Argyle.
*Fl.* May and June.    H. ♄.


3. I. foliis ovatis cum acumine inermibus fubintegris.    *Perado.*
Thick-leav'd Smooth Holly.
*Nat.* of Madeira.
*Introd.* 1760, by Mr. James Gordon.
*Fl.* April and May.    G. H. ♄.


4. I. foliis elliptico-lanceolatis acutis deciduis ferratis :    *Prinoi-*
ferraturis muticis.    *des.*
Deciduous Holly.
*Nat.* of Carolina and Virginia.
*Cult.* before 1760, by Archibald Duke of Argyle.
*Fl.* July.    H. ♄.


5. I.

*Caſſine.*   5. I. foliis alternis diſtantibus ſempervirentibus lanceolatis ſerratis : ſerraturis acuminatis.

Ilex Caſſine.   *Sp. pl.* 181.

latifolia.   α foliis lanceolato-oblongis ſerratis.
Broad-leav'd Dahoon Holly.

anguſti-   β foliis lanceolatis ſubintegerrimis.
folia.    Narrow-leav'd Dahoon Holly.

*Nat.* of Carolina and Florida.

*Introd.* about 1726, by Mr. Mark Cateſby.   *Mill.
dict. edit.* 8. *n.* 3.

*Fl.* Auguſt.                                    H. ♄.

vomito-   6. I. foliis alternis diſtantibus oblongis obtuſiuſculis cre-
ria.       nato-ſerratis : ſerraturis muticis.

Caſſine *Paragua* foliis lanceolatis alternis ſempervirentibus, floribus axillaribus.   *Mill. dict. ic. t.* 83.
*f.* 2.

Caſſine vera Floridanorum arbuſcula baccifera, alater-
ni ferme facie, foliis alternatim ſitis, tetrapyrene.
*Pluk. mant.* 40. *t.* 376. *f.* 2.   *Cateſb. car.* 2. *p.* 57.
*t.* 57.

South-Sea Tea, or Ever-green Caſſine.

*Nat.* of Weſt Florida.

*Cult.* 1700.   *Pluk. mant. loc. cit.*

*Fl.*                                            H. ♄.

## COLDENIA.   *Gen. pl.* 173.

*Cal.* 4-phyllus.   *Cor.* infundibuliformis.   *Styli* 4.   *Sem.* 2,
bilocularia.

procum-   1. COLDENIA.   *Sp. pl.* 182.
bens.     Trailing Coldenia.

*Nat.* of the Eaſt Indies.

*Cult.* 1759, by Mr. Philip Miller.   *Mill. dict. edit.* 7.

*Fl.* July and Auguſt.                          S. ☉.

POTAMO-

# POTAMOGETON. *Gen. pl.* 174.

*Cal.* o.　*Petala* 4.　*Stylus* o.　*Sem.* 4.

1. P. foliis oblongo-ovatis petiolatis natantibus. *Sp. pl.*   *natans.*
　182.
　Broad-leav'd Pondweed.
　*Nat.* of Britain.
　*Fl.* Auguft.                     H. ♃.

2. P. foliis cordatis amplexicaulibus. *Sp. pl.* 182.   *perfolia-*
　Perfoliate Pondweed.                              *tum.*
　*Nat.* of Britain.
　*Fl.* June and July.              H. ♃.

3. P. foliis ovatis acuminatis oppofitis confertis, caulibus   *denfum.*
　dichotomis, fpica quadriflora. *Sp. pl.* 182.
　Forked Pondweed, or Leffer Water-caltrops.
　*Nat.* of Britain.
　*Fl.* June.                        H. ♃.

4. P. foliis lanceolatis planis in petiolos definentibus.   *lucens.*
　*Sp. pl.* 183.
　Shining Pondweed.
　*Nat.* of Britain.
　*Fl.* June.                        H. ♃.

5. P. foliis lanceolatis alternis oppofitifve undulatis ferratis.   *crifpum.*
　*Syft. veget.* 169.  *Curtis lond.*
　Curl'd Pondweed, or Greater Water-caltrops.
　*Nat.* of Britain.
　*Fl.* May and June.             H. ♃.

6. P. foliis linearibus obtufis, caule compreffo. *Sp. pl.* 183.   *compref-*
　Flat-ftalk'd Pondweed.                          *fum.*

*Nat.*

*Nat.* of Britain.
*Fl.* June and July.     H. ♃.

*pectina-*
*tum.*

7. P. foliis setaceis parallelis approximatis distichis. *Sp.*
    *pl.* 183.
Fine leav'd Pondweed.
*Nat.* of Britain.
*Fl.* June and July.     H. ♃.

*gramine-*
*um.*

8. P. foliis lineari-lanceolatis alternis sessilibus stipula la-
    tioribus. *Sp. pl.* 184.
Grassy Pondweed.
*Nat.* of Britain.
*Fl.* June.     H. ♃.

*pusillum.*

9. P. foliis linearibus oppositis alternisque distinctis
    basi patentibus, caule tereti. *Sp. pl.* 184.
Small Grafs-leav'd Pondweed.
*Nat.* of Britain.
*Fl.* July and August.     H. ♃.

## S A G I N A. *Gen. pl.* 176.

*Cal.* 4-phyllus. *Petala* 4. *Capf.* 1-locularis, 4-valvis,
polyfperma.

*procum-*
*bens.*

1. S. ramis procumbentibus. *Syft. veget.* 169. *Curtis*
    *lond.*
Procumbent Pearl-wort.
*Nat.* of Britain.
*Fl.* June.     H. ☉. ♃.

*apetala.*

2. S. caule erectiufculo pubefcente, floribus alternis ape-
    talis. *Linn. mant.* 559. *Curt. lond.*

                                      Annual.

Annual Pearl-wort.
*Nat.* of England.
*Fl.* May and June.                               H. ☉.

3. S. caule erecto fubunifloro. *Syft. veget.* 169.  *Curtis*   *erecta.*
   *lond.*
   Upright Pearl-wort.
   *Nat.* of Britain.
   *Fl.* April and May.                           H. ☉.


## T I L L Æ A.  *Gen. pl.* 177.

*Cal.* 3- f. 4-partitus.  *Petala* 3 f. 4, æqualia.  *Capf.*
         3 f. 4, polyfpermæ.

1. T. procumbens, floribus trifidis.  *Syft. veget.* 170.      *mufcofa.*
   Procumbent Tillæa.
   *Nat.* of England.
   *Fl.* June——October.                          H. ☉.

*Claffis*

*Claſſis V.*

# PENTANDRIA

## *MONOGYNIA.*

### HELIOTROPIUM. *Gen. pl.* 179.

*Cor.* hypocrateriformis, 5-fida, interjectis dentibus : fauce clauſa fornicibus.

*peruvia-*
*num.*

1. H. foliis lanceolato-ovatis, caule fruticoſo, ſpicis nu-
   meroſis aggregato-corymboſis. *Sp. pl.* 187.
   Peruvian Turnſole, or Heliotrope.
   *Nat.* of Peru.
   *Cult.* 1757. *Mill. ic.* 96. *t.* 144.
   *Fl.* moſt part of the year.                    S. ♄.

*indicum.*

2. H. foliis cordato-ovatis acutis ſcabriuſculis, ſpicis ſo-
   litariis, fructibus bifidis. *Sp. pl.* 187.
   Indian Turnſole, or Heliotrope.
   *Nat.* of the Weſt Indies.
   *Cult.* 1713. *Philoſoph. tranſ. n.* 337. *p.* 60. *no.* 95.
   *Fl.* July and Auguſt.                         S. ♂.

*parviflo-*
*rum.*

3. H. foliis ovatis rugoſis ſcabris oppoſitis alterniſque.
   *Linn. mant.* 201.
   Small-flower'd Turnſole, or Heliotrope.
   *Nat.* of the Weſt Indies.
   *Cult.* 1732, by James Sherard, M. D. *Dill. elth.* 178.
   *t.* 146. *f.* 175.
   *Fl.* July and Auguſt.                         S. ☉.

                                                 4. H.

4. H. foliis ovatis integerrimis tomentofis rugofis, fpicis *europæ-*
conjugatis. *Sp. pl.* 187. *Jacqu. auftr.* 3. *p.* 4. *um.*
*t.* 207.
European Turnfole, or Heliotrope.
*Nat.* of Italy and France.
*Cult.* 1562, by William Turner, M. D. *Turn. herb.*
*part.* 2. *fol.* 13 *verfo.*
*Fl.* June——October. H. ⊙.

5. H. foliis ovatis integerrimis tomentofis plicatis, fpicis *fupinum.*
folitariis. *Sp. pl.* 187.
Trailing Turnfole, or Heliotrope.
*Nat.* of the South of Europe.
*Cult.* 1640. *Park. theat.* 438. *no.* 2.
*Fl.* June and July. H. ⊙.

6. H. foliis lanceolato-linearibus glabris aveniis, fpicis *curaſſa-*
conjugatis. *Sp. pl.* 188. *vicum.*
Glaucous Turnfole, or Heliotrope.
*Nat.* of the Weft Indies.
*Cult.* 1759, by Mr. Philip Miller. *Mill. dict. edit.* 7.
*no.* 7.
*Fl.* June and July. S. ⊙.

### MYOSOTIS. *Gen. pl.* 180.

*Cor.* hypocrateriformis, 5-fida, emarginata : fauce
claufa fornicibus.

1. M. feminibus lævibus, foliorum apicibus callofis. *Syft.* *fcorpi-*
*veget.* 184. *Curtis lond.* *oides.*
α Myofotis foliis hirfutis. *Hort. cliff.* 45. arvenfis.
Hairy Moufe-ear Scorpion-grafs.
β Myofotis foliis glabris. *Hort. cliff.* 46. paluftris.

Marfh

Marſh Mouſe-ear Scorpion-graſs.
*Nat.* of Britain.
*Fl.* April——Auguſt.　　　　　　　　　　H. ♃.

*Lappula.*　2. M. ſeminibus aculeatis glochidibus, foliis lanceolatis
　　　　　　piloſis. *Sp. pl.* 189.
　　　　Prickly-ſeeded Scorpion-graſs.
　　　　*Nat.* of Europe.
　　　　*Cult.* 1683, by Mr. James Sutherland. *Sutherl. hort.*
　　　　*edin.* 100. *no.* 3.
　　　　*Fl.* April——Auguſt.　　　　　　　　H. ☉.

*apula.*　3. M. ſeminibus nudis, foliis hiſpidis, racemis folioſis.
　　　　　　*Sp. pl.* 189.
　　　　Small Scorpion-graſs.
　　　　*Nat.* of the South of Europe.
　　　　*Cult.* 1768, by Mr. Philip Miller. *Mill. dict. edit.* 8.
　　　　*Fl.* June and July.　　　　　　　　H. ☉.

## LITHOSPERMUM. *Gen. pl.* 181.

*Cor.* infundibuliformis : fauce perforata, nuda. *Cal.*
5-partitus.

*officinale.*　1. L. ſeminibus lævibus, corollis calycem vix ſuperanti-
　　　　　　bus, foliis lanceolatis. *Sp. pl.* 189.
　　　　Officinal Gromwell.
　　　　*Nat.* of Britain.
　　　　*Fl.* May——Auguſt.　　　　　　　　H. ♃.

*arvenſe.*　2. L. ſeminibus rugoſis, corollis vix calycem ſuperanti-
　　　　　　bus. *Sp. pl.* 190.
　　　　Alkanet Gromwell, or Baſtard Alkanet.
　　　　*Nat.* of Britain.
　　　　*Fl.* May and June.　　　　　　　　　H. ☉.

3. L.

3. L. ramis floriferis lateralibus, bracteis cordatis am-    *orien-*
      plexicaulibus. *Syft. veget.* 185.                          *tale.*
   Anchufa orientalis.  *Sp. pl.* 191.
   Yellow Gromwell, or Buglofs.
   *Nat.* of the Levant.
   *Cult.* 1713, in Chelfea Garden. *Philofoph. tranf. n.*
      337. *p.* 60. *no.* 94.
   *Fl.* May and June.                          G. H. ♃.


4. L. feminibus lævibus, corollis calycem multoties fu-    *purpuro-*
      perantibus. *Sp. pl.* 190. *Jacqu. auftr.* 1. *p.* 11.   *cæruleum.*
      *t.* 14.
   Creeping Gromwell.
   *Nat.* of England.
   *Fl.* June.                                   H. ♃.


5. L. fruticofum, foliis linearibus hifpidis, ftaminibus    *frutico-*
      corollam fubæquantibus. *Sp. pl.* 190.                  *fum.*
   Shrubby Gromwell.
   *Nat.* of the South of Europe.
   *Cult.* 1683, by Mr. James Sutherland.  *Sutherl. hort.*
      *edin.* 24. *no.* 5.
   *Fl.* May and June.                          H. ♃.


### A N C H U S A.  *Gen. pl.* 182.

*Cor.* infundibuliformis: fauce claufa fornicibus. *Sem.*
                    bafi infculpta.


1. A. foliis lanceolatis ftrigofis integerrimis, panicula    *panicula-*
      dichotoma divaricata, floribus pedunculatis, calyci-   *ta.*
      bus quinquepartitis: laciniis fubulatis.
   Panicled Buglofs.
   *Nat.* of Madeira.  Mr. *Francis Maffon.*

                    N                          *Introd.*

178  PENTANDRIA MONOGYNIA. Anchusa.

*Introd.* 1777.

*Fl.* May and June.                                    H. ♂.

Obs. Facile dignofcitur calycibus ufque ad bafin quinquepartitis, et quod reliquis magis ftrigofa.

*officinalis.*  2. A. foliis lanceolatis ftrigofis, fpicis fecundis imbrica-
tis, calycibus quinquepartitis.
Anchufa officinalis.  *Sp. pl.* 191.
Officinal Buglofs.
*Nat.* of Europe.
*Cult.* 1748, by Mr. Philip Miller.  *Mill. dict. edit.* 5.
Bugloffum 7.
*Fl.* June——October.                               H. ♃.

*anguftifo-*  3. A. foliis oblongo-lanceolatis integris, floribus fpica-
*lia.*        tis, calycibus quinquefidis.
Anchufa anguftifolia.  *Sp. pl.* 191.
α Bugloffum creticum majus, flore albo.  *Boerh. lugdb.*
1. *p.* 189.
White-flower'd narrow-leav'd Buglofs.
β Bugloffum creticum majus, flore purpurafcente.
*Boerh. lugdb.* 1. *p.* 189.
Purple-flower'd narrow-leav'd Buglofs.
*Nat.* of the South of Europe.
*Cult.* 1759, by Mr. Ph. Miller.  *Mill. dict. edit.* 7. *x.* 2.
*Fl.* May and June.                                H. ♃.

*undulata.*  4. A. ftrigofa, foliis linearibus dentatis, pedicellis bractea
minoribus, calycibus fructiferis inflatis. *Sp. pl.* 191.
Waved-leav'd Buglofs.
*Nat.* of Spain and Portugal.
*Cult.* 1739, by Mr. Philip Miller.  *Mill. dict. vol.* 2.
Bugloffum 6.
*Fl.* July and Auguft.                             H. ♃.

5. A.

5. A. tomentofa, foliis lanceolatis obtufis, ftaminibus co-  *tinctoria.*
rolla brevioribus. *Sp. pl.* 192.
Dyers Buglofs.
*Nat.* of Montpellier.
*Cult.* 1683, by Mr. James Sutherland. *Sutherl. hort.*
*edin.* 24. *no.* 7.
*Fl.* June——October. H. ♃.

6. A. pedunculis diphyllis capitatis. *Sp. pl.* 192.  *fempervi-*
Ever-green Buglofs, or Alkanet.  *rens.*
*Nat.* of Britain.
*Fl.* March——July. H. ♃.

## CYNOGLOSSUM. *Gen. pl.* 183.

*Cor.* infundibuliformis : fauce claufa fornicibus. *Semina*
deprefla, interiore tantum latere ftylo affixa.

1. C. ftaminibus corolla brevioribus, foliis lato-lanceo-  *officinale.*
latis bafi attenuatis tomentofis feffilibus, laciniis ca-
lycinis oblongis.
Cynogloffum officinale. *Sp. pl.* 192. *Curtis lond.*
α Cynogloffum majus vulgare. *Bauh. pin.* 257.
Officinal Hound's-tongue.
β Cynogloffum fempervirens. *Bauh. pin.* 257.
Ever-green Hound's-tongue.
*Nat.* of Britain.
*Fl.* α, May——September. β, May and June. H. ☉.

2. C. corollis calyci fubæqualibus : laciniis fubrotundo-  *pictum.*
dilatatis, foliis lanceolatis tomentofis : fuperioribus
bafi cordatis.
Cynogloffum folio molli incano, flore cæruleo ftriis
rubris variegato. *Morif. blæf.* 258. *hift.* 3. *p.* 449.
Madeira Hound's-tongue.

N 2  *Nat.*

*Nat.* of Madeira.   Mr. *Francis Maſſon.*
*Introd.* 1777.
*Fl.* Auguſt.                                          H. 4.
Obs. Flores pallide cærulei vel pallide purpuraſcen-
tes, venis ſaturate coloratis pulchre picti.

*cheirifoli-*      3. C. corollis calyce duplo longioribus, foliis lanceola-
*um.*                    tis.  *Sp. pl.* 193.
Silvery-leav'd Hound's-tongue.
*Nat.* of Spain and the Levant.
*Cult.* 1596, by Mr. John Gerard.   *Hort. Ger.*
*Fl.* June and July.                             H. ♂.

*apenni-*      4. C. ſtaminibus corollam æquantibus.  *Sp. pl.* 193.
*num.*            Apennine Hound's-tongue.
*Nat.* of the Alps of Italy.
*Cult.* 1731, by Mr. Philip Miller.   *Mill. dict. edit.* 1.
*no.* 3.
*Fl.* April——June.                               H. ♂.

*linifoli-*      5. C. foliis lineari-lanceolatis glabris.  *Sp. pl.* 193.
*um.*             Flax-leav'd Hound's-tongue, or Venus's Navel-wort.
*Nat.* of Portugal.
*Cult.* 1731, by Mr. Philip Miller.   *Mill. dict. edit.* 1.
Omphalodes 1.
*Fl.* June——Auguſt.                             H. ☉.

*Omphalo-*   6. C. repens, foliis radicalibus cordatis.  *Sp. pl.* 193.
*des.*                *Curtis magaz.* 7.
Comfrey-leav'd Hound's-tongue.
*Nat.* of the South of Europe.
*Cult.* 1633.   *Ger. emac.* 806. *f.* 4.
*Fl.* March——May.                               H. 4.

PULMO-

## PULMONARIA. *Gen. pl.* 184.

*Cor.* infundibuliformis: fauce pervia. *Cal.* prifmatico-pentagonus.

1. P. foliis hirfutis : caulinis oblongo-lanceolatis amplexicaulibus ; radicalibus ellipticis. *anguftifolia.*
Pulmonaria anguftifolia. *Sp. pl.* 194.
a foliis non maculatis.
Pulmonaria III. auftriaca. *Cluf. hift.* 2. *p.* 169. *cum fig.*
Narrow-leav'd Lung-wort.
β foliis maculatis.
Pulmonaria faccharata. *Mill. dict.*
Pulmonaria V. pannonica. *Cluf. hift.* 2. *p.* 170. *cum fig.*
Spotted narrow-leav'd Lung-wort.
*Nat.* of Sweden, Germany, and Switzerland.
*Cult.* 1731. *Mill. dict. ed.* 1. *n.* 4.
*Fl.* April and May. H. ♃.

2. P. foliis hirfutis : caulinis ovato-oblongis ; radicalibus fubcordatis. *officinalis.*
Pulmonaria officinalis. *Sp. pl.* 194.
a Symphytum maculofum f. Pulmonaria latifolia. *Bauh. pin.* 259.
Common Lung-wort.
β Pulmonaria vulgaris latifolia, flore albo. *Tournef. inft.* 136.
White-flower'd Common Lung-wort.
*Nat.* of Britain.
*Fl.* March——May. H. ♃.

3. P. calycibus abbreviatis quinquepartitis hifpidis, foliis ovato-oblongis acuminatis pilofiufculis. *paniculata.*

N 3 floribus

α floribus cæruleis.

Blue-flower'd panicl'd Lung-wort.

β floribus albis.

White-flower'd panicl'd Lung-wort.

*Nat.* of Hudson's Bay.

*Introd.* 1778, by Daniel Charles Solander, LL.D.

*Fl.* May and June.                             H. ♃.

*virginica.*  4. P. calycibus abbreviatis glaberrimis, foliis lanceolatis
obtusiusculis.

Pulmonaria virginica.  *Sp. pl.* 194.

Virginian Lung-wort.

*Nat.* of Virginia and Maryland.

*Cult.* 1699, in Chelsea Garden.  *Morif. hift.* 3. *p.* 444.
*no.* 6.

*Fl.* March——May.                             H. ♃.

*maritima.*  5. P. calycibus abbreviatis, foliis ovatis, caule ramofo
procumbente.  *Sp. pl.* 195.

Sea Lung-wort.

*Nat.* of Britain.

*Fl.* July.                             H. ☉.


S Y M P H Y T U M.  *Gen. pl.* 185.

*Corollæ* limbus tubulato-ventricofus : fauce claufa radiis fubulatis.

*officinale.*  1. S. foliis ovato-lanceolatis decurrentibus.  *Sp. pl.* 195.
*Curtis lond.*

Common Comfrey.

*Nat.* of Britain.

*Fl.* May——October.                             H. ♃.

*tubera-*  2. S. foliis femidecurrentibus: fummis oppofitis.  *Syft.*
*fum.*       veget. 187.  *Jacqu. auftr.* 3. *p.* 14. *t.* 225.

Tuberous

Tuberous-rooted Comfrey.
*Nat.* of Germany, France, and Spain.
*Cult.* 1596, by Mr. John Gerard. *Hort. Ger.*
*Fl.* May and October. H. ♃.

C E R I N T H E. *Gen. pl.* 186.

*Corollæ* limbus tubulato-ventricofus: fauce pervia.
*Semina* 2, bilocularia.

1. C. foliis amplexicaulibus, fructibus geminis, corollis   *major.*
obtufiufculis patulis. *Sp. pl.* 195.
α Cerinthe flore ex rubro purpurafcente. *Bauh. pin.* 258.
Great Purple Honey-wort.
β Cerinthe flore flavo afperior. *Bauh. pin.* 258.
Great Yellow Honey-wort.
*Nat.* of the South of Europe.
*Cult.* 1596, by Mr. John Gerard. *Hort. Ger.*
*Fl.* July and Auguft. H. ⊙.

2. C. foliis amplexicaulibus integris, fructibus geminis,   *minor.*
corollis acutis claufis. *Sp. pl.* 196. *Jacqu. auftr.* 2.
*p.* 15. *t.* 124.
Small Honey-wort.
*Nat.* of Auftria.
*Cult.* 1570, by Mr. Hugh Morgan. *Lobel. adv.* 172.
*Fl.* June——October. H. ♂.

O N O S M A. *Gen. pl.* 187.

*Cor.* campanulata: fauce pervia. *Semina* 4.

1. O. foliis lanceolatis hifpidis, fructibus erectis. *Sp. pl.*   *echioides.*
196. *Jacqu. auftr.* 3. *p.* 52. *t.* 295.
Hairy Onofma.

N 4        *Nat.*

*Nat.* of the South of Europe.
*Cult.* 1683, by Mr. James Sutherland.   *Sutherl. hort.*
    *edin.* 24. *n.* 6.
*Fl.* March——June                H. ♃.

## BORAGO. *Gen. pl.* 188.

*Corolla* rotata : fauce radiis clausa.

*officinalis.*    1. B. foliis omnibus alternis, calycibus patentibus.   *Sp.*
         *pl.* 197.
         Common Borage.
         *Nat.* of England.
         *Fl.* June——September.             H. ☉.

*indica.*    2. B. foliis ramificationum oppositis amplexicaulibus, pe-
         dunculis unifloris.   *Syst. veget.* 188.
         Indian Borage.
         *Nat.* of the East Indies.
         *Cult.* 1759, by Mr. Philip Miller.   *Mill. dict. edit.* 7.
         *no.* 4.
         *Fl.* June——October.           G. H. ☉.

*africana.*    3. B. foliis oppositis petiolatis ovatis, pedunculis multi-
         floris.   *Syst. veget.* 188.
         African Borage.
         *Nat.* of the Cape of Good Hope.
         *Cult.* 1759, by Mr. Philip Miller.   *Mill. dict. edit.* 7.
         *no.* 3.
         *Fl.* July and August.           G. H. ☉.

*orientalis.*    4. B. calycibus tubo corollæ brevioribus, foliis cordatis.
         *Sp. pl.* 197.
         Oriental Borage.
         *Nat.* of the neighbourhood of Constantinople.

                                     *Cult.*

*Cult.* 1752, by Mr. Philip Miller.   *Mill. dict. edit.* 6.
*no.* 5.
*Fl.* March——May.                              H. ♃.


A S P E R U G O.   *Gen. pl.* 189.
*Cal.* fructus compreffus : lamellis plano-parallelis,
            finuatis.

1. A. calycibus fructus compreffis.   *Sp. pl.* 198.          *procum-*
Procumbent Afperugo.                                         *bens.*
*Nat.* of Britain.
*Fl.* April and May.                              H. ☉.


L Y C O P S I S.   *Gen. pl.* 190.
            *Corolla* tubo incurvato.

1. L. foliis integerrimis, caule proftrato, calycibus fruc-   *vefica-*
    tefcentibus inflatis pendulis.   *Sp. pl.* 198.            *ria.*
Bladder-podded wild Buglofs.
*Nat.* of the South of Europe.
*Introd.* 1770, by Monf. Richard.
*Fl.* June and July.                              H. ☉.


2. L. foliis integerrimis, caule erecto, calycibus fructef-   *pulla.*
    centibus inflatis pendulis.   *Sp. pl.* 198. *Jacqu. auftr.*
    2. *p.* 53. *t.* 188.
Dark-flower'd wild Buglofs.
*Nat.* of Germany.
*Introd.* 1785, by William Pitcairn, M. D.
*Fl.* June and July.                              H. ♂.


3. L. foliis lanceolatis hifpidis, calycibus florefcentibus   *arvenfis.*
    erectis.   *Sp. pl.* 199.   *Curtis lond.*
Small wild Buglofs.
*Nat.* of Britain.
*Fl.* June and July,                              H. ☉.
                                        ECHIUM.

# ECHIUM. *Gen. pl.* 191.

*Cor.* irregularis : fauce nuda.

*frutico-*
*sum.*

1. E. caule fruticoso, foliis lanceolatis basi attenuatis villoso-strigosis aveniis, foliolis calycinis lanceolatis acutis.

Echium fruticosum. *Sp. pl.* 199.
Shrubby Viper's Bugloss.
*Nat.* of the Cape of Good Hope.
*Cult.* 1759, by Mr. Philip Miller. *Mill. dict. edit.* 7. *no.* 7.
*Fl.* May and June.     G. H. ♄.

*candi-*
*cans.*

2. E. caule fruticoso, foliis lanceolatis nervosis ramisque hirsutis, foliolis calycinis oblongis lanceolatisque acutis, stylis hirtis.

Echium candicans. *Linn. suppl.* 131. *Jacqu. ic. collect.* 1. *p.* 44.
Hoary Tree Viper's Bugloss.
*Nat.* of Madeira. Mr. *Francis Masson.*
*Introd.* 1777.
*Fl.* May.     G. H. ♄.

*gigante-*
*um.*

3. E. caule fruticoso, foliis lanceolatis basi attenuatis pilosis : pilis brevissimis, bracteis calycibusque strigosis, staminibus corolla longioribus.

Echium giganteum. *Linn. suppl.* 131.
Gigantic Viper's Bugloss.
*Nat.* of the Canary Islands. Mr. *Francis Masson.*
*Introd.* 1779.
*Fl.* July——September.     G. H. ♄.

*strictum.*

4. E. caule fruticoso stricto ramoso, foliis oblongo-lanceolatis pilosis, corollis subcampanulatis, staminibus corolla longioribus.

Echium

Echium ftri&um. *Linn. fuppl.* 131.
Upright Viper's Buglofs.
*Nat.* of the Canary Iflands. Mr. *Francis Maffon.*
*Introd.* 1779.
*Fl.* Moft part of the year. G. H. ♂.

5. E. foliis radicalibus ovatis lineatis petiolatis. *Linn.* *plantagi-*
   mant. 202. *neum.*
   Plantaln-leav'd Viper's Buglofs.
   *Nat.* of Italy.
   *Introd.* 1776, by Monf. Thouin.
   *Fl.* July——Octtober. H. ☉.

6. E. caule lævi, foliis lanceolatis nudis margine carina *læviga-*
   apiceque fcabris. *Sp. pl.* 199. *tum.*
   Smooth-ftalk'd Viper's Buglofs.
   *Nat.* of the Cape of Good Hope.
   *Introd.* 1774, by Mr. Francis Maffon.
   *Fl.* June and July. G. H. ♄.

7. E. caule herbaceo pilofo, foliis lineari-lanceolatis ftri- *italicum.*
   gofo-hirfutis: inferioribus nervofis, corollis fubæ-
   qualibus, ftaminibus corolla longioribus.
   Echium italicum. *Sp. pl.* 200.
   Wall Viper's Buglofs.
   *Nat.* of England.
   *Fl.* July and Auguft. H. ♃.

8. E. caule tuberculato-hifpido, foliis caulinis lanceo- *vulgare.*
   latis hifpidis, floribus fpicatis lateralibus. *Sp. pl.*200.
   Common Viper's Buglofs.
   *Nat.* of Britain.
   *Fl.* July and Auguft. H. ♂.

9. E.

*violace-*
*um.*

9. E. corollis ſtamina æquantibus : tubo calycibus bre-
viore. *Linn. mant.* 42.
Violet-flower'd Viper's Bugloſs.
*Nat.* of Auſtria.
*Introd.* 1780, by Caſimir Gomez Ortega, M. D.
*Fl.* July.                                                    H. ⊙.

*creticum.*

10. E. calycibus fructeſcentibus diſtantibus, caule pro-
cumbente. *Sp. pl.* 200.
Cretan Viper's Bugloſs.
*Nat.* of the Levant.
*Cult.* 1683, by Mr. James Sutherland. *Sutherl. hort.*
*edin.* 108. *no.* 2.
*Fl.* July——September.                                        H. ♂.

*orientale.*

11. E. caule ramoſo, foliis caulinis ovatis, floribus ſoli-
tariis lateralibus. *Sp. pl.* 200.
Oriental Viper's Bugloſs.
*Nat.* of the Levant.
*Introd.* 1780, by Monſ. Thouin.
*Fl.* July.                                                    H. ⊙.

*luſitani-*
*cum.*

12. E. corollis ſtamine longioribus.  *Sp. pl.* 200.
Portugal Viper's Bugloſs.
*Nat.* of the South of Europe.
*Cult.* 1731, by Mr. Philip Miller. *Mill. dict. edit.* 1.
*no.* 4.
*Fl.* July and Auguſt.                                        H. ⊙.

MESSERSCHMIDIA. *Linn. mant.* 5.
*Bacca* ſuberoſa, bipartibilis : ſingulo diſpermo.

*fruticoſa.*

1. M. caule fruticoſo, foliis petiolatis, corollis hypocrateriformibus. *Linn. ſuppl.* 132.  *L'Her. ſtirp. nov.*
*tom.* 2. *tab.* 1.
Shrubby Meſſerſchmidia.

                                                              *Nat.*

*Nat.* of the Canary Iflands.  Mr. *Francis Maffon.*
*Introd.* 1779.
*Fl.* June——October.                       G. H. ♄.


2. M. caule herbaceo, foliis feffilibus, corollis infundi-    *Arguzia.*
   buliformibus.  *Linn. fuppl.* 132.
   Tournefortia fibirica.  *Sp. pl.* 202.
   Herbaceous Mefferfchmidia.
   *Nat.* of Siberia.
   *Introd.* 1780, by *Peter Simon Pallas*, M. D.
   *Fl.*                                     H. ♃.


## TOURNEFORTIA. *Gen. pl.* 192.

*Bacca* 2-locularis, difperma, fupera, apice duobus
poris perforata.

1. T. foliis ovatis acuminatis glabris, petiolis reflexis,    *volubilis.*
   caule volubili.  *Sp. pl.* 201.
   Climbing Tournefortia.
   *Nat.* of Jamaica.
   *Cult.* 1739, by Mr. Philip Miller.  *Mill. dict. vol.* 2.
   Pittonia 4.
   *Fl.* July and Auguft.                    S. ♄.


2. T. foliis lanceolatis feffibilibus, fpicis fimplicibus re-   *humilis.*
   curvis lateralibus.  *Sp. pl.* 202.
   Dwarf Tournefortia.
   *Nat.* of South America.
   *Cult.* 1739, by Mr. Philip Miller.  *Mill. dict. vol.* 2.
   Pittonia 3.
   *Fl.* May and June.                        S. ♄.


3. T. foliis ovatis integerrimis nudis, fpicis cymofis.   *cymofa.*
   *Sp. pl.* 202.  *Jacqu. ic. collect.* 1. *p.* 96.
   Broad-leav'd Tournefortia.
                                              *Nat.*

*Nat.* of Jamaica.
*Cult.* 1777, by Mr. William Malcolm.
*Fl.* July.      S. ♄.

*suffruti-*
*cosa.*    4. T. foliis sublanceolatis incanis, caule suffruticoso. *Sp.*
     *pl.* 202.
     Hoary-leav'd Tournefortia.
     *Nat.* of Jamaica.
     *Cult.* 1768, by Mr. Philip Miller. *Mill. dict. edit.* 8.
     *no.* 7.
     *Fl.*      S. ♄.

## N O L A N A.    *Gen. pl.* 193.

*Cor.* campanulata. *Stylus* inter germina. *Sem.* 5, bac‑
cata, bilocularia.

*prostrata.*    1. NOLANA.    *Sp. pl.* 202.
     Trailing Nolana.
     *Nat.* of Peru.
     *Cult.* 1761, by Mr. Philip Miller. *Philos. transact.*
     *vol.* 53. *p.* 130.
     *Fl.* July——September.      H. ☉.

## A R E T I A.    *Gen. pl.* 195.

*Cor.* hypocraterif. 5-fida : tubo ovato. *Stigma* de‑
presso-capitatum. *Caps.* 1-locularis, globosa, sub-
pentasperma.

*helvetica.*    1. A. foliis imbricatis, floribus subsessilibus. *Syst. veget.*
     192.
     Imbricated Aretia.
     *Nat.* of Switzerland.
     *Introd.* 1775, by the Doctors Pitcairn and Fothergill.
     *Fl.*      H. ♃.

2. A.

2. **A.** foliis linearibus patentibus, floribus pedunculatis. *alpina.*
    *Syft. veget.* 192. *Jacqu. auftr.* 5. *p.* 36. *t. app.* 18.
    Linear-leav'd Aretia.
    *Nat.* of Switzerland.
    *Introd.* 1775, by the Doctors Pitcairn and Fothergill.
    *Fl.*                                H. ♃.

3. **A.** foliis linearibus recurvatis, floribus fubfeffilibus. *Vitalia-*
    *Syft. veget.* 192.                                     *na.*
    Primula Vitaliana. *Sp. pl.* 206.
    Grafs-leav'd Aretia.
    *Nat.* of the Pyrenees.
    *Introd.* 1787, by Monf. Cels.
    *Fl.*                                H. ♃.

## ANDROSACE. *Gen. pl.* 196.

*Involucrum* umbellulæ. *Corollæ* tubus ovatus : ore
    glandulofo. *Capf.* 1-locularis, globofa.

1. **A.** perianthiis fructuum maximis. *Sp. pl.* 203. *Jacqu.* *maxima.*
    *auftr.* 4. *p.* 16. *t.* 331.
    Oval-leav'd Androface.
    *Nat.* of Auftria.
    *Cult.* 1596, by Mr. John Gerard. *Hort. Ger.*
    *Fl.* March——June.                        H. ☉.

2. **A.** foliis fubdentatis, pedicellis longiffimis, corollis ca- *elongata.*
    lyce brevioribus. *Syft. veget.* 192. *Jacqu. auftr.* 4.
    *p.* 16. *t.* 330.
    Clufter-flower'd Androface.
    *Nat.* of Auftria.
    *Introd.* 1776, by Monf. Thouin.
    *Fl.* April and May.                     H. ☉.

3. **A.** foliis lanceolatis dentatis glabris, perianthiis angu- *fepten-*
    latis corolla brevioribus. *Sp. pl.* 203.           *trionalis.*
                                    Toothed-

Toothed-leav'd Androsace.
*Nat.* of Ruffia and Lapland.
*Cult.* 1755, by Mr. Philip Miller. *Mill. ic.* 20. *t.* 30.
*f.* 2.
*Fl.* April and May.                                    H. ☉.

*villosa.*    4. A. foliis pilosis, perianthiis hirsutis. *Sp. pl.* 203.
*Jacqu. austr.* 4. *p.* 16. *t.* 332.
Hairy Androsace.
*Nat.* of Austria and Switzerland.
*Introd.* 1768, by Professor de Saussure.
*Fl.* June——Auguft.                                    H. ♃.

*lactea.*    5. A. foliis lanceolatis glabris, umbella involucris mul-
toties longiore. *Syst. veget.* 192. *Jacqu. austr.* 4.
*p.* 17. *t.* 333.
Grafs-leav'd Androsace.
*Nat.* of Austria.
*Introd.* 1768, by Professor de Saussure.
*Fl.* June.                                    H. ♃.

*carnea.*    6. A. foliis fubulatis glabris, umbella involucrum æquante.
*Sp. pl.* 204.
Awl-leav'd Androsace.
*Nat.* of Switzerland.
*Introd.* 1768, by Professor de Saussure.
*Fl.* July and Auguft.                                    H. ♃.

## PRIMULA. *Gen. pl.* 197.

*Involucr.* umbellulæ. *Corollæ* tubus cylindricus: ore
patulo.

*acaulis.*    1. P. foliis rugosis dentatis subtus hirsutis, scapis unifloris.
*Jacqu. miscell.* 1. *p.* 158. *n.* 1.

Primula

Primula veris γ. acaulis. *Sp. pl.* 205.
Primula vulgaris α. *Hudf. angl.* 83.
Common Primrofe.
*Nat.* of Britain.
*Fl.* April and May.                                   H. ♃.

2. P. foliis rugofis dentatis hirfutis, fcapo multifloro : flo-   *elatior.*
ribus exterioribus nutantibus ; medio erecto. *Jacqu.*
*mifcell.* I. *p.* 158. *n.* 2.
Primula veris β. elatior. *Sp. pl.* 204.
Primula vulgaris β. *Hudf. angl.* 84.
Oxflip.
*Nat.* of Britain.
*Fl.* April and May.                                   H. ♃.

3. P. foliis rugofis dentatis fubtus hirfutis, fcapo multi-   *officinalis.*
floro : floribus omnibus nutantibus. *Jacqu. mifcell.*
I. *p.* 159. *n.* 3.
Primula veris α. officinalis. *Sp. pl.* 204.
Primula veris. *Hudf. angl.* 84.
Common Cowflip.
*Nat.* of Britain.
*Fl.* April and May.                                   H. ♃.

4. P. foliis oblongis denticulatis undulatis fubtus ꞏfari-   *farinofa.*
nofis, umbella erecta faftigiata, corollæ limbo plano.
Primula farinofa. *Sp. pl.* 205. *Jacqu. mifcell.* I.
*p.* 159. *n.* 4.
Bird's-eye Primrofe, or Cowflip.
*Nat.* of Britain.
*Fl.* May and June.                                   H. ♃.

5. P. foliis ferratis glabris. *Sp. pl.* 205. *Jacqu. mifcell.*   *Auricula.*
I. *p.* 160. *n.* 9. *auftr.* 5. *p.* 7. *t.* 415.
α Sanicula alpina lutea. *Bauh. pin.* 242.

O                                   Yellow

Yellow Auricula, or Bear's-ear.

β Sanicula alpina purpurea. *Bauh. pin.* 242.

Purple Auricula, or Bear's-ear.

γ Sanicula alpina flore variegato. *Bauh. pin.* 242.

Variegated Auricula, or Bear's-ear.

*Nat.* of the Alps of Switzerland and Auſtria.

*Cult.* 1597. *Ger. herb.* 640.

*Fl.* April and May. H. ♃.

*villoſa.* 6. P. foliis ovato-cuneiformibus dentatis carnoſis pubeſcentibus, ſcapo umbellato, corollis glabris.

Primula villoſa β. pubeſcens. *Jacqu. miſcell.* 1. *p.* 159. *n.* 6. *tab.* 18. *fig.* 2. *Curtis magaz.* 14.

Hairy Auricula, or Bear's-ear.

*Nat.* of Carinthia and Switzerland.

*Introd.* 1768, by Profeſſor de Sauſſure.

*Fl.* April and May. H. ♃.

*glutinoſa.* 7. P. involucro longitudine florum ſeſſilium, foliis lanceolatis glutinoſis. *Linn. ſuppl.* 133. *Jacqu. miſcell.* 1. *p.* 159. *n.* 8. *auſtr.* 5. *p.* 41. *tab. app.* 26.

Clammy Primroſe.

*Nat.* of the Alps of Carinthia and Tyrol.

*Introd.* 1777, by Meſſrs. Kennedy and Lee.

*Fl.* April and May. H. ♃.

## CORTUSA. *Gen. pl.* 198.

*Corolla* rotata: fauce annulo elevato. *Capſ.* 1-locularis, ovalis, apice 5-valvi.

*Matthioli.* 1. C. calycibus corolla brevioribus. *Sp. pl.* 206.

Bear's-ear Sanicle.

*Nat.* of Siberia and Auſtria.

*Cult.*

*Cult.* 1596, by Mr. John Gerard. *Hort. Ger.*
*Fl.* April——June.  H. ♃.


## SOLDANELLA. *Gen. pl.* 199.

*Corolla* campanulata, lacero-multifida. *Capf.* 1-locularis,
apice multidentata.

1. SOLDANELLA. *Sp. pl.* 206. *Jacqu. auftr.* 1. *p.* 11.  *alpina.*
t. 13. *Curtis magaz.* 49.
Alpine Soldanella.
*Nat.* of the Alps of Switzerland and Auftria.
*Cult.* 1656, by Mr. John Tradefcant. *Muf. Tradefc.*
169.
*Fl.* April.  H. ♃.


## DODECATHEON. *Gen. pl.* 200.

*Cor.* rotata, reflexa. *Stam.* tubo infidentia. *Capf.* 1-
locularis, oblonga.

1. DODECATHEON. *Sp. pl.* 207. *Curtis magaz.* 12.  *Meadia.*
Virginian Cowflip.
*Nat.* of North America.
*Cult.* 1744, by Peter Collinfon, Efq. *Catefb. carol.*
append. 1.
*Fl.* April——June.  H. ♃.


## CYCLAMEN. *Gen. pl.* 201.

*Cor.* rotata, reflexa, tubo breviffimo : fauce prominente.
*Bacca* tecta capfula.

1. C. foliis orbiculatis cordatis integerrimis.  *coum.*
Cyclamen coum. *Mill. dict. Curtis magaz.* 4.
Round-leav'd Cyclamen.
*Nat.* of the South of Europe.

O 2  *Cult.*

Cult. 1731. *Mill. dict. edit.* 1. *n.* 4.
Fl. February.                          H. ♃.

*europæum.*    2. C. foliis orbiculatis cordatis crenatis.
Cyclamen europæum. *Sp. pl.* 207. *Jacqu. austr.* 5.
*p.* 1. *t.* 401.
Common european Cyclamen.
*Nat.* of Austria.
Cult. 1596, by Mr. John Gerard. *Hort. Ger.*
Fl. April.                             H. ♃.

*persicum.*    3. C. foliis oblongo-ovatis cordatis crenatis.
Cyclamen persicum. *Mill. dict. Curtis magaz.* 44.
Persian Cyclamen.
*Nat.* of the Island of Cyprus. *John Sibthorp,* M. D.
Cult. 1731. *Mill. dict. edit.* 1. *n.* 5.
Fl. February——April.               G. H. ♃.

*hederæfo-*    4. C. foliis cordatis angulatis denticulatis.
*lium.*        Cyclamen europæum. *Mill. dict. Regn. botan.*
Ivy-leav'd Cyclamen.
*Nat.* of Italy.
Cult. 1596, by Mr. John Gerard. *Hort. Ger.*
Fl. April.                             H. ♃.

## MENYANTHES. *Gen. pl.* 202.

*Corolla* hirsuta. *Stigma* 2-fidum. *Caps.* 1-locularis.

*Nymphoi-*    1. M. foliis cordatis integerrimis, corollis ciliatis. *Sp.*
*des.*        *pl.* 207.
Fringed Buck-bean, or Water-lily.
*Nat.* of Britain.
Fl. June and July.                   H. ♃.

*ovata.*    2. M. foliis ovatis petiolatis, caule paniculato. *Linn.*
*suppl.* 133.

                                    Renealmia

Renealmia capenfis. *Houtt. nat. hiſt.* 8. *p.* 335. *t.* 47.
*f.* 1.
Oval-leav'd Buck-bean.
*Nat.* of the Cape of Good Hope.
*Introd.* 1786, by Mr. Francis Maſſon.
*Fl.* May and June.                          G. H. ♃.


3. M. foliis ternatis. *Sp. pl.* 208. *Curtis lond.*          *trifoliata.*
Common Buck-bean, or Marſh Trefoil.
*Nat.* of Britain.
*Fl.* July.                                    H. ♃.


## HOTTONIA. *Gen. pl.* 203.

*Corolla* hypocrateriformis. *Stamina* tubo corollæ impoſita. *Capſ.* 1-locularis.

1. H. pedunculo verticillato-multifloro. *Sp. pl.* 208. *paluſtris.*
   *Curtis lond.*
Water Violet.
*Nat.* of England.
*Fl.* July and Auguſt.                         H. ♃.


## HYDROPHYLLUM. *Gen. pl.* 204.

*Cor.* campanulata, interne ſtriis 5, melliferis, longitudinalibus. *Stigm.* 2-fidum. *Capſ.* globoſa, 2-valvis.

1. H. foliis pinnatifidis. *Sp. pl.* 208.                      *virgini-*
Virginian Water-leaf.                                          *cum.*
*Nat.* of Virginia and Carolina.
*Cult.* 1739, by Mr. Philip Miller. *Rand. chel.*
*Fl.* May and June.                            H. ♃.


2. H. foliis lobato-angulatis. *Sp. pl.* 208.                  *canadenſe.*
Canadian Water-leaf.
*Nat.* of Canada.

*Cult.*

*Cult.* 1759, by Mr. Philip Miller.
*Fl.* May. H. ♃.


ELLISIA. *Gen. pl.* 244.

*Cor.* infundibuliformis, angufta. *Bacca* ficca, bilocularis,
bivalvis. *Sem.* 2, punctata : altero fupra alterum.

*Nyctelea.*  1. ELLISIA. *Syft. veget.* 195.
Cut-leav'd Ellifia.
*Nat.* of Virginia.
*Cult.* 1755, by Peter Collinfon, Efq. *Ehret nov. act.
nat. curiof.* 2. *p.* 330. *t.* 7. *f.* 1.
*Fl.* July and Auguft. H. ☉.


LYSIMACHIA. *Gen. pl.* 205.

*Cor.* rotata. *Capf.* globofa, mucronata, 10-valvis.

* *Pedunculis multifloris.*

*vulgaris.*  1. L. paniculata, racemis terminalibus. *Sp. pl.* 209.
*Curtis lond.*
Common Loofe-ftrife.
*Nat.* of Britain.
*Fl.* July——September. H. ♃.


*Epheme-*  2. L. racemis terminalibus, petalis obovatis patulis, fo-
*rum.*      liis lineari-lanceolatis feffilibus.
Lyfimachia Ephemerum. *Sp. pl.* 209. (exclufo fyno-
nymo Buxbaumii) *Murray in commentat. gotting.*
1782. *p.* 9. *t.* 2.
Lyfimachia Otani. *D'Affo arag.* 22. *tab.* 2. *f.* 1.
(qui vero folia inpunctata effe dicit.)
Lyfimachia falicifolia. *Mill. dict.*
Ephemerum fpurium Lobelii. *Rob. ic.*
Willow-leav'd Loofe-ftrife.

*Nat.*

*Nat.* of Spain.
*Cult.* 1731, by Mr. Philip Miller. *Mill. dict. ed.* 1.
*n.* 5.
*Fl.* July——September. H. ♃.

3. L. racemis terminalibus, petalis lanceolatis patulis, *ſtricta.*
foliis lanceolatis ſeſſilibus.
Upright Looſe-ſtrife.
*Nat.* of North America.
*Introd.* about 1781, by Mr. William Curtis.
*Fl.* July and Auguſt. H. ♃.
DESCR. *Caulis* erectus, tetragonus, glaber. *Folia* integerrima, acuta, glabra, punctata. *Racemi* ſimpli-
ces. *Pedicelli* ſubverticillati, filiformes, unciales.
*Bracteæ* lanceolatæ, breviſſimæ. *Calycis* laciniæ
lanceolatæ, glabræ, rubro maculatæ. *Petala* calyce triplo longiora, lutea, punctis et lituris rubris,
maculiſque duabus ſaturate rubris. *Stamina* corolla breviora.

4. L. racemis terminalibus, petalis conniventibus, ſta- *dubia.*
minibus corolla brevioribus, foliis lanceolatis pe-
tiolatis.
Lyſimachia atropurpurea. *Murray in commentat.*
*gotting.* 1782. *p.* 6. *tab.* 1.
Lyſimachia Ephemerum. *Mill. dict.*
Lyſimachia ſpicata purpurea minor. *Buxb. cent.* 1. *p.*
*t.* 33.
Purple-flower'd Looſe-ſtrife.
*Nat.* of the Levant.
*Cult.* 1759, by Mr. Philip Miller. *Mill. dict. edit.* 7.
*n.* 4.
*Fl.* July and Auguſt. H. ♂.

5. L. racemis lateralibus pedunculatis. *Sp. pl.* 209. *thyrſiflo-*
O 4 Tufted *ra.*

Tufted Looſe-ſtrife.

*Nat.* of England.

*Fl.* May——July.                                        H. ♃.

** *Pedunculis unifloris.*

*punctata.*  6. L. foliis ſubquaternis, pedunculis verticillatis unifloris.  *Sp. pl.* 210.

Four-leav'd Looſe-ſtrife.

*Nat.* of Holland.

*Cult.* 1759, by Mr. Philip Miller.   *Mill. dict. edit.* 7.
*no.* 10.

*Fl.* July and Auguſt.                                   H. ♃.


*ciliata.*  7. L. petiolis ciliatis, floribus cernuis.  *Sp. pl.* 210.

Ciliated Looſe-ſtrife.

*Nat.* of Virginia and Canada.

*Cult.* 1759, by Mr. Philip Miller.   *Mill. dict. edit.* 7.
*no.* 5.

*Fl.* July and Auguſt.                                   H. ♃.


*Linum*      8. L. calycibus corollam ſuperantibus, caule erecto ra-
*ſtellatum.*      moſiſſimo.  *Sp. pl.* 211.

Small Looſe-ſtrife.

*Nat.* of Italy.

*Introd.* 1776, by Monſ. Thouin.

*Fl.* June.                                               H. ☉.


*nemorum.*  9. L. foliis ovatis acutis, floribus ſolitariis, caule procumbente.  *Sp. pl.* 211.  *Curtis lond.*

Wood Looſe-ſtrife, or Pimpernel.

*Nat.* of Britain.

*Fl.* May——July.                                        H. ♃.


*Nummu-*  10. L. foliis ſubcordatis, floribus ſolitariis, caule repente.
*laria.*        *Sp. pl.* 211.  *Curtis lond.*

Creeping

Creeping Loose-strife, or Money-wort.

*Nat.* of Britain.

*Fl.* June and July.         H. ♃.

ANAGALLIS. *Gen. pl.* 206.

*Cor.* rotata. *Capf.* circumfcifla.

1. A. foliis indivifis, caule procumbente. *Sp. pl.* 211. *arvenfis.*
  Curtis lond.

α Anagallis phœniceo flore. *Bauh. pin.* 252.
  Red Pimpernel.

β Anagallis cæruleo flore. *Bauh. pin.* 252.
  Blue Pimpernel.

γ Anagallis terreftris flore albo. *Raj. fyn.* 282.
  White Pimpernel.
  *Nat.* of Britain.
  *Fl.* July——September.        H. ☉.

2. A. foliis indivifis, caule erecto. *Sp. pl.* 211.     *monelli.*
  Italian blue Pimpernel.
  *Nat.* of Italy.
  *Cult.* 1648, in Oxford Garden. *Hort. oxon. edit.* 1.
  *p.* 4. *no.* 8.
  *Fl.* May——September.        G. H. ♃.

3. A. foliis cordatis amplexicaulibus, caulibus com-  *latifolia.*
  preffis. *Sp. pl.* 212.
  Broad-leav'd Pimpernel.
  *Nat.* of Spain.
  *Cult.* 1759, by Mr. Philip Miller. *Mill. dict. edit.* 7.
  *Fl.* July.        H. ☉.

4. A. foliis ovatis acutiufculis, caule repente. *Syft. veget. tenella.*
  196. *Curtis lond.*

Lyfimachia

Lyſimachia tenella.   *Sp. pl.* 211.
Creeping Pimpernel, or Purple Money-wort.
*Nat.* of Britain.
*Fl.* Auguſt and September.                                   H. ♃.


S P I G E L I A.   *Gen. pl.* 209.

*Cor.* infundibulif.   *Capſ.* didyma, 2-locularis, polyſperma.

*Anthel-*     1. S. caule herbaceo, foliis ſummis quaternis. *Syſt. veget.*
*mia.*             197.
              Annual Worm-graſs.
              *Nat.* of the Weſt Indies.
              *Cult.* 1759, by Mr. Philip Miller.   *Mill. diɛt. edit.* 7.
              *no.* 1.
              *Fl.* July.                                      S. ☉.


*marilan-*    2. S. caule tetragono, foliis omnibus oppoſitis.  *Syſt.*
*dica.*            *veget.* 197.
              Lonicera marilandica.   *Sp. pl.* 249.
              Perennial Worm-graſs.
              *Nat.* of North America.
              *Cult.* 1694, by Mr. Bobart. *Br. muſ. Sloan. mſs.* 3343.
              *Fl.* July and Auguſt.                          H. ♃.


A Z A L E A.   *Gen. pl.* 212.

*Cor.* ſubcampanulata.   *Stam.* receptaculo inſerta.  *Capſ.*
5-locularis.

*nudiflora.*   1. A. foliis ovatis, corollis piloſis, ſtaminibus longiſſimis.
                   *Sp. pl.* 214.
coccinea.      α floribus coccineis.
                   Deep ſcarlet Azalea.

                                                  β floribus

β floribus faturate rubris, calycibus minutis.                     rutilans.
   Deep red Azalea.

γ floribus pallide rubicundis : tubo bafi rubro, calyci-    carnea.
   bus foliaceis.
   Pale red Azalea.

δ floribus albidis, calycibus mediocribus.                 alba.
   Early white Azalea.

ε florum limbo pallido; tubo rubro, calyce parvo, ramu-    bicolor.
   lis pilofis.
   Red and white Azalea.

ζ floribus rubicundis : lacinia infima alba, calycibus     papilio-
   foliaceis.                                              nacea.
   Variegated Azalea.

η floribus carneis ufque ad bafin quinquepartitis.         partita,
   Downy Azalea.
   *Nat.* of North America.
   *Introd.* 1734, by Peter Collinfon, Efq. *Coll. mfs.*
   *Fl.* May and June.                          H. ♄.


2. A. foliis margine fcabris, corollis pilofo-glutinofis.   *vifcofa.*
   *Sp. pl.* 214.

α floribus albis, ramis diffufis, foliis faturate viridibus   odorata.
   lucidis.
   Common white Azalea.

β floribus albis : carinis carneis, ftylis elongatis apice   vittata.
   rubris, foliis pallidis ovato-oblongis.
   White-ftrip'd flower'd Azalea.

γ floribus albis ad bafin ufque divifis, foliis faturate vi-  fiffa.
   ridibus lucidis.
   Narrow-petal'd white Azalea.

δ floribus albis, foliis fubtus glaucis, ftylis corolla lon-  floribun-
   gioribus.                                               da.
   Clufter-flower'd white Azalea.

                                              ε floribus

glauca.  ε floribus albis, foliis utrinque glaucis; junioribus fu-
pra pilis adfperfis.
Glaucous Azalea.
*Nat.* of North America.
*Introd.* 1734, by Peter Collinfon, Efq.  *Coll. mff.*
*Fl.* δ, and ε, June, and α, β, and γ, July and Auguft.
H. ♄.

procum-
bens.
3. A. ramis diffufo-procumbentibus.  *Sp. pl.* 215.
Procumbent Azalea.
*Nat.* of Scotland.
*Fl.* April and May.                                H. ♄.

### PLUMBAGO.  *Gen. pl.* 213.

*Cor.* infundibulif.  *Stamina* fquamis bafin corollæ clau-
dentibus inferta.  *Stigma* 5-fidum.  *Sem.* 1, oblongum, tunicatum.

europeæa.  1. P. foliis amplexicaulibus lanceolatis fcabris.  *Sp. pl.*
215.
European Lead-wort.
*Nat.* of the South of Europe.
*Cult.* 1597, by Mr. John Gerard.  *Ger. herb.* 1069.
*Fl.* September and October.                        H. ♃.

zeylanica.  2. P. foliis petiolatis ovatis glabris, caule filiformi.  *Sp.
pl.* 215.
Ceylon Lead-wort.
*Nat.* of the Eaft Indies.
*Cult.* 1731, by Mr. Philip Miller.  *Mill. dict. edit.* 1.
*no.* 2.
*Fl.* April——September.                            S. ♄.

rofea.  3. P. foliis petiolatis ovatis glabris fubdenticulatis, caule
geniculis gibbofis.  *Syft. veget.* 199.
Rofe-

Rofe-colour'd Lead-wort.
*Nat.* of the Eaft Indies.
*Introd.* 1777, by John Fothergill, M. D.
*Fl.* July.                                         S. ♄.

4. P. foliis petiolatis ovatis glabris, caule flexuofo-fcan-   *fcandens.*
   dente. *Sp. pl.* 215.
Climbing Lead-wort.
*Nat.* of South America.
*Introd.* 1778, by Henry de Ponthieu, Efq.
*Fl.* July and Auguft.                             S. ♄.


P H L O X.   *Gen. pl.* 214.

*Cor.* hypocraterif. *Filamenta* inæqualia. *Stigma*
   3-fidum. *Cal.* prifmaticus. *Capf.* 3-locularis, 1-
   fperma.

1. P. foliis lanceolatis planis margine fcabris, caule lævi,   *panicula-*
   corymbis paniculatis, corollæ laciniis rotundatis.         *ta.*
Phlox paniculata. *Sp. pl.* 216.
Panicl'd Lychnidea.
*Nat.* of North America.
*Cult.* 1732, by James Sherard, M. D. *Dill. elth.* 205.
   *t.* 166. *f.* 203.
*Fl.* Auguft and September.                        H. ♃.

2. P. foliis oblongo-lanceolatis fubundulatis margine       *undulata.*
   fcabris, caule lævi, corymbis paniculatis, corollæ
   laciniis fubretufis.
Waved-leav'd Phlox.
*Nat.* of North America.
*Cult.* 1759, by Mr. Philip Miller.
*Fl.* July and Auguft.                             H. ♃.
OBS. *Flores* cærulei.

                                                   3. P.

*suaveo-*
*lens.*

3. P. foliis ovato-lanceolatis undique lævibus, caule
glaberrimo, racemo pañiculato.
White-flower'd Phlox.
*Nat.* of North America.
*Introd.* about 1766, by Peter Collinſon, Eſq.
*Fl.* July and Auguſt.　　　　　　　　　　　H. ♃.
Obs. *Flores* albi, leviter ſuaveolentes.

*maculata.*

4. P. foliis oblongo-lanceolatis glabris, caule ſcabriuſ-
culo, racemo corymboſo.
Phlox maculata. *Sp. pl.* 216. *Jacqu. hort.* 2. *p.* 58.
*t.* 127.
Spotted-ſtalk'd Lychnidea.
*Nat.* of North America.
*Cult.* 1759, by Mr. Philip Miller. *Mill. dict. edit.* 7.
*no.* 3.
*Fl.* Auguſt.　　　　　　　　　　　　　　　H. ♃.

*carolina.*

5. P. foliis lanceolatis lævibus, caule ſcabro, corymbis
ſubfaſtigiatis. *Sp. pl.* 216.
Carolina Lychnidea.
*Nat.* of Carolina.
*Cult.* before 1728. *Mart. dec.* 1. *p.* 10.
*Fl.* July——September.　　　　　　　　　　H. ♃.

*glaberri-*
*ma.*

6. P. foliis lineari-lanceolatis glabris, caule erecto, co-
rymbo terminali. *Sp. pl.* 217.
Smooth Lychnidea.
*Nat.* of North America.
*Cult.* 1731, by Mr. Philip Miller. *Mill. dict. edit.* 1.
Lychnidea 1.
*Fl.* June——Auguſt.　　　　　　　　　　　H. ♃.

*divarica-*
*ta.*

7. P. foliis lato-lanceolatis: ſuperioribus alternis, caule
bifido, pedunculis geminis. *Sp. pl.* 217.

Early-

Early-flowering Lychnidea.
*Nat.* of North America.
*Cult.* 1758, by Mr. Philip Miller. *Mill. ic.* 137.
t. 205. f. 1.
*Fl.* April——June.                                    H. ♃.


## CONVOLVULUS. *Gen. pl.* 215.

*Cor.* campanulata, plicata. *Stigmata* 2. *Capſ.* 2-lo-
     cularis : loculis diſpermis.

### * *Caule volubili.*

1. C. foliis ſagittatis utrinque acutis, pedunculis ſubuni-   *arvenſis.*
     floris. *Syſt. veget.* 200. *Curtis lond.*
Small Bind-weed.
*Nat.* of Britain.
*Fl.* June——September.                                H. ♃.


2. C. foliis ſagittatis poſtice truncatis, pedunculis tetra-   *ſepium.*
     gonis unifloris. *Sp. pl.* 218. *Curtis lond.*
Great Bind-weed.
*Nat.* of Britain.
*Fl.* June——September.                                H. ♃.


3. C. foliis ſagittatis poſtice truncatis, pedunculis tereti-   *Scammo-*
     bus ſubtrifloris. *Sp. pl.* 218.                          *nia.*
Scammony Bind-weed.
*Nat.* of the Levant.
*Cult.* 1596, by Mr. John Gerard. *Hort. Ger.*
*Fl.* July and Auguſt.                                H. ♃.


4. C. foliis cordatis acuminatis lævibus, pedunculis uni-   *ſibiricus.*
     floris. *Linn. mant.* 203.
Sibirian Bind-weed.
*Nat.* of Sibiria.

                                            *Introd.*

*Introd.* 1779, by Meſſrs. Kennedy and Lee.
*Fl.* July and Auguſt.                                    H. ☉.

*farino-*      5. C. foliis cordatis acuminatis repandis, pedunculis tri-
*ſus.*            floris, caule farinoſo.  *Linn. mant.* 203.  *Jacqu.*
                  *hort.* 1. *p.* 13. *t.* 35.
              Mealy-ſtalk'd Bind-weed.
              *Nat.* of Madeira.
              *Introd.* 1777, by Mr. Francis Maſſon.
              *Fl.* May and June.                         G. H. ♃.

*Medium.*     6. C. fol. linearibus haſtato-acuminatis : auriculis den-
                  tatis, pedunculis unifloris, calycibus ſagittatis, caule
                  volubili.  *Syſt. veget.* 200.
              Arrow-headed Bind-weed.
              *Nat.* of the Eaſt Indies.
              *Introd.* 1778, by Mr. William Roxburgh.
              *Fl.* July and Auguſt.                      S. ☉.

*tridenta-*   7. C. foliis cuneiformibus tricuſpidatis, baſi dilatata den-
*tus.*            tatis, pedunculis unifloris.  *Sp. pl. ed.* 1. *p.* 157.
              Evolvulus tridentatus.  *Sp. pl. ed.* 2. *p.* 392.  *Burm.*
                  *ind.* 77. *t.* 16. *f.* 3.
              Convolvulus indicus barbatus minor, foliorum apici-
                  bus lunulatis.  *Pluk. alm.* 117. *t.* 276. *f.* 5.
              Sendera-clandi.  *Rheed. mal.* 11. *p.* 133. *t.* 65.
              Trifid Bind-weed.
              *Nat.* of the Eaſt Indies.
              *Introd.* 1778, by Sir Joſeph Banks, Bart.
              *Fl.* July and Auguſt.                      S. ☉.

*pandura-*    8. C. foliis cordatis integris panduriformibus, calycibus
*tus.*            lævibus.  *Sp. pl.* 219.
              Virginian Bind-weed.
              *Nat.* of Carolina and Virginia.

                                                          *Cult.*

*Cult.* 1732, by James Sherard, M. D. *Dill. elth.* 101.
*t.* 85. *f.* 99.
*Fl.* June——September. G. H. ♃ .

9. C. foliis cordatis integris trilobifque, corollis indi- *hederace-*
vifis, fructibus erectis. *Sp. pl.* 219. *us.*
Ivy-leav'd Bind-weed.
*Nat.* of Afia, Africa, and America.
*Cult.* 1732, by James Sherard, M. D. *Dill. elth.* 97.
*t.* 81. *f.* 93.
*Fl.* June and July. H. ☉ .

10. C. foliis cordatis trilobis, corollis femiquinquefidis, *Nil.*
pedunculis petiolo brevioribus. *Sp. pl.* 219.
Blue Bind-weed.
*Nat.* of America.
*Cult.* before 1597, by Mr. John Gerard. *Ger. herb.* 715.
*Fl.* July——September. S. ☉ .

11. C. foliis cordatis indivifis, fructibus cernuis, pedi- *purpure-*
cellis incraffatis. *Sp. pl.* 219. *us.*
α Convolvulus purpureus folio fubrotundo. *Bauh. pin.*
295.
Great purple Bind-weed, or Convolvulus major.
β Convolvulus cæruleus minor, folio fubrotundo. *Dill.*
*elth.* 97. *t.* 82. *f.* 94.
Small purple Bind-weed.
*Nat.* of America.
*Cult.* 1732, by James Sherard, M. D. *Dill. elth.*
*loc. cit.*
*Fl.* June——September. H. ☉ .

12. C. foliis cordatis indivifis, caule fubpubefcente, pe- *obfcurus.*
dunculis incraffatis unifloris, calycibus glabris.
*Sp. pl.* 220.

P                                   Hairy

Hairy Bind-weed.
*Nat.* of the Eaſt Indies.
*Cult.* 1732, by James Sherard, M. D.   *Dill. elth.* 98.
t. 83. f. 95.
*Fl.* July and Auguſt.                          S. ☉.

*Ḅatatas.*  13. C. foliis cordatis haſtatis quinquenerviis, caule re-
pente hiſpido tuberifero.   *Sp. pl.* 220.
Tuberous-rooted Bind-weed.
*Nat.* of both Indies.
*Cult.* 1597, by Mr. John Gerard.   *Ger. herb.* 780.
*Fl.*                                           S. ♃.

*umbella-*  14. C. foliis cordatis, pedunculis umbellatis, caule volu-
*tus.*          bili.   *Sp. pl.* 221.
Umbel'd Bind-weed.
*Nat.* of the Weſt Indies.
*Introd.* 1774, by George Young, M. D.
*Fl.* June and July.                            S. ♃.

*canarien-*  15. C. foliis cordatis pubeſcentibus, caule perenni villo-
*ſis.*          ſo, pedunculis multifloris.   *Sp. pl.* 221.
Canary Bind-weed.
*Nat.* of the Canary Iſlands.
*Cult.* 1690, by the Dutcheſs of Beaufort.   *Br. Muſ.*
H. S. 139. fol. 63.
*Fl.* May——September.                           G. H. ♄.

*murica-*  16. C. foliis cordatis, pedunculis incraſſatis calycibuſ-
*tus.*          que lævibus, caule muricato.   *Linn. mant.* 44.
Rough-ſtalk'd Bind-weed.
*Nat.* of the Eaſt Indies.
*Introd.* 1777, by Daniel Charles Solander, LL.D.
*Fl.* July and Auguſt.                          S. ☉.

17. C.

17. C. foliis cordatis angulatis, caule membranaceo-qua-   *Turpe-*
drangulari, pedunculis multifloris. *Sp. pl.* 221.   *thum.*
Square-ftalk'd Bind-weed.
*Nat.* of Ceylon.
*Cult.* 1759, by Mr. Philip Miller. *Mill. dict. edit.* 7.
*no.* 31.
*Fl.*                                    S. ♃.

18. C. foliis cordatis fubtus tomentofo-fericeis, pedunculis   *fpeciofus.*
petiolo longioribus umbelliferis, calycibus acutis,
caule volubili.
Cónvolvulus fpeciofus. *Linn. fuppl.* 137.
Convolvulus nervofus. *Burm. ind.* 48. *t.* 20. *f.* 1.
Broad-leav'd Bind-weed.
*Nat.* of the Eaft Indies.
*Introd.* 1778, by Sir Jofeph Banks, Bart.
*Fl.*                                    S. ♄.

19. C. caule volubili, foliis ovatis fubcordatis obtufis ob-   *Jalapa.*
folete repandis fubtus villofis, pedunculis unifloris.
Convolvulus Jalapa. *Linn. mant.* 43.
Jalap Bind-weed.
*Nat.* of South America.
*Introd.* 1778, by Monf. Thouin.
*Fl.* Auguft and September.               S. ♄.

20. C. foliis cordatis finuatis fericeis: lobis repandis,   *althæ-*
pedunculis bifloris. *Syft. veget.* 202.   *oides.*
α Convolvulus argenteus Altheæ folio. *Bauh. pin.* 295.
Mallow-leav'd Bind-weed.
β Convolvulus argenteus elegantiffimus, foliis tenuiter
incifis. *Tournef. inft.* 85.
Silky-leav'd Bind-weed.
*Nat.* α. of the Levant; β. of Sicily.

Cult. 1656, by Mr. John Tradefcant, Jun. *Trad.*
*muf.* 104.

*Fl.* June —— September.                          G. H. ♃.

cairicus.   21. C. foliis pinnato-palmatis ferratis, pedunculis filifor-
mibus paniculatis, calycibus lævibus.   *Syft. veget.*
202.
Jagged-leav'd Bind-weed.
*Nat.* of Egypt.
*Introd.* 1770, by Monf. Richard.
*Fl.* June and July.                          G. H. ♃.

macro-      22. C. foliis palmato-pedatis quinquepartitis, peduncu-
carpos.         lis unifloris.   *Sp. pl.* 222.
Long-fruited Bind-weed.
*Nat.* of South America.
*Cult.* 1752, by Mr. Ph. Miller. *Mill. dict. edit.* 6. *n.* 27.
*Fl.* July and Auguft.                          S. ☉.

penta-      23. C. foliis digitatis quinis pilofis integerrimis, caule
phyllus.        pilofo.   *Sp. pl.* 223.
Five-leav'd Bind-weed.
*Nat.* of the Weft Indies.
*Cult.* 1759, by Mr. Philip Miller.   *Mill. dict. edit.* 7.
*no.* 12.
*Fl.* Auguft and September.                          S. ☉.

** *Caule non volubili.*

ficulus.    24. C. foliis cordato-ovatis, pedunculis unifloris: brac-
teis lanceolatis, flore feffili.   *Sp. pl.* 223.
Small-flower'd Bind-weed.
*Nat.* of the South of Europe.
*Cult.* 1739, by Mr. Ph. Miller.   *Rand. chel. n.* 14.
*Fl.* June —— Auguft.                          H. ☉.

25. C.

25. C. foliis lanceolatis fericeis lineatis petiolatis, pedun-     *lineatus.*
culis bifloris, calycibus fericeis fubfoliaceis. *Sp.*
*pl.* 224.
Dwarf Bind-weed.
*Nat.* of France and Spain.
*Cult.* 1714, by the Dutchefs of Beaufort. *Br. Muf.*
*H. S.* 139. *fol.* 27.
*Fl.* June.                                        H. ♃.

26. C. foliis lanceolatis tomentofis, floribus umbella-     *Cneorum.*
tis, calycibus hirfutis, caule erecto. *Syft. veget.*
203.
Silvery-leav'd Bind-weed.
*Nat.* of Spain and the Levant.
*Cult.* 1739. *Rand. chelf. n.* 6.
*Fl.* May——September.                             G. H. ♄.

27. C. foliis lineari-lanceolatis acutis pilofis, pedunculis     *Canta-*
inferioribus foliis longioribus fubbifloris, calycibus     *brica.*
oblongo-lanceolatis hirfutis.
*α* caulibus decumbentibus.
Convolvulus Cantabrica. *Sp. pl. ed.* 2. *p.* 225. *Jacq.*
*auftr.* 3. *p.* 53. *t.* 296.
Decumbent flax-leav'd Bind-weed.
*β* caule erecto ramofo.
Convolvulus Cantabrica. *Sp. pl. ed.* 1. *p.* 158.
Upright flax-leav'd Bind-weed.
*Nat.* of the South of Europe.
*Cult.* 1743, by Mr. Philip Miller. *Mill. dict. edit.* 4.
*no.* 16.
*Fl.* moft part of the Summer.                    G. H. ♄.

28. C. foliis linearibus pilofiufculis, pedunculis fubtri-     *fcoparius.*
floris, calycibus fericeis ovatis acutis, caule fruti-
cofo, ramis virgatis.

P 3                         Convolvulus

Convolvulus fcoparius. *Linn. fuppl.* 135.
Convolvulus linarifolius. *Mill. dict.*
Broom Bind-weed.
*Nat.* of the Canary Iflands.
*Cult.* 1768, by Mr. Philip Miller. *Mill. dict. edit.* 8.
*Fl.* Auguft and September.      G. H. ♄.

*floridus.*    29. C. foliis oblongo-lanceolatis bafi attenuatis fubpilo-
fis, ramis floriferis pedunculifque paniculatis.
Convolvulus floridus. *Linn. fuppl.* 136. *Jacqu. ic.*
*collect.* 1. *p.* 62.
Many-flower'd Bind-weed.
*Nat.* of the Canary Iflands.   Mr. *Francis Maffon.*
*Introd.* 1779.
*Fl.* Auguft and September.      G. H. ♄.

*tricolor.*    30. C. foliis lanceolato-ovatis glabris, caule declinato,
floribus folitariis. *Sp. pl.* 225. *Curtis magaz.* 27.
Trailing Bind-weed, or Convolvulus minor.
*Nat.* of Barbary, Spain, and Sicily.
*Cult.* 1629, by Mr. John Parkinfon.   *Park. parad.*
361. *f.* 3.
*Fl.* June —— Auguft.      H. ☉.

*Soldanel-*   31. C. foliis reniformibus, pedunculis unifloris. *Sp.*
*la.*          *pl.* 226.
Sea Bind-weed.
*Nat.* of Britain.
*Fl.* June and July.      H. ☉.

*Pes ca-*   32. C. foliis bilobis, pedunculis unifloris.   *Sp. pl.* 226.
*præ.*       Thick-leav'd Bind-weed.
*Nat.* of the Eaft Indies.
*Introd.* 1770, by Monf. Richard.
*Fl.* June and July.      S. ☉.
                                       33. C.

33. C. foliis emarginatis bafi biglandulofis, pedunculis *brafilien-* trifloris. *Sp. pl.* 226.     *fis.*
Broad-leav'd Bind-weed.
*Nat.* of South America.
*Introd.* before 1726, by Mr. Mark Catefby. *Mill. dict. edit.* 1. *no.* 4.
*Fl.*          S. ♃.

I P O M Œ A. *Gen. pl.* 216.
*Cor.* infundibulif. *Stigma* capitato-globofum. *Capf.* 3-locularis.

1. I. foliis pinnatifidis linearibus, floribus fubfolitariis. *Quamo-* *Sp. pl.* 227.     *clit.*
Winged-leav'd Ipomœa.
*Nat.* of the Eaft Indies.
*Cult.* 1629. *Park. parad.* 358. *no.* 3.
*Fl.* July——September.     S. ☉.

2. I. foliis pinnatifidis linearibus, floribus racemofis pen- *rubra.* dulis. *Syft. veget.* 204.
Polemonium rubrum. *Sp. pl.* 231.
Upright Ipomœa.
*Nat.* of Carolina.
*Cult.* 1732, by James Sherard, M. D. *Dill. elth.* 321. *t.* 241. *f.* 312.
*Fl.* September.     S. ♄.

3. I. foliis cordatis acuminatis bafi angulatis, pedun- *coccinea.* culis multifloris. *Sp. pl.* 228.
Scarlet-flower'd Ipomœa, or Convolvulus.
*Nat.* of the Weft Indies.
*Cult.* 1759, by Mr. Philip Miller. *Mill. dict. edit.* 7. *no.* 2.
*Fl.* June——September.     H. ☉.

*lacunofa.*    4. I. foliis cordatis acuminatis fcrobiculatis bafi angula-
tis, pedunculis fubunifloris flore brevioribus. *Sp.
pl.* 228.
Starry Ipomœa.
*Nat.* of Virginia and Carolina.
*Cult.* 1640. *Park. theat.* 164. *no.* 5.
*Fl.* July.                        H. ☉.

*tuberofa.*    5. I. foliis palmatis: lobis feptenis lanceolatis acutis in-
tegerrimis, pedunculis trifloris. *Syft. veget.* 204.
Tuberous-rooted Ipomœa.
*Nat.* of the Weft Indies.
*Introd.* 1780, by Mr. Alexander Anderfon.
*Fl.*                            S. ♃.

*bona nox.*    6. I. foliis cordatis acutis integerrimis, caule aculeato,
floribus ternis, corollis indivifis. *Sp. pl.* 228.
Prickly Ipomœa.
*Nat.* of the Weft Indies.
*Introd.* 1773, by John Earl of Bute.
*Fl.* July and Auguft.                 S. ☉.

*triloba.*    7. I. foliis trilobis cordatis, pedunculis trifloris. *Sp. pl.*
229.
Three-lobed Ipomœa.
*Nat.* of the Weft Indies.
*Cult.* 1759, by Mr. Philip Miller. *Mill. dict. edit.* 7.
*no.* 6.
*Fl.* June and July.                 H. ☉.

*hederifo-
lia.*    8. I. foliis trilobis cordatis, pedunculis multifloris race-
mofis. *Sp. pl.* 229.
Ivy-leav'd Ipomœa.
*Nat.* of South America.
*Introd.* 1773, by Jofeph Nicholas de Jacquin, M.D.
*Fl.* July.                         H. ☉.

9. I.

9. I. foliis cordatis acuminatis pilofis, floribus aggregatis. *tamnifo-*
     *Sp. pl.* 230.                                                    *lia.*
   Tamnus-leav'd Ipomœa.
   *Nat.* of Carolina.
   *Cult.* 1732, by James Sherard, M. D. *Dill. elth.*
     428.
   *Fl.* July.                                             H. ☉.

10. I. foliis palmatis, floribus aggregatis.  *Sp. pl.* 230.   *Pes ti-*
    Palmated Ipomœa.                                           *gridis.*
    *Nat.* of the Eaft Indies.
    *Cult.* 1732, by James Sherard, M. D. *Dill. elth.* 429.
      *t.* 318. *f.* 411.
    *Fl.* Auguft.                                         S. ☉.

## LIGHTFOOTIA. *L'Heritier fert. angl.*

*Cor.* 5-petala, fundo claufo valvis ftaminiferis. *Cal.*
    5-phyllus. *Stigma* 3-5-fidum. *Capf.* 3-5-locula-
    ris, 3-5-valvis, femifupera.

1. L. foliis petalifque lanceolatis.  *L'Herit. fert. angl.*   *oxycoccoi-*
     *tab.* 4.                                                 *des.*
   Lobelia tenella.  *Linn. mant.* 120.
   Lance-leav'd Lightfootia.
   *Nat.* of the Cape of Good Hope.
   *Introd.* 1787, by Mr. Francis Maffon.
   *Fl.* July.                                         G. H. ♄.

2. L. foliis fubulatis, petalis linearibus.  *L'Herit. fert.*  *fubulata.*
     *angl. tab.* 5.
   Awl-leav'd Lightfootia.
   *Nat.* of the Cape of Good Hope.  Mr. *Fr. Maffon.*
   *Introd.* 1787.
   *Fl.* Auguft,                                       G. H. ♃.

POLEMO-

## POLEMONIUM. *Gen. pl.* 217.

*Cor.* 5-partita, fundo clauſo valvis ſtaminiferis. *Stigma* 3-fidum. *Capſ.* 3-locularis, ſupera.

*cæruleum.*

1. P. foliis pinnatis, floribus erectis, calycibus corollæ tubo longioribus. *Sp. pl.* 230.

α flore cæruleo.
　　Blue-flower'd Greek Valerian.

β flore albo.
　　White-flower'd Greek Valerian.
　　*Nat.* of Britain.
　　*Fl.* May——July. 　　　　　　　　　　H. ♃.

*reptans.*

2. P. foliis pinnatis ſeptenis, floribus terminalibus nutantibus. *Syſt. veget.* 205.
　　Creeping Greek Valerian.
　　*Nat.* of North America.
　　*Cult.* 1758, by Mr. Ph. Miller. *Mill. ic.* 140. *t.* 209.
　　*Fl.* April and May. 　　　　　　　　　H. ♃.

## CAMPANULA. *Gen. pl.* 218.

*Cor.* campanulata, fundo clauſo valvis ſtaminiferis. *Stigma* 3-fidum. *Capſ.* infera, poris lateralibus dehiſcens.

### * Foliis lævioribus anguſtioribus.

*ceniſia.*

1. C. caulibus unifloris, foliis ovatis glabris integerrimis ſubciliatis. *Syſt. veget.* 206.
　　Ciliated Bell-flower.
　　*Nat.* of Switzerland.
　　*Introd.* 1775, by the Doctors Pitcairn and Fothergill.
　　*Fl.* June and July. 　　　　　　　　　H. ♃.

*grandiflora.*

2. C. foliis ternis oblongis ſerratis, caule unifloro, flore patulo. *Linn. ſuppl.* 140. *Jacqu. hort.* 3. *p.* 4. *t.* 2.
　　　　　　　　　　　　　　　　　　Great-

Great-flower'd Bell-flower.
*Nat.* of Siberia.
*Introd.* 1782, by Mr. John Bell.
*Fl.* July. H. ♃.

3. C. foliis radicalibus reniformibus; caulinis lineari- *rotundi-*
    bus. *Sp. pl.* 232. *Curtis lond.* *folia.*
α Campanula minor rotundifolia vulgaris. *Bauh. pin.* 93.
    Round-leav'd Bell-flower, or Hare Bells.
β Campanula minor rotundifolia alpina. *Bauh. pin.* 93.
    Small Round-leav'd Bell-flower.
*Nat.* of Britain.
*Fl.* June and July. H. ♃.

4. C. foliis omnibus cordatis ferratis petiolatis glabris, *carpati-*
    ramis filiformibus unifloris. *ca.*
    Campanula carpatica. *Jacqu. hort.* 1. *p.* 22. *t.* 57.
      *Linn. fuppl.* 140.
    Heart-leav'd Bell-flower.
*Nat.* of the Carpatian Alps.
*Introd.* 1774, by Jofeph Nicholas de Jacquin, M. D.
*Fl.* June and July. H. ♃.

5. C. cauliculis teretibus ftrictis glabris, foliis lineari- *Lobelioi-*
    lanceolatis denticulatis, corollis fubinfundibulifor- *des.*
    mibus trifidis quadrifidifque.
    Campanula Lobelioides. *Linn. fuppl.* 140.
    Small-flower'd Bell-flower.
*Nat.* of Madeira. Mr. *Francis Maffon.*
*Introd.* 1777.
*Fl.* July and Auguft. H. ☉.

6. C. foliis ftrictis: radicalibus lanceolato-ovalibus, pa- *patula.*
    nicula patula. *Sp. pl.* 232.
                              Spreading

Spreading Bell-flower.
*Nat.* of England.
*Fl.* July and Auguſt.                                        H. ♂.

*Rapun-*      7. C. foliis undulatis: radicalibus lanceolato-ovalibus, pa-
*culus.*         nicula coarctata. *Sp. pl.* 232.
              Eſculent Bell-flower, or Rampions.
              *Nat.* of England.
              *Fl.* July——September.                          H. ♂.

*perſicifo-*   8. C. foliis radicalibus obovatis, caulinis lanceolato-
*lia.*            linearibus ſubſerratis ſeſſilibus remotis. *Syſt. veget.*
              206.
              Peach-leav'd Bell-flower.
              *Nat.* of the Northern parts of Europe.
              *Cult.* 1596, by Mr. John Gerard. *Hort. Ger.*
              *Fl.* July——September.                          H. ♃.

*pyrami-*      9. C. foliis lævibus ſerratis cordatis: caulinis lanceola-
*dalis.*          tis, caulibus junceis ſimplicibus, umbellis ſeſſilibus
              lateralibus. *Syſt. veget.* 206.
              Pyramidal Bell-flower.
              *Nat.*
              *Cult.* 1596, by Mr. John Gerard. *Hort. Ger.*
              *Fl.* July——October.                            H. ♃.

*america-*    10. C. foliis cordatis lanceolatiſque ſerratis, petiolis in-
*na.*             feriorum ciliatis, floribus axillaribus ſeſſilibus, co-
              rollis quinquepartitis planis, ſtylis corolla longio-
              ribus.
              Campanula americana. *Sp. pl.* 233. (excluſis ſynony-
              mis, quæ ſequentis)
              American Bell-flower.
              *Nat.* of Penſylvania.

                                                           *Introd.*

*Introd.* 1763, by Mr. John Bartram.
*Fl.* July.                                             H. ☉.

11. C. foliis oblongis crenatis lævigatis: caulinis lan-   *nitida.*
ceolatis fubintegris, corollis campanulato-rotatis.
*L'Herit. fert. angl.*
Campanula americana minor, flore cæruleo patulo.
*Rob. ic.*
Trachelium americanum minus, flore cæruleo pa-
tulo. *Dod. mem.* 119. *tab.* 118.
Smooth-leav'd Bell-flower.
*Nat.* of North America.
*Cult.* 1743, by Mr. Philip Miller. *Mill. dict. edit.* 4.
*no.* 13.
*Fl.* July.                                             H. ♃.

12. C. foliis lanceolatis: caulinis acute ferratis, floribus   *lilifolia.*
paniculatis nutantibus. *Sp. pl.* 233.
Lily Bell-flower.
*Nat.* of Siberia.
*Introd.* 1784, by Monf. Thouin.
*Fl.* moft part of the Summer.                         H. ♃.

13. C. foliis rhomboidibus ferratis, fpica fecunda, caly-   *rhomboi-*
cibus dentatis. *Syft. veget.* 206.                       *dea.*
Germander-leav'd Bell-flower.
*Nat.* of the Alps of Switzerland.
*Introd.* 1775, by the Doctors Pitcairn and Fothergill.
*Fl.* July.                                             H. ♃.

** *Foliis fcabris latioribus.*
14. C. foliis ovato-lanceolatis, caule fimpliciffimo tereti,   *latifolia.*
floribus folitariis pedunculatis, fructibus cernuis.
*Sp. pl.* 234.
Broad-leav'd Bell-flower.
                                                        *Nat.*

*Nat.* of Britain.

*Fl.* July.                                              H. ♃.

*rapuncu-*      15. C. foliis cordato-lanceolatis, caule ramofo, floribus
*loides.*              fecundis fparfis, calycibus reflexis. *Syft. veget.* 207.
                Nettle-leav'd Bell-flower.
                *Nat.* of the South of Europe.
                *Cult.* 1683, by Mr. James Sutherland. *Sutherl. hort.*
                      *edin.* 64. *no.* 3.
                *Fl.* June and July.                           H. ♃.

*bononien-*     16. C. foliis ovato-lanceolatis fubtus fcabris feffilibus,
*fis.*                caule paniculato. *Sp. pl.* 234.
                Panicled Bell-flower.
                *Nat.* of Italy.
                *Introd.* 1773, by John Earl of Bute.
                *Fl.* Auguft and September.                    H. ♃.

*Tracheli-*     17. C. caule angulato, foliis petiolatis, calycibus ciliatis,
*um.*                 pedunculis trifidis. *Sp. pl.* 235.
                Great Bell-flower.
                *Nat.* of Britain.
                *Fl.* July and Auguft.                         H. ♃.

*glomera-*      18. C. caule angulato fimplici, floribus feffilibus, ca-
*ta.*                 pitulo terminali. *Sp. pl.* 235.
                Clufter'd Bell-flower.
                *Nat.* of Britain.
                *Fl.* May——September.                          H. ♃.

*Cervica-*      19. C. hifpida, floribus feffilibus, capitulo terminali, fo-
*ria.*                liis lanceolato-linearibus undulatis. *Sp. pl.* 235.
                Waved-leav'd Bell-flower.
                *Nat.* of Sweden and Germany.

                                                          *Introd.*

*Introd.* 1783, by William Pitcairn, M. D.
*Fl.* July.                                        H. ♂ .

\*\*\* *Capſulis obtectis calycis ſinubus reflexis.*

20. C. capſulis quinquelocularibus obtectis, caule ſim-    *Medium.*
plici erecto folioſo, floribus erectis. *Sp. pl.* 236.
Canterbury Bell-flower.
*Nat.* of Germany and Italy.
*Cult.* 1597. *Ger. herb.* 362.
*Fl.* June——September.                         H. ♂ .

21. C. capſulis quinquelocularibus obtectis, caule ſim-    *barbata.*
pliciſſimo unifolio, foliis lanceolatis, corollis bar-
batis. *Sp. pl.* 236.
One-leav'd Bell-flower.
*Nat.* of Italy and Switzerland.
*Introd.* 1775, by the Doctors Pitcairn and Fothergill.
*Fl.* June and July.                            H. ♃ .

22. C. hiſpida, ſpica laxa, floribus alternis, foliis lineari-    *ſpicata.*
bus integerrimis. *Syſt. veget.* 208.
Spiked Bell-flower.
*Nat.* of Switzerland.
*Introd.* 1786, by William Pitcairn, M. D.
*Fl.* July.                                        H. ♂ .

23. C. capſulis trilocularibus obtectis, caule paniculato.    *ſibirica.*
*Sp. pl.* 236. *Jacqu. auſtr.* 2. *p.* 60. *t.* 200.
Siberian Bell-flower.
*Nat.* of Siberia and Auſtria.
*Introd.* 1783, by William Pitcairn, M. D.
*Fl.* July and September.                       H. ♂ .

24. C. capſulis quinquelocularibus, foliis ellipticis ſer-    *aurea.*
ratis glabris, floribus ſubpaniculatis quinqueparti-
tis, caulibus fruticoſis carnoſis.

Campanula

Campanula aurea. *Linn. fuppl.* 141.
Golden Bell-flower.
*Nat.* of Madeira.    Mr. *Francis Maſſon.*
*Introd.* 1777.
*Fl.* Auguſt and September.                    G. H. ♃.

*Speculum.*    25. C. caule ramoſiſſimo diffuſo, foliis oblongis ſubcrena-
tis, floribus ſolitariis, capſulis priſmaticis. *Syſt.
veget.* 209.
Venus's Looking-glaſs, or Bell-flower.
*Nat.* of the South of Europe.
Cult. 1683, by Mr. James Sutherland. *Sutherl. hort.
edin.* 250. *no.* 2.
*Fl.* May——July.                    H. ☉.

*hybrida.*    26. C. caule baſi ſubramoſo ſtricto, foliis oblongis crena-
tis, calycibus aggregatis corolla longioribus, cap-
ſulis priſmaticis. *Sp. pl.* 239.
Corn Bell-flower, or Codded Corn-violet.
*Nat.* of England.
*Fl.* May——July.                    H. ☉.

*Priſma-*    27. C. capſulis linearibus bilocularibus, foliis lanceola-
*tocarpus.*        tis laxe ſerratis glaberrimis, caule decumbente.
*L'Herit. ſert. angl. t.* 3.
Long-capſuled Bell-flower.
*Nat.* of the Cape of Good Hope.   Mr. *Fr. Maſſon.*
*Introd.* 1787.
*Fl.* September.                    G. H. ☉.

*perfolia-*    28. C. caule ſimplici, foliis cordatis dentatis amplexi-
*ta.*        caulibus, floribus ſeſſilibus aggregatis. *Sp. pl.* 239.
Perfoliate Bell-flower.
*Nat.* of North America.

                                        *Cult.*

*Cult.* 1680. *Morif. hift.* 2. *p.* 457. *no.* 23. *f.* 5. *t.* 2. *f.* 23.
*Fl.* June.                                                    H. ☉.

29. C. foliis cordatis quinquelobis petiolatis glabris, *hedera-cea.*
    caule laxo. *Sp. pl.* 240.
    Ivy-leav'd Bell-flower.
    *Nat.* of England.
    *Fl.* May——Auguft.                                H. ♃.

30. C. caule dichotomo, foliis feffilibus : fuperioribus op- *Erinus.*
    pofitis tridentatis. *Syft. veget.* 210.
    Fork'd Bell-flower.
    *Nat.* of Spain, Italy, and the South of France.
    *Cult.* 1768, by Mr. Philip Miller. *Mill. dict. edit.* 8.
    *Fl.* July and Auguft.                             H. ☉.

### ROELLA. *Gen. pl.* 219.

*Cor.* infundibulif. fundo claufo valvis ftaminiferis.
    *Stigma* 2-fidum. *Capf.* 2-locularis, cylindrica,
    infera.

1. R. foliis ciliatis : mucrone recto. *Sp. pl.* 241.          *ciliata.*
    Ciliated Roella.
    *Nat.* of the Cape of Good Hope.
    *Introd.* 1774, by Mr. Francis Maffon.
    *Fl.* July——September.                           G. H. ♄.

2. R. herbacea diffufa, foliis ovatis recurvatis dentatis, *fquarro-*
    floribus terminalibus aggregatis. *Linn. fuppl.* 143. *fa.*
    Trailing Roella.
    *Nat.* of the Cape of Good Hope.
    *Introd.* 1787, by Mr. Francis Maffon.
    *Fl.* June.                                       G. H. ♃.

*decur-*
*rens.*

3. R. foliis lanceolatis ciliatis decurrentibus.   *L'Herit.*
      *fert. angl. tab.* 6.
   Decurrent Roella.
   *Nat.* of the Cape of Good Hope.   Mr. *Fr. Maſſon.*
   *Introd.* 1787.
   *Fl.* September.                                    G. H. ☉

### PHYTEUMA.   *Gen. pl.* 220.

   *Cor.* rotata, 5-partita :  laciniis linearibus.   *Stigma*
      2. ſ. 3-fidum.  *Capſ.* 2. ſ. 3-locularis, infera.

*orbicula-*
*ris.*

1. P. capitulo ſubrotundo, foliis ſerratis :  radicalibus cor-
      datis.  *Sp. pl.* 242. *Jacqu. auſtr.* 5. *p.* 18. *t.* 437.
   Round-headed horn'd Rampion.
   *Nat.* of England.
   *Fl.* June——Auguſt.                                H. ♃

*ſpicata.*

2. P. ſpica oblonga, capſulis bilocularibus, foliis radica-
      libus cordatis.  *Sp. pl.* 242.
   Spiked horn'd Rampion.
   *Nat.* of Europe.
   *Cult.* 1683, by Mr. James Sutherland.   *Sutherl. hort.*
      *edin.* 290. *no.* 3.
   *Fl.* June.                                         H. ♃

### TRACHELIUM.   *Gen. pl.* 221.

   *Cor.* infundibuliformis.  *Stigma* globoſum.  *Capſ.* tri-
      locularis, infera.

*cærule-*
*um.*

1. TRACHELIUM.   *Sp. pl.* 243.
   Blue Throat-wort.
   *Nat.* of Italy, and the Levant.
   *Cult.* 1739.  *Mill. dict. vol.* 2. *n.* 1.
   *Fl* July——September.                              H. ♂

                                       SAMOLUS.

## S A M O L U S. *Gen. pl.* 222.

*Cor*, hypocrateriformis. *Stamina* munita fquamulis corollæ. *Capf.* 1-locularis, infera.

1. SAMOLUS. *Sp. pl.* 243. *Curtis lond.*                    *valeran-*
   Water Pimpernel.                                           *di.*
   *Nat.* of Britain.
   *Fl.* June——Auguft.                           H. ♃.


## R O N D E L E T I A. *Gen. pl.* 224.

*Cor.* infundibuliformis. *Capfula* 2-locularis, infera, polyfperma, fubrotunda, coronata.

1. R. foliis feffilibus, panicula dichotoma. *Sp. pl.* 243.    *america-*
   American Rondeletia.                                        *na.*
   *Nat.* of the Weft Indies.
   *Cult.* 1748, by Mr. Philip Miller.  *Mill. dict. edit.* 5.
   *Fl.* Auguft.                                  S. ♄.

2. R. foliis petiolatis ovato-oblongis acutis pilofis, pa-    *hirta.*
   niculis trichotomis axillaribus.
   Hairy Rondeletia.
   *Nat.* of Jamaica.
   *Introd.* about 1776, by John Blackburne, Efq.
   *Fl.* June——Auguft.
   OBS. Valde affinis *R. odoratæ*, differt vero infloref-
   centia axillari et foliis majoribus, acutioribus, fupra
   non fcabris, bafi vix cordatis, petiolifque longioribus. *Tubus* corollæ huic calyce duplo tantum longior. *Stylus* extra faucem exfertus. *Stigmata* erecta, conniventia. *Corolla* e luteo rufefcens.


Q 2                          SOLAN-

## SOLANDRA. *Swartz.*

*Cor.* infundibuliformis, maxima. *Cal.* tandem rum-
pens. *Stam.* inclinata. *Bacca* fupera, 4-locularis,
polyfperma.

*grandi-*
*flora.*

1. SOLANDRA. *Swartz in act. stockholm.* 1787. *pag.*
300. *tab.* 11.
Great-flower'd Solandra.
*Nat.* of Jamaica. Mr. *Francis Maffon.*
*Introd.* 1781.
*Fl.* March. S. ♄.

## PORTLANDIA. *Gen. pl.* 227.

*Cor.* clavato-infundibuliformis. *Antheræ* longitudi-
nales. *Capf.* pentagona, retufa, bilocularis, polyfperma, coronata calyce 5-phyllo.

*grandi-*
*flora.*

1. P. floribus pentandris. *Syst. veget.* 213.
Great-flower'd Portlandia.
*Nat.* of Jamaica.
*Introd.* 1775, by —— Ellis, Efq.
*Fl.* July and Auguft. S. ♄.

## CINCHONA. *Gen. pl.* 228.

*Cor.* hypocrateriformis. *Capf.* infera, bilocularis:
diffepimento intus dehifcente. *Sem.* alata, imbricata.

*caribæa.*

1. C. pedunculis unifloris. *Sp. pl.* 245.
Caribean Cinchona.
*Nat.* of the Weft Indies.
*Introd.* 1780, by William Philp Perrin, Efq.
*Fl.* S. ♄.

COFFEA.

C O F F E A. *Gen. pl.* 230.

*Cor.* hypocrateriformis. *Stamina* fupra tubum. *Bacca* infera, difperma. *Sem.* arillata.

1. C. floribus quinquefidis difpermis. *Sp. pl.* 245.     *arabica.*
   Coffee-tree.
   *Nat.* of Yemen.
   *Cult.* 1696, by Bifhop Compton. *Douglas hiftory of*
     *the Coffee-tree,* p. 21.
   *Fl.* October and November.      S. ♄.

C H I O C O C C A. *Gen. pl.* 231.

*Cor.* infundibuliformis, æqualis. *Bacca* 1-locularis, 2-fperma, infera.

1. C. foliis oppofitis. *Sp. pl.* 246.     *racemofa.*
   Oppofite-leav'd Chiococca, or Snow-berry.
   *Nat.* of Jamaica.
   *Cult.* 1729, by James Sherard, M.D. *Dill. elth.* 306.
     *t.* 228. *f.* 295.
   *Fl.* September.      S. ♄.

H A M E L L I A. *Gen. pl.* 232.

*Cor.* 5-fida. *Bacca* 5-locularis, infera, polyfperma.

1. H. foliis lævigatis, tubo corollæ ventricofo. *L'Herit.*   *grandi-*
   *fert. angl. tab.* 7.      *flora.*
   Great-flower'd Hamellia.
   *Nat.* of the Weft Indies.
   *Introd.* 1778, by Thomas Clark, M.D.
   *Fl.* September——November.      S. ♄.

Q 3      LONI-

## LONICERA. *Gen. pl.* 233.

*Cor.* monopetala, irregularis. *Bacca* polyſperma, 2-
locularis, infera.

\* Periclymena *caule volubili.*

*Caprifo-*      1. L. floribus ringentibus verticillatis terminalibus, fo-
*lium.*              liis deciduis : ſummis connato-perfoliatis.
                    Lonicera Caprifolium. *Sp. pl.* 246. *Jacqu. auſtr.* 4.
                    *p.* 30. *t.* 357.

alba.           α Caprifolium italicum perfoliatum præcox. *Hort.*
                    *angl. p.* 14. *no.* 3. *tab.* 5.
                    Italian early White Honey-ſuckle.

rubra.          β Caprifolium italicum. *Hort. angl. p.* 14. *no.* 1. *tab.* 5.
                    Italian early Red Honey-ſuckle.
                    *Nat.* of the South of Europe.
                    *Cult.* 1596, by Mr. John Gerard. *Hort. Ger.*
                    *Fl.* May and June.                                    H. ♄.

*dioica.*       2. L. verticillis ſubcapitatis bracteolatis, foliis deciduis
                    ſubtus glaucis : ſummis connato-perfoliatis, corollis
                    ringentibus baſi gibbis.
                    Lonicera dioica. *Syſt. veget.* 215.
                    Lonicera media. *Murray in nov. comm. gotting.* 1776.
                    *p.* 28. *t.* 3. (excluſis ſynonymis Milleri)
                    Glaucous Honey-ſuckle.
                    *Nat.* of North America.
                    *Introd.* 1766, by Peter Collinſon, Eſq.
                    *Fl.* June and July.                                   H. ♄.

*ſempervi-*    3. L. ſpicis ſubnudis terminalibus, foliis oblongis : ſum-
*rens.*             mis connato-perfoliatis, corollis ſubæqualibus : tubo
                    ſuperne ventricoſo.
                    Lonicera ſempervirens. *Sp. pl.* 247.
major.          α foliis ſubrotundis.
                    Great Trumpet Honey-ſuckle.

                                                            β foliis

β foliis oblongis.                                             minor.
  Small Trumpet Honey-fuckle.
  *Nat.* of North America.
  *Cult.* 1656, by Mr. John Tradefcant, Jun. *Muf.*
    *Trad.* 95.
  *Fl.* May——Auguft.                                  H. ♄.

4. L. floribus verticillatis terminalibus, foliis perennan-    *grata.*
     tibus obovatis fubtus glaucis, fummis connato-fub-
     perfoliatis, corollis ringentibus.
  Periclymenum americanum.    *Mill. dict.*
  Ever-green Honey-fuckle.
  *Nat.* of North America.
  *Cult.* 1739, by Mr. Philip Miller.    *Rand. chelf.* Ca-
     prifolium 9.
  *Fl.* June——October.                                H. ♄.

5. L. floribus ringentibus verticillatis, bracteis lævibus,    *implexa.*
     foliis perennantibus glabris oblongis: fuperioribus
     connato-perfoliatis : fummis dilatatis.
  Minorca Honey-fuckle.
  *Nat.* of Minorca.
  *Introd.* about 1772, by Monf. Richard.
  *Fl.* June——September.                              H. ♄.
  Obs. *Folia* minora et anguftiora quam in plerifque
     fpeciebus hujus generis.

6. L. floribus ringentibus capitatis terminalibus, foliis     *Pericly-*
     deciduis : omnibus diftinctis.                            *menum.*
  Lonicera Periclymenum.    *Sp. pl.* 247.    *Curtis lond.*
  α Periclymenum non perfoliatum germanicum.    *Bauh.*    vulgaris.
     *pin.* 302.
  Common Wood-bine, or Honey-fuckle.
  β Caprifolium germanicum flore rubello ferotinum.    ferotina.
     *Hort. angl. p.* 14. *n.* 4. *tab.* 7.
  Late Red Honey-fuckle.

                    Q 4                     γ Caprifo-

belgica.    γ Caprifolium germanicum floribus fpeciofis.  *Hort.*
           *angl. p.* 15. *n.* 5. *tab.* 6.
        Dutch Honey-fuckle.

quercifo-    δ Periclymenum foliis quercinis.  *Merr. pin.* 92.
lia.       Oak-leav'd Honey-fúckle.
        *Nat.* α, and δ, of Britain.
        *Fl.* May——July.    H. ♄.

           ** Chamæcerafa *pedunculis bifloris.*

nigra.    7. L. pedunculis bifloris, baccis diftinctis, foliis ellipti-
        cis integerrimis.  *Sp. pl.* 247.  *Jacqu. auftr.* 4. *p.* 7.
        *t.* 314.
        Black-berried upright Honey-fuckle.
        *Nat.* of Switzerland and France.
        *Cult.* 1683.  *Sutherl. hort. edinb.* 77. *n.* 2.
        *Fl.* March and April.    H. ♄.

tatarica.    8. L. pedunculis bifloris, baccis diftinctis, foliis cordatis
        obtufis.  *Sp. pl.* 247.  *Jacqu. ic. collect.* 1. *p.* 34.
        Tartarian upright Honey-fuckle.
        *Nat.* of Ruffia.
        *Cult.* 1752, by Mr. Ph. Miller.  *Mill. dict. edit.* 6. *n.* 7.
        *Fl.* April and May.    H. ♄.

Xylofte-    9. L. pedunculis bifloris, baccis diftinctis, foliis integer-
um.       rimis pubefcentibus.  *Sp. pl.* 248.
        Fly Honey-fuckle.
        *Nat.* of the North of Europe.
        *Cult.* 1683.  *Sutherl. hort. edinb.* 77. *n.* 3.
        *Fl.* May.    H. ♄.

pyrenai-    10. L. pedunculis bifloris, baccis diftinctis, foliis oblon-
ca.       gis glabris.  *Sp. pl.* 248.
        Pyrenean upright Honey-fuckle.
        *Nat.* of the Pyrenean Mountains.
                            *Cult.*

*Cult.* 1739, by Mr. Philip Miller. *Rand. chel.* Xylofteon.
*Fl.* May.                                              H. ♄.

11. L. pedunculis bifloris, baccis coadunatis didymis, foliis ovali-lanceolatis. *Syft. veget.* 216. *Jacqu. auftr.* 3. *p.* 40. *t.* 274.   *alpigena.*
Red-berried upright **Honey-fuckle.**
*Nat.* of Switzerland and Auftria.
*Cult.* 1596, by Mr. John Gerard. *Hort. Ger.*
*Fl.* April and May.                                    H. ♄.

12. L. pedunculis bifloris, baccis coadunato-globofis, ftylis indivifis. *Sp. pl.* 249. *Jacqu. auftr.* 5. *p.* 35. *tab. app.* 17.   *cærulea.*
Blue-berried upright Honey-fuckle.
*Nat.* of Switzerland.
*Cult.* 1724. *Furber's catal.*
*Fl.* March and April.                                  H. ♄.

\*\*\* *Caule erecto, pedunculis multifloris.*

13. L. capitulis lateralibus pedunculatis, foliis petiolatis. *Sp. pl.* 249.   *Symphoricarpos.*
Shrubby St. Peter's Wort.
*Nat.* of Virginia and Carolina.
*Cult.* 1730. *Hort. angl. t.* 20.
*Fl.* Aug.ft and September.                              H. ♄.

14. L. racemis terminalibus, foliis ferratis. *Sp. pl.* 249.   *Diervilla.*
Yellow-flower'd upright Honey-fuckle.
*Nat.* of North America.
*Cult.* 1739, by Mr. Philip Miller. *Rand. chel. p.* 68.
*Fl.* June.                                             H. ♄.

TRIOS-

## TRIOSTEUM. *Gen. pl.* 234.

*Cor.* monopetala, fubæqualis. *Cal.* longitudine corollæ.
*Bacca* 3-locularis, monofperma, infera.

*perfolia-*  1. T. floribus verticillatis feffilibus. *Sp. pl.* 250.
*tum.*  Fever-root.
*Nat.* of North America.
*Cult.* 1732, by James Sherard, M.D. *Dill. elth.* 394.
t. 293.
*Fl.* June and July. H. ♃.

## CONOCARPUS. *Gen. pl.* 236.

*Cor.* 5-petala, aut nulla. *Semina* nuda, folitaria, infera.
*Flores* aggregati.

*erecta.*  1. C. erecta, foliis lanceolatis. *Sp. pl.* 250.
Jamaica Button-tree.
*Nat.* of Jamaica.
*Cult.* 1752, by Mr. Philip Miller. *Mill. dict. edit.* 6.
n. 1.
*Fl.* S. ♄.

## MIRABILIS. *Gen. pl.* 242.

*Cor.* infundibul. fupera. *Cal.* inferus. *Nectarium*
globofum, germen includens.

*Jalapa.*  1. M. floribus congeftis terminalibus erectis. *Sp. pl.* 252.
α Jalapa flore purpureo. *Tournef. inft.* 129.
Common Marvel of Peru.
β Jalapa flore flavo. *Tournef. inft.* 129.
Yellow Marvel of Peru.
γ Jalapa flore exalbido. *Tournef. inft.* 129.
White Marvel of Peru.
δ Jalapa flore ex purpureo & luteo mixto. *Tournef.*
*inft.* 129.
Yellow-ftriped Marvel of Peru.
*Nat.*

*Nat.* of both Indies.
*Cult.* 1596, by Mr. John Gerard. *Hort. Ger.*
*Fl.* June——September. H. ♃.

2. M; floribus feffilibus axillaribus erectis folitaris. *Sp.* *dichoto*
   *pl.* 252. *ma;*
Forked Marvel of Peru.
*Nat.* of Mexico.
*Introd.* before 1728, by Charles Du Bois, Efq. *Mart.*
   *dec.* 1. *p.* 2.
*Fl.* July. S. ♃.

3. M. floribus congeftis longiffimis fubnutantibus ter- *longiflora.*
   minalibus, foliis fubvillofis. *Syft. veget.* 219.
Sweet-fcented Marvel of Peru.
*Nat.* of Mexico.
*Cult.* 1759, by Mr. Philip Miller. *Mill. dict. edit.* 7.
*Fl.* June——September. H. ♃.

### C O R I S. *Gen. pl.* 243.

*Cor.* monopetala, irregularis. *Cal.* fpinofus. *Capf.*
   5-valvis, fupera.

1. Coris. *Sp. pl.* 252. *monfpe*
Montpelier Coris. *lienfis.*
*Nat.* of the South of Europe.
*Cult.* 1739. *Mill. dict. vol.* 2.
*Fl.* June and July. G. H. ♂.

### V E R B A S C U M. *Gen. pl.* 245.

*Cor.* rotata, fubinæqualis. *Capf.* 2-locularis, 2-valvis.

1. V. foliis decurrentibus utrinque tomentofis, caule *Thapfus.*
   fimplici. *Syft. veget.* 219.

Great

Great broad-leav'd Mullein.
*Nat.* of Britain.
*Fl.* July and Auguſt.     H. ♂.

*Thapſo-ides.*     2. V. foliis decurrentibus, caule ramoſo. *Syſt. veget.* 219.
Baſtard Mullein.
*Nat.* of England.
*Fl.* July and Auguſt.     H. ♂.

*Boerhaa-vii.*     3. V. foliis ſublyratis, floribus ſeſſilibus. *Linn. mant.* 45.
Annual Mullein.
*Nat.* of the South of Europe.
*Cult.* 1758, by Mr. Ph. Miller. *Mill. ic.* 182. *t.* 273.
*Fl.* July.     H. ☉.

*hæmor-rhoidale.*     4. V. foliis ovato-oblongis baſi attenuatis tomentoſis ob-
ſolete crenulatis, racemis ſpiciformibus elongatis,
faſciculis florum ebracteatis.
Madeira Mullein.
*Nat.* of Madeira.    Mr. *Francis Maſſon.*
*Introd.* 1777.
*Fl.* June——Auguſt.     G. H. ♂.

*Lychni-tis.*     5. V. foliis cuneiformi-oblongis. *Sp. pl.* 253.
α Verbaſcum mas, anguſtioribus foliis, floribus pallidis.
*Bauh. pin.* 239.
White Mullein.
β Verbaſcum lychnitis flore albo parvo. *Bauh. pin.* 240.
Small-flower'd white Mullein.
*Nat.* of Britain.
*Fl.* June —— Auguſt.     H. ♂.

*phlomo-ides.*     6. V. foliis ovatis utrinque tomentoſis : inferioribus petiolatis. *Sp. pl.* 253.
Woolly Mullein.

                                               *Nat.*

*Nat.* of Italy.
*Cult.* 1739, by Mr. Philip Miller. *Rand. chelſ. n.* 2.
*Fl.* June and July.                                    H. ♂.

7. V. foliis ſubvilloſis rugoſis: caulinis ſubſeſſilibus æqua-    *ferrugi-*
    liter crenatis; radicalibus oblongis cordatis du-    *neum.*
    plicato crenatis.
  Verbaſcum ferrugineum. *Mill. dict.*
  Ruſty Mullein.
  *Nat.* of the South of Europe.
  *Cult.* 1683, by Mr. James Sutherland. *Sutherl. hort.*
    *edin.* 51.
  *Fl.* May——Auguſt.                                  H. ♃.

8. V. foliis oblongo-cordatis petiolatis. *Sp. pl.* 253.    *nigrum.*
  Black Mullein.
  *Nat.* of England.
  *Fl.* June——Auguſt.                                 H. ♃.

9. V. foliis ovatis nudis crenatis radicalibus, caule ſubnu-    *phœnice-*
    do racemoſo. *Syſt. veget.* 219. *Jacqu. auſtr.* 2. *p.* 15.    *um.*
    *t.* 125.
  Purple Mullein.
  *Nat.* of the South of Europe.
  *Cult.* 1597, by Mr. John Gerard. *Ger. herb.* 633. *f.* 2.
  *Fl.* May——July.                                    H. ♂.

10. V. foliis amplexicaulibus oblongis glabris, peduncu-    *Blatta-*
     lis ſolitariis. *Sp. pl.* 254.    *ria.*
  α Blattaria lutea, folio longo laciniato. *Bauh. pin.* 240.
    Yellow-flower'd Moth Mullein.
  β Blattaria alba. *Bauh. pin.* 241.
    White-flower'd Moth Mullein.
  *Nat.* of England.
  *Fl.* July and Auguſt.                               H. ♂.
                                                   11. V.

*finuatum.* 11. V. foliis radicalibus pinnatifido-repandis tomentofis, caulinis amplexicaulibus nudiufculis, rameis primis oppofitis. *Sp. pl.* 254.
Scollop-leav'd Mullein.
*Nat.* of Montpellier and Florence.
*Cult.* 1731, by Mr. Philip Miller. *Mill. dict. edit.* 1. *n.* 7.
*Fl.* July and Auguft. H. ♂.

*Ofbeckii.* 12. V. foliis incifis nudis, caule foliofo, calycibus lanatis, pedunculis bifloris. *Sp. pl.* 255.
Great-flower'd Mullein.
*Nat.* of Spain.
*Introd.* 1775, by Monf. Thouin.
*Fl.* July and Auguft. H. ♂.

*myconi.* 13. V. foliis lanatis radicalibus, fcapo nudo. *Sp. pl.* 255.
Borage-leav'd Mullein.
*Nat.* of the Pyrenees.
*Cult.* 1731. *Mill. dict. edit.* 1. *n.* 8.
*Fl.* May. H. ♃.

### D A T U R A. *Gen. pl.* 246.

*Cor.* infundib. plicata. *Cal.* tubulofus, angulatus, deciduus. *Capf.* 4-valvis.

*ferox.* 1. D. pericarpiis fpinofis erectis ovatis : fpinis fupremis maximis convergentibus. *Sp. pl.* 255.
Rough Thorn-apple.
*Nat.* of China.
*Cult.* 1731, by Mr. Philip Miller. *Mill. dict. edit.* 1. Stramonium 3.
*Fl.* July——September. S. ☉.

2. D.

2. D. pericarpiis fpinofis erectis ovatis, foliis ovatis gla- *Stramo-*
bris. *Sp. pl.* 255. *nium.*
Common Thorn-apple.
*Nat.* of England.
*Fl.* July. H. ☉.

3. D. pericarpiis fpinofis erectis ovatis, foliis cordatis *Tatula.*
glabris dentatis. *Sp. pl.* 256.
Blue Thorn-apple.
*Nat.*
*Cult.* 1686, by Mr. John Ray. *Raj. hift.* 1. *p.* 748.
*Fl.* July. H. ☉.

4. D. pericarpiis tuberculatis nutantibus globofis, foliis *faftuofa.*
ovatis angulatis. *Syft. veget.* 220.
Purple Thorn-apple.
*Nat.* of Egypt.
*Cult.* 1731, by Mr. Philip Miller. *Mill. dict. edit.* 1.
Stramonium 6.
*Fl.* July——November. S. ☉.

5. D. pericarpiis fpinofis nutantibus globofis, foliis cor- *Metel.*
datis fubintegris pubefcentibus. *Sp. pl.* 256.
Hairy Thorn-apple.
*Nat.* of Afia, Africa, and the Canary Iflands.
*Cult.* 1759, by Mr. Philip Miller. *Mill. dict. edit.* 7.
*Fl.* June——September. S. ☉.

6. D. pericarpiis glabris inermibus erectis, foliis glabris, *lævis.*
caule fiftulofo herbaceo.
Datura lævis. *Linn. fuppl.* 146.
Datura inermis. *Jacqu. hort.* 3. *p.* 44. *t.* 82.
Smooth-capful'd Thorn-apple.
*Nat.* of Africa.
*Introd.* 1780, by Monf. Thouin.
*Fl.* July——September. S. ☉.

7. D.

*arborea.*    7. D. pericarpiis glabris inermibus nutantibus, caule ar-
                 boreo.  *Syſt. veget.* 220.
             Tree Thorn-apple.
             *Nat.* of Peru.
             *Introd.* 1783, by Monſ. Thouin.
             *Fl.* Auguſt.                                    S. ♄.


             H Y O S C Y A M U S.  *Gen. pl.* 247.

         *Cor.* infundibul. obtuſa.  *Stam.* inclinata.  *Capſ.* oper-
                         culata, 2-locularis.

*niger.*      1. H. foliis amplexicaulibus ſinuatis, floribus feſſilibus.
                 *Sp. pl.* 257.
             Black Henbane.
             *Nat.* of Britain.
             *Fl.* June.                                       H. ♂.


*albus.*      2. H. foliis petiolatis ſinuatis obtuſis, floribus ſubſeſſili-
                 bus.  *Sp. pl.* 257.
             White Henbane.
             *Nat.* of the South of Europe.
             *Cult.* 1570.   *Lobel. adv.* 107.
             *Fl.* Auguſt.                                     H. ☉.


*aureus.*     3. H. foliis petiolatis eroſo-dentatis acutis, floribus pe-
                 dunculatis, fructibus pendulis.  *Sp. pl.* 257.
             Shrubby Henbane.
             *Nat.* of the Levant.
             *Cult.* 1640, by Mr. John Parkinſon.  *Park. theat.*
                 362. *f.* 3.
             *Fl.* March——October.                            G. H. ♂.

*puſillus.*   4. H. foliis lanceolatis dentatis : floralibus inferioribus
                 binis, calycibus ſpinoſis.  *Sp. pl.* 258.
                                                            Dwarf

Dwarf Henbane.
*Nat.* of Perſia.
*Cult.* 1691, by Biſhop Compton.- *Pluk. phyt. t.* 37. *f.* 5.
*Fl.* July. H. ☉.

5. H. foliis ovatis integerrimis, calycibus inflatis ſub-  *phyſaloi-*
   globoſis. *Sp. pl.* 258.  *des.*
   Purple-flower'd Henbane.
   *Nat.* of Siberia.
   *Introd.* 1777, by Meſſrs. Gordon and Græffer.
   *Fl.* March and April. H. ♃.

6. H. foliis ovatis integris, calycibus inflatis campanu-  *Scopolia.*
   latis lævibus. *Syſt. veget.* 221.
   Nightſhade-leav'd Henbane.
   *Nat.* of Carniola.
   *Introd.* 1780, by Mr. Daniel Grimwood.
   *Fl.* May. H. ♃.


N I C O T I A N A. *Gen. pl.* 248.
*Cor.* infundibul. limbo plicato. *Stam.* inclinata. *Capſ.*
      2-valvis, 2-locularis.

1. N. foliis lanceolatis ſubpetiolatis amplexicaulibus,  *fruticoſa.*
   floribus acutis, caule fruteſcente. *Syſt. veget.* 221.
   Shrubby Tobacco.
   *Nat.* of China.
   *Cult.* 1699, by the Dutcheſs of Beaufort. *Br. muſ.*
      *Sloan. mſs.* 525, and 3349.
   *Fl.* July and Auguſt. G. H. ♄.

2. N. foliis lanceolato-ovatis ſeſſilibus decurrentibus, flo-  *Tabacum.*
   ribus acutis. *Sp. pl.* 258.
α Nicotiana major latifolia. *Bauh. pin.* 169.
   Broad-leav'd Virginian Tobacco.
                    R           β Nicotiana

angusti-
folia.

β Nicotiana foliis lanceolatis acutis sessilibus, calycibus
acutis, tubo floris longissimo. *Mill. dict.*
Narrow-leav'd Virginian Tobacco.
*Nat.* of America.
*Cult.* before 1570. *Lobel. adv.* 251.
*Fl.* July and August.                                    H. ⊙.

rustica.

3. N. foliis petiolatis ovatis integerrimis, floribus obtu-
sis. *Sp. pl.* 258.
Common Tobacco.
*Nat.* of America.
*Cult.* 1731, by Mr. Philip Miller. *Mill. dict. edit.* 1.
*n.* 4.
*Fl.* July——September.                                    H. ⊙.

panicula-
ta.

4. N. foliis petiolatis cordatis integerrimis, floribus pa-
niculatis obtusis clavatis. *Sp. pl.* 259.
Panicled Tobacco.
*Nat.* of Peru.
*Cult.* 1739, by Mr. Ph. Miller. *Mill. dict. vol.* 2. *n.* 4.
*Fl.* July——September.                                    S. ⊙.

glutinosa.

5. N. foliis petiolatis cordatis integerrimis, floribus race-
mosis secundis ringentibus. *Sp. pl.* 259.
Clammy-leav'd Tobacco.
*Nat.* of Peru.
*Cult.* 1759, by Mr. Philip Miller. *Mill. dict. edit.* 7.
*Fl.* July——September.                                    S. ⊙.

### A T R O P A. *Gen. pl.* 249.

*Cor.* campanulata. *Stam.* distantia. *Bacca* globosa,
2-locularis.

Mandra-
gora.

1. A. acaulis, scapis unifloris. *Sp. pl.* 259.
Mandrake.

*Nat.*

*Nat.* of the South of Europe and the Levant.
*Cult.* 1562. *Turn. herb. part* 2. *fol.* 46.
*Fl.* March and April. H. ♃.

2. A. caule herbaceo, foliis ovatis integris. *Sp. pl.* 260. *Belladon-na.*
 Curtis lond. *Jacqu. auſtr.* 4. *p.* 5. *t.* 309.
 Deadly Nightſhade.
 *Nat.* of Britain.
 *Fl.* June and July. H. ♃.

3. A. caule herbaceo, foliis ſinuato-angulatis, calycibus *phyſaloi-des.*
 clauſis acutangulis. *Sp. pl.* 260.
 Blue-flower'd Atropa.
 *Nat.* of Peru.
 *Cult.* 1759, by Mr. Philip Miller. *Mill. dict. edit.* 7.
 Phyſalis 18.
 *Fl.* July——September. S. ☉.

4. A. caule fruticoſo, pedunculis confertis, foliis cordato- *fruteſcens.*
 ovatis obtuſis. *Sp. pl.* 260.
 Shrubby Atropa.
 *Nat.* of Spain.
 *Cult.* 1739, by Mr. Philip Miller. *Mill. dict. vol.* 2.
 Belladona 3.
 *Fl.* January——March. S. ♄.

## P H Y S A L I S. *Gen. pl.* 250.

*Cor.* rotata. *Stam.* conniventia. *Bacca* intra calycem
 inflatum, bilocularis.

\* *Perennes.*

1. P. caule fruticoſo, ramis rectis, floribus confertis. *Sp.* *ſomnifera.*
 *pl.* 261.
 Cluſter'd Winter Cherry.

R 2 *Nat.*

*Nat.* of Spain and Mexico.
*Cult.* 1596, by Mr. John Gerard.  *Hort. Ger.*
*Fl.* July and Auguft.                              G. H. ♄.

*ariftata.*  2. P. caule fruticofo, foliis oblongis integris glabris, ra-
mis petiolis pedunculifque lanuginofis, denticulis
calycinis ariftatis.
Bearded Winter Cherry.
*Nat.* of the Canary Iflands.   Mr. *Francis Maffon.*
*Introd.* 1779.
*Fl.*                                               G. H. ♄.

*curaffa-*  3. P. caule fruticofo, foliis ovatis tomentofis. *Sp. pl.* 261.
*vica.*     Curaffavian Winter Cherry.
*Nat.* of South America.
*Cult.* 1699, by Vifcount Falkland. *Morif. hift.* 3.
*p.* 527. *n.* 26.
*Fl.* June——September.                              S. ♃.

*vifcofa.*  4. P. foliis geminis repandis obtufis fubtomentofis, caule
herbaceo fuperne paniculato. *Sp. pl.* 261. *Jacqu.*
*hort.* 2. *p.* 64. *t.* 136.
Clammy Winter Cherry.
*Nat.* of America.
*Cult.* 1732, by James Sherard, M. D. *Dill. elth.* 11.
*t.* 10. *f.* 10.
*Fl.* July.                                         S. ♃.

*penfylva-*  5. P. foliis ovatis fubrepandis obtufis nudiufculis; flora-
*nica.*     libus geminis, caule herbaceo. *Sp. pl.* 1670.
Penfylvanian Winter Cherry.
*Nat.* of North America.
*Cult.* 1726, by Mr. Philip Miller.  *R. S. n.* 207.
*Fl.* July——September.                              H. ♃.

6. P.

6. P. foliis geminis integris acutis, caule herbaceo in- *Alkeken-*
 ferne ſubramoſo. *Sp. pl.* 262. *gi.*
 Common Winter Cherry.
 *Nat.* of the South of Europe.
 *Cult.* 1597. *Ger. herb.* 271. *f.* 1.
 *Fl.* July——September. H. ♃.

7. P. pubeſcens, foliis cordatis integerrimis. *Sp. pl.* 1670. *peruvia-*
 Peruvian Winter Cherry. *na.*
 *Nat.* of South America.
 *Introd.* about 1772.
 *Fl.* April——October. S. ♄.

\*\* *Annuæ.*

8. P. ramoſiſſima, ramis angulatis glabris, foliis ovatis *angulata.*
 dentatis. *Sp. pl.* 262.
 Tooth'd-leav'd Winter Cherry.
 *Nat.* of both Indies.
 *Cult.* 1732, by James Sherard, M. D. *Dill. elth.* 13.
 *t.* 12. *f.* 12.
 *Fl.* June——September. H. ☉.

9. P. ramoſiſſima, foliis villoſo-viſcoſis, floribus pendu- *pubeſcens.*
 lis. *Sp. pl.* 262.
 Woolly Winter Cherry.
 *Nat.* of America.
 *Cult.* 1739, by Mr. Philip Miller. *Rand. chelſ.* Al-
 kekengi 7.
 *Fl.* July and Auguſt. S. ☉.

10. P. ramoſiſſima, caule procumbente tereti hirſuto, fo- *proſtrata.*
 liis ſubcarnoſis. *L'Herit. ſtirp. nov. p.* 43. *t.* 22.
 *Jacqu. ic. collect.* 1. *p.* 99.
 Trailing blue-flower'd Winter Cherry.
 *Nat.* of Peru.

R 3 *Introd.*

*Introd.* 1782, by Monſ. Thouin.
*Fl.* Auguſt and September.                              S. ☉.

*minima.*  11. P. ramoſiſſima, pedunculis fructiferis folio villoſo
longioribus. *Sp. pl.* 263.
Small Winter Cherry.
*Nat.* of the Eaſt Indies.
*Cult.* 1759, by Mr. Philip Miller. *Mill. dict. edit.* 7.
*Fl.* July and Auguſt.                                   S. ☉.

*pruinoſa.*  12. P. ramoſiſſima, foliis villoſis, pedunculis ſtrictis. *Sp.*
*pl.* 263.
Hairy annual Winter Cherry.
*Nat.* of America.
*Cult.* 1726, by Mr. Philip Miller. *R. S. no.* 203.
*Fl.* July and Auguſt.                                   H. ☉.

## SOLANUM.  *Gen. pl.* 251.

*Cor.* rotata. *Antheræ* ſubcoalitæ, apice poro gemino
dehiſcentes. *Bacca* 2-locularis.

### * *Inermia.*

*auricula-*  1. S. caule inermi fruticoſo, foliis ovatis integerrimis
*tum.*       tomentoſis, ſtipulis ſemicircularibus. *L'Herit.*
*ſolan. tab.* 1.
Ear-leav'd Nightſhade.
*Nat.* of the Iſlands of Madagaſcar, Mauritius, and
Bourbon.
*Introd.* 1773, by Monſ. Richard.
*Fl.*                                                    S. ♄.

*Pſeudo-*   2. S. caule inermi fruticoſo, foliis lanceolatis repandis,
*capſicum.*     umbellis feſſilibus. *Sp. pl.* 263.
Shrubby Winter Cherry.
*Nat.* of Madeira.

*Cult.*

*Cult.* 1596, by Mr. John Gerard.   *Hort. Ger.*
*Fl.* June——September.                     G. H. ♄.

3. S. caule inermi fruticofo, foliis geminis: altero mi-   *diphyllum.*
nore, floribus cymofis.  *Sp. pl.* 264.
Two-leav'd Nightſhade.
*Nat.* of the Weſt Indies.
*Cult.* 1759, by Mr. Philip Miller.
*Fl.* June and July.                          S. ♄.

4. S. caule inermi frutefcente flexuofo, foliis fuperiori-   *Dulca-*
bus haſtatis, racemis cymofis. *Sp. pl.* 264. *Curtis*   *mara.*
*lond.*
α Solanum fcandens feu Dulcamara. *Bauh. pin.* 167.
Common Woody Nightſhade.
β Solanum dulcamarum africanum, foliis craffis hirfutis. *Dill. elth.* 365. *t.* 273. *f.* 352.
African Woody Nightſhade.
*Nat.* α. of Britain; β. of Africa.
*Fl.* June and July.              α. H. β. G. H. ♄.

5. S. caule inermi fubherbaceo angulato flexuofo fcabro,   *quercifo-*
foliis pinnatifidis, racemis cymofis. *Syſt. veget.* 223.   *lium.*
Oak leav'd Nightſhade.
*Nat.* of Peru.
*Introd.* 1787, by Monf. Vare.
*Fl.* July.                                    H. ♃.

6. S. caule fuffruticofo inermi glaberrimo, foliis pinnati-   *lacinia-*
fidis: laciniis lanceolatis acutis, paniculis axillari-   *tum.*
bus binis ternifve.
Cut-leav'd Nightſhade.
*Nat.* of New Zealand.  Sir *Jofeph Banks*, Bart.
*Introd.* 1772.
*Fl.* July and Auguſt.                        S. ♃.

*radicans.*    7. S. caule inermi herbaceo laevi teretiusculo prostrato
           radicante, foliis pinnatifidis, racemis cymosis. *Sp.*
           *pl.* 264.
           Climbing Nightshade.
           *Nat.* of Peru.
           *Introd.* 1771, by Monf. Richard.
           *Fl.* July and Augufl.                        G. H. ♃

*racemo-*    8. S. caule inermi frutescente, foliis lanceolatis repandis
*fum.*         undulatis, racemis longis rectis. *Linn. mant.* 47.
           Wave-leav'd Nightshade.
           *Nat.* of the Weft Indies.
           *Introd.* 1781, by Mr. Francis Masson.
           *Fl.*                                        S. ♄.

*corymbo-*   9. S. caule inermi suffruticoso, foliis ovato-lanceolatis
*fum.*         integris basi acuminatis, floribus corymbosis. *L'He-*
           *rit. solan. tab.* 3. *Jacqu. ic. collect.* 1. *p.* 78.
           Oval-leav'd Nightshade.
           *Nat.* of Peru.
           *Introd.* 1786, by Monf. Thouin.
           *Fl.* July.                                  S. ♃.

*bonari-*    10. S. caule subinermi fruticoso, foliis ovato-oblongis si-
*enfe.*        nuato-repandis scabris. *Syft. veget.* 223.
           Tree Nightshade.
           *Nat.* of Buenos Ayres.
           *Cult.* 1727, by James Sherard, M. D. *Dill. elth.* 364.
           t. 272. *f.* 351.
           *Fl.* June——September.                       G. H. ♄.

*macro-*     11. S. caule inerme suffruticoso, foliis cuneatis repandis
*carpon.*       glabris. *Linn. mant.* 205. (excluso synonymo
           Feuillei).
           Smooth flefhy-leav'd Nightshade.

                                                        *Nat.*

*Nat.* of Peru.

*Cult.* 1759, by Mr. Philip Miller. *Mill. ic.* 196.
t. 294.

*Fl.* moſt part of the Summer. S. ♄.

12. S. caule inermi herbaceo, foliis pinnatis integerrimis, *tubero-*
pedunculis ſubdiviſis. *Sp. pl.* 265. *ſum.*

Tuberous-rooted Nightſhade, or **Common** Potato.

*Nat.* of Peru.

*Cult.* 1597, by Mr. John Gerard. *Ger. herb.* 781.

*Fl.* July and Auguſt. H. ☉.

13. S. caule inermi herbaceo, foliis pinnatis inciſis, race- *Lycoper-*
mis ſimplicibus. *Sp. pl.* 265. *ſicum.*

Love-apple.

*Nat.* of South America.

*Cult.* 1596, by Mr. John Gerard. *Hort. Ger.*

*Fl.* July——September. H. ☉.

14. S. caule inermi herbaceo, foliis ovatis dentato-an- *nigrum.*
gulatis, racemis diſtichis nutantibus. *Syſt. veget.*
224. *Curtis lond.*

α Solanum officinarum. *Bauh. pin.* 166. vulga-
Common Nightſhade.. tum.

β Solanum ramis teretibus villoſis, foliis angulatis ſub- villo-
villoſis. *Sp. pl.* 266. ſum.

Yellow-berried Nightſhade.

γ Solanum ramis angulatis dentatis, foliis integerrimis guine-
glabris. *Sp. pl.* 266. enſe. -

Large black-berried Nightſhade.

δ Solanum ramis angulatis dentatis, foliis repandis virgini-
glabris. *Sp. pl.* 266. cum.

Small black-berried Nightſhade.

ε Solanum caule inermi herbaceo glabro, foliis ob- rubrum.
longo-ovatis acuminatis dentatis glabris, umbellis
nutantibus. *Mill. dict.*

Red-

Red-berried Nightſhade.

*Nat.* α. of Britain, β. of Barbadoes, γ. of Guinea, δ. of Virginia, and ε. of the Weſt Indies.

*Fl.* July.　　　　　　　　　　　　H. ☉.

*æthiopi-*
*cum.*

15. S. caule inermi herbaceo, foliis ovatis repando-an-
gulatis, pedunculis fertilibus unifloris cernuis. *Sp.*
*pl.* 265. *Jacqu. hort.* I. *p.* 4. *t.* 12.

Ethiopian Nightſhade.

*Nat.* of China.

*Cult.* before 1597, by Mr. John Gerard. *Ger. herb.*
276.

*Fl.* July.　　　　　　　　　　　　G. H. ☉.

*Melonge-*
*na.*

16. S. caule inermi herbaceo, foliis ovatis tomentoſis,
pedunculis pendulis incraſſatis, calycibus inermibus. *Syſt. veget.* 224.

Large-fruited Nightſhade, or Egg-plant.

*Nat.* of Aſia, Africa, and America.

*Cult.* before 1597. *Ger. herb.* 274.

*Fl.* June and July.　　　　　　　　G. H. ☉.

*ſubin-*
*erme.*

17. S. caule ſubinermi fruticoſo, foliis lanceolato-ellipticis integerrimis ſupra glabris; ſubtus tomentoſis,
cymis farinoſis.

Solanum ſubinerme. *Jacqu. hiſt.* 50.

Spear-leav'd Nightſhade.

*Nat.* of the Weſt Indies.

*Introd.* 1778, by Mr. Gilbert Alexander.

*Fl.* July and Auguſt.　　　　　　　S. ♄.

*murica-*
*tum.*

18. S. caule ſubinermi ſuffruticoſo radicante, turionibus
muricatis, foliis oblongo-lanceolatis integris pubeſcentibus. *L'Herit. ſolan. tab.* 6.

Melongena

Melongena laurifolia, fructu turbinato, variegato.
*Feuillée it.* 2. *p.* 735. *t.* 26.
Warted Nightshade.
*Nat.* of Peru.
*Introd.* 1785, by Monf. Thouin.
*Fl.*                                                              S. ♄.

\*\* *Aculeata.*

19. S. caule aculeato hirto, foliis cordato-oblongis quin-    *campe-*
     quelobis : finubus obtufis elevatis. *Syft. veget.* 225.    *chienfe.*
     Purple-fpin'd Nightshade.
     *Nat.* of America.
     *Cult.* 1732, by James Sherard, M.D. *Dill. elth.*
         361. *t.* 268. *f.* 347.
     *Fl.* July.                                                G. H. ♄.

20. S. caule aculeato fruticofo, foliis cuneiformibus an-    *indicum.*
     gulatis fubvillofis integerrimis : aculeis utrinque
     rectis. *Sp. pl.* 268.
     Indian Nightshade.
     *Nat.* of both Indies.
     *Cult.* 1732, by James Sherard, M.D. *Dill. elth.*
         362. *t.* 270. *f.* 349.
     *Fl.* July.                                                S. ♄.

21. S. caule aculeato annuo, foliis haftato-angulatis :    *caroli-*
     aculeis utrinque rectis, racemis laxis. *Sp. pl.* 268.    *nenfe.*
     Carolina Nightshade.
     *Nat.* of Carolina.
     *Cult.* 1732, by James Sherard, M.D. *Dill. elth.*
         361. *t.* 269. *f.* 348.
     *Fl.* July——September.                                  G. H. ♃.

22. S. caule aculeato fruticofo tereti, foliis pinnatifido-    *fodomeum.*
     finuatis fparfe aculeatis nudis, calycibus aculeatis.
     *Syft. veget.* 225.
                                                         Black-

Black-fpin'd Nightſhade.
*Nat.* of Africa.
*Cult.* 1731, by Mr. Philip Miller. *Mill. dict. edit.* 1.
*n.* 12.
*Fl.* June and July.                    G. H. ♄.

*margina-*    23. S. caule aculeato fruticoſo, foliis cordatis repandis
*tum.*              margine albis. *Syſt. veget.* 226.  *L'Herit. ſolan.*
                    *tab.* 11. *Murray in commentat. gotting.* 1783. *p.* 11.
                    *tab.* 4. *Jacqu. ic. collect.* 1. *p.* 50.
              White Nightſhade.
              *Nat.* of Africa. *James Bruce*, Eſq.
              *Introd.* 1775.
              *Fl.* moſt part of the Summer.          G. H. ♄.

*ſtramoni-*   24. S. caule aculeato fruticoſo, foliis cordatis angulato-
*folium.*           lobatis integris ſubinermibus ſubtus tomentoſiuſcu-
                    lis.  *L'Herit. ſolan. tab.* 9.
              Solanum ſtramonifolium. *Jacqu. ic. miſcell.* 2. *p.* 298.
              Broad-leav'd Nightſhade.
              *Nat.* of the Weſt Indies.
              *Introd.* 1778, by Meſſrs. Kennedy and Lee.
              *Fl.* June——September.                S. ♄.

*Veſperti-*   25. S. caule aculeato fruticoſo, foliis cordatis integris,
*lio.*              corollis ſubirregularibus, anthera ima productiore.
                    *L'Herit. ſolan.*
              Canary Nightſhade.
              *Nat.* of the Canary Iſlands.   Mr. *Francis Maſſon.*
              *Introd.* 1779.
              *Fl.* March and April.                 G. H. ♄.

*tomento-*    26. S. caule aculeato fruticoſo, aculeis aceroſis, foliis cor-
*ſum.*              datis inermibus ſubrepandis : tenellis purpureo-
                    pulverulentis. *Sp. pl.* 269.
                                            Woolly

Woolly Nightſhade.
*Nat.* of the Cape of Good Hope.
*Cult.* 1731, by Mr. Philip Miller. *Mill. dict. edit.* 1.
*n.* 10.
*Fl.* June and July.                                    G. H. ♄.

27. S. caule aculeato fruticoſo, foliis lanceolatis acumina-   *igneum.*
tis baſi utrinque revolutis, racemis ſimplicibus. *Sp.*
*pl.* 270. *Jacqu. hort.* 1. *p.* 5. *t.* 14.
Red-ſpin'd Nightſhade.
*Nat.* of South America.
*Cult.* 1714, by the Dutcheſs of Beaufort. *Br. Muſ.*
*H. S.* 133. *fol.* 28.
*Fl.* March——November.                          S. ♄.

### C A P S I C U M. *Gen. pl.* 252.

*Cor.* rotata. *Bacca* exſucca.

1. C. caule herbaceo, pedunculis ſolitariis. *Sp. pl.* 270.   *annuum.*
α Capſicum ſiliquis longis propendentibus. *Tournef.*
*inſt.* 152.
Long-podded Capſicum.
β Capſicum caule herbaceo, fructu rotundo glabro. *Mill.*
*dict.*
Cherry Capſicum.
γ Capſicum caule herbaceo, fructu ovato. *Mill. dict.*
Olive Capſicum.
*Nat.* of both Indies.
*Cult.* 1596, by Mr. John Gerard. *Hort. Ger.*
*Fl.* June and July.                              S. ☉.

2. C. caule fruticoſo lævi, pedunculis geminis. *Linn.*   *baccatum.*
*mant.* 47.
Small-fruited Capſicum, or Bird Pepper.
*Nat.* of both Indies.
                                                *Cult.*

Cult. 1731, by Mr. Philip Miller.  Mill. dict. edit. 1.
n. 15.
Fl. June——September.                              S. ♄.

grossum.    3. C. caule suffrutescente, fructibus incrassatis variis.
Linn. mant. 47.
Heart-shap'd Capsicum, or Bell Pepper.
Nat. of both Indies.
Cult. 1759, by Mr. Ph. Miller. Mill. dict. edit. 7. n. 3.
Fl. July.                                        S. ♃.

frutes-    4. C. caule fruticoso scabriusculo, pedunculis solitariis.
cens.          Syst. veget. 227.
Shrubby Capsicum.
Nat. of both Indies.
Cult. 1656, by Mr. John Tradescant, Jun. Trad.
muf. 95.
Fl. June——September.                              S. ♄.


S T R Y C H N O S.  Gen. pl. 253.

Cor. 5-fida.  Bacca 1-locularis, cortice lignoso.

Nux vo-    1. S. foliis ovatis, caule inermi.  Sp. pl. 271.
mica.      Poison-nut.
Nat. of the East I dies.
Introd. 1778, by Patrick Russell, M. D.
Fl.                                              S. ♄.


C E S T R U M.  Gen. pl. 261.

Cor. infundibuliformis.  Stam. denticulo in medio.
Bacca 1-locularis, polysperma.

nocturn-    1. C. filamentis dentatis, pedunculis subracemosis folio
num.            æqualibus. L'Herit. stirp. nov. 70.
Cestrum nocturnum.  Sp. pl. 277.

                                                Night-

Night-fmelling Ceftrum.
*Nat.* of the Weft Indies.
*Cult.* 1732, by James Sherard, M.D. *Dill. elth.* 183.
*t.* 153. *f.* 185.
*Fl.* November.                                                     S. ♄.

2. C. filamentis denticulatis nudifve, foliis ellipticis co-          *laurifoli-*
     riaceis nitidiffimis, pedunculis petiolo brevioribus.            *um.*
     *L'Herit. ftirp. nov.* 69. *t.* 34.
Laureola fempervirens americana, latioribus foliis, flo-
     ribus albicantibus odoris. *Pluk. alm.* 209. *t.* 95. *f.* 1.
Lawrel-leav'd Ceftrum.
*Nat.* of the Weft Indies.
*Cult.* 1691, in the Royal Garden at Hampton-court.
     *Pluk. l. c.*
*Fl.* Auguft.                                                       S. ♄.

3. C. filamentis edentatis, tubo filiformi, pedunculis           *vefperti-*
     breviffimis. *L'Herit. ftirp. nov.* 72.                          *num.*
Ceftrum vefpertinum. *Linn. mant.* 206.
Clufter-flower'd Ceftrum.
*Nat.* of the Weft Indies.
*Cult.* 1759, by Mr. Ph. Miller. *Mill. dict. edit.* 7. *n.* 5.
*Fl.* May——July.                                                   S. ♄.

4. C. filamentis edentatis, laciniis corollæ fubrotundis          *diurnum.*
     reflexis, foliis lanceolatis. *L'Herit. ftirp. nov.* 74.
Ceftrum diurnum. *Sp. pl.* 277. (exclufis fynonymis
     Plukenetii et Feuillei).
Day-fmelling Ceftrum.
*Nat.* of the Weft Indies.
*Cult.* 1732, by James Sherard, M.D. *Dill. elth.* 186.
     *t.* 154. *f.* 186.
*Fl.* November.                                                     S. ♄.

5. C. filamentis edentatis, ftipulis lunatis. *L'Herit.*          *auricu-*
     *ftirp. nov.* 71. *t.* 35.                                         *latum.*

                                    Hediunda

Hediunda jafmiano flore.　*Feuill. it.* 3. *p.* 25. *t.* 20.
*f.* 3.
Ear-leav'd Ceftrum.
*Nat.* of Peru.
*Introd.* about 1774, by Monf. Richard.
*Fl.* June.　　　　　　　　　　　　　　　　S. ♄.

## L Y C I U M.　*Gen. pl.* 262.

*Cor.* tubulofa, fauce claufa filamentorum barba.　*Bacca*
2-locularis, polyfperma.

*japoni-*
*cum.*　　1. L. inerme, foliis ovatis nervofis planis, floribus feffili-
　　　　　bus.　*Thunb. jap.* 93. *tab.* 17.　*Syft. veget.* 228.
　　　　Lycium fœtidum.　*Linn. fuppl.* 150.　*Syft. veget.* 228.
　　　　Lycium indicum.　*Retz. obf.* 2. *p.* 12. *n.* 21.
　　　　Japan Box-thorn.
　　　　*Nat.* of Japan.
　　　　*Introd.* 1787, by Monf. Cels.
　　　　*Fl.* moft part of the Summer.　　　　　G. H. ♄.

*afrum.*　　2. L. foliis linearibus.　*Sp. pl.* 277.
　　　　African Box-thorn.
　　　　*Nat.* of the Cape of Good Hope.
　　　　*Cult.* 1712, by the Dutchefs of Beaufort.　*Br. Muf.*
　　　　　*H. S.* 137. *fol.* 58.
　　　　*Fl.* June and July.　　　　　　　　　　G. H. ♄.

*boerha-*
*viæfoli-*
*um.*　　3. L. fpinofum, foliis ovatis integerrimis acutis glaucis,
　　　　floribus paniculatis.　*Linn. fuppl.* 150.　*Syft. veget.*
　　　　228.
　　　　Lycium heterophyllum.　*Murray commentat. gotting.*
　　　　　1783. *p.* 6. *t.* 2.　*Syft. veget.* 228.
　　　　Ehretia halimifolia.　*L'Herit. ftirp. nov.* 45. *t.* 23.
　　　　Glaucous-leav'd Box-thorn.
　　　　*Nat.* of Peru.
　　　　*Introd.* 1780, by Monf. Thouin.
　　　　*Fl.* April.　　　　　　　　　　　　　　S. ♄.

4. L.

4. L. foliis oblongo-lanceolatis, ramis angulatis.          *barbarum.*

α ſtylo longitudine ſtaminum.          vulgare.

Lycium barbarum. *Sp. pl.* 277.

Willow-leav'd Box-thorn.

β ſtylo ſtaminibus longiore.          chinenſe.

Lycium chinenſe. *Mill. dict.*

Chineſe Box-thorn.

*Nat.* of Europe, Aſia, and Africa.

*Cult.* 1709, by the Dutcheſs of Beaufort. *Br. Muſ.*

H. S. 137. *fol.* 54.

*Fl.* May——October.          H. ♄.

5. L. foliis obliquis, ramulis flexuoſis teretibus. *Linn.*   *europæ-*
     *mant.* 47.          *um.*

European Box-thorn.

*Nat.* of the South of Europe.

*Introd.* 1780, by Peter Simon Pallas, M. D.

*Fl.*          H. ♄.

JACQUINIA. *Gen. pl.* 254.

*Cor.* 10-fida.   *Stam.* receptaculo inſerta. *Bacca*
                    1-ſperma.

1. J. foliis obtuſis cum acumine. *Sp. pl.* 272.          *armilla-*
Obtuſe-leav'd Jacquinia.          *ris.*

*Nat.* of the Weſt Indies.

*Cult.* 1768, by Mr. Philip Miller. *Mill. dict. edit.* 8.

*Fl.*          S. ♄.

2. J. foliis lanceolatis acuminatis. *Sp. pl.* 271.          *ruſcifolia.*
Prickly Jacquinia.

*Nat.* of South America.

*Cult.* 1729, by James Sherard, M. D. *Dill. elth.* 148.

t. 123. *f.* 149.

*Fl.*          S. ♄.

S          CHIRO-

## CHIRONIA. *Gen. pl.* 255.

*Cor.* rotata. *Piſtillum* declinatum. *Stam.* tubo corollæ inſidentia. *Antheræ* demum ſpirales. *Peric.*
2-loculare.

*baccife-*    1. C. fruteſcens baccifera. *Sp. pl.* 273.
*ra.*           Berry-bearing Chironia.
              *Nat.* of the Cape of Good Hope.
              *Cult.* 1759, by Mr. Ph. Miller. *Mill. dict. edit.* 7. *n.* 2.
              *Fl.* June and July.                    G. H. ♄.

*fruteſcens.*  2. C. fruteſcens, foliis lanceolatis ſubtomentoſis, calyci-
              bus campanulatis. *Syſt. veget.* 229. *Curtis magaz.*
              37.
              Shrubby Chironia.
              *Nat.* of the Cape of Good Hope.
              *Cult.* 1756, by Mr. Philip Miller. *Mill. ic.* 65. *t.* 97.
              *Fl.* June——September.                 G. H. ♄.

## CORDIA. *Gen. pl.* 256.

*Cor.* infundibuliformis. *Stylus* dichotomus. *Drupa*
nucleis 2-locularibus.

*Myxa.*    1. C. foliis ovatis ſupra glabris, corymbis lateralibus, ca-
            lycibus decemſtriatis. *Syſt. veget.* 230.
            Smooth-leav'd Cordia.
            *Nat.* of Egypt and the Eaſt Indies.
            *Cult.* 1640. *Park. theat.* 252. *n.* 1.
            *Fl.*                                     S. ♄.

*Sebeſtena.*  2. C. foliis oblongo-ovatis repandis ſcabris. *Sp. pl.* 274.
             Rough-leav'd Cordia.
             *Nat.* of both Indies.

                                                    *Cult.*

*Cult.* 1728, by James Sherard, M.D. *Dill. elth.* 340.
*t.* 255. *f.* 331.
*Fl.*                                               S. ♄.

3. C. foliis cordato-ovatis integerrimis, floribus co-      *Collococ-*
   rymbofis, calycibus interne tomenfis. *Syft. veget.*    *ca.*
   230.
Long-leav'd Cordia.
*Nat.* of Jamaica.
*Cult.* 1759, by Mr. Philip Miller. *Mill. dict. edit.* 7.
Addenda. Collococcus 1.
*Fl.*                                               S. ♄.

4. C. foliis oblongo-lanceolatis utrinque glabris: fupe-   *Patago-*
   rioribus ferratis, ramulis pilofis.                     *nula.*
Patagonula americana. *Sp. pl.* 212.
Spear-leav'd Cordia, or Patagonula.
*Nat.* of South America.
*Cult.* 1732, by James Sherard, M.D. *Dill. elth.* 304.
*t.* 226. *f.* 293.
*Fl.* June——Auguft.                                 S. ♄.

E H R E T I A. *Gen. pl.* 257.

*Bacca* 2-locularis. *Sem.* folitaria, bilocularia. *Stigma*
emarginatum.

1. E. foliis oblongo-ovatis integerrimis glabris, floribus  *tinifolia.*
   paniculatis. *Sp. pl.* 274.
Tinus-leav'd Ehretia.
*Nat.* of Jamaica.
*Introd.* 1734. *Mill. dict. edit.* 8.
*Fl.* June and July.                                S. ♄.

2. E. foliis ovatis integerrimis lævibus, floribus fubco-   *Bourre-*
   rymbofis, calycibus glabris. *Sp. pl.* 275.             *ria.*

                         S 2                        Oval-

Oval-leav'd Ehretia.
*Nat.* of the Weſt Indies.
*Cult.* 1758, by Mr. Philip Miller.
*Fl.*                                                          S. ♄.

## TECTONA. *Linn. ſuppl.* 20.

*Cal.* campanulatus. *Cor.* campanulata, 5-fida. *Drupa*
ſicca, calyce inflato ſcarioſo tecta : *Nux* quadrilocu-
laris.

*grandis.* 1. TECTONA. *Linn. ſuppl.* 151.
Theka. *Rheed. mal.* 4. *p.* 57. *t.* 27.
Jatus. *Rumph. amb.* 3. *p.* 34. *t.* 18.
Indian Oak, or Teak-wood.
*Nat.* of India.
*Introd.* 1777, by John Walſh, Eſq.
*Fl.*                                                          S. ♄.

## VARRONIA. *Gen. pl.* 258.

*Cor.* 5-fida. *Drupa* nucleo 4-loculari.

*curaſſa-* 1. V. foliis lanceolatis, ſpicis oblongis. *Sp. pl.* 276.
*vica.* Long-ſpiked Varronia.
*Nat.* of South America.
*Introd.* about 1778, by John Hope, M.D.
*Fl.*                                                          S. ♄.

## CHRYSOPHYLLUM. *Gen. pl.* 263.

*Cor.* campanulata, 10-fida : laciniis alternis patulis.
*Bacca* 10-ſperma.

*Cainito.* 1. C. foliis ovatis parallele ſtriatis ſubtus tomentoſo-
nitidis. *Sp. pl.* 278.
Broad-leav'd Star Apple.

*Nat.*

*Nat.* of the Weſt Indies.

*Cult.* 1739, by Mr. Philip Miller. *Rand. chel.* Cainito 1.

*Fl.* S. ♄.

2. C. foliis falcato-ovatis ſubtus tomentoſo-nitidis. *Syſt.* *veget.* 232.        *argenteum.*

Narrow-leav'd Star Apple.

*Nat.* of Martinico.

*Cult.* 1758, by Mr. Philip Miller.

*Fl.* S. ♄.


# A R D I S I A.

*Cor.* 5-partita. *Cal.* 5-phyllus. *Antheræ* magnæ, erectæ. *Stigma* ſimplex. *Drupa* ſupera.

1. A. racemis axillaribus ſimplicibus, foliis obovatis margine cartilagineo-ſerratis.       *excelſa.*

Laurel-leav'd Ardiſia, or Aderno.

*Nat.* of Madeira. Mr. *Francis Maſſon.*

*Introd.* 1784.

*Fl.* G. H. ♄.


# S I D E R O X Y L O N. *Gen. pl.* 264.

*Cor.* 10-fida : laciniis alternis incurvis. *Stigma* ſimplex. *Bacca* 5-ſperma.

1. S. inerme, foliis perennantibus obovatis, pedunculis teretibus. *Syſt. veget.* 232.       *inerme.*

Smooth Iron-wood.

*Nat.* of the Cape of Good Hope.

*Cult.* 1729, by James Sherard, M. D. *Dill. eltb.* 357. *t.* 265. *f.* 344.

*Fl.* July. G. H. ♄.

S 3       2. S.

*melano-phleos.*

2. S. inerme, foliis perennantibus lanceolatis, pedunculis angulatis. *Linn. mant.* 48. *Jacqu. hort.* 1. *p.* 29. *t.* 71.

Laurel-leav'd Iron-wood.

*Nat.* of the Cape of Good Hope.

*Introd.* 1783, by Mr. John Græffer.

*Fl.*                    G. H. ♄.

*sericeum.*

3. S. inerme, foliis ovatis subtus tomentoso-sericeis.

Silky Iron-wood.

*Nat.* of New South Wales.    Sir *Joseph Banks,* Bart.

*Introd.* 1772.

*Fl.*                      S. ♄.

*tenax.*

4. S. subinerme, foliis deciduis lanceolatis subtus tomentosis, pedunculis filiformibus. *Syst. veget.* 232.

Silvery-leav'd Iron-wood.

*Nat.* of Carolina.

*Introd.* 1765, by Mr. John Cree.

*Fl.* July and August.            H. ♄.

*lycioides.*

5. S. spinosum, foliis deciduis lanceolatis. *Syst. veget.* 232.

Willow-leav'd Iron-wood.

*Nat.* of North America.

*Cult.* 1758, by Mr. Philip Miller.

*Fl.* August.                H. ♄.

*spinosum.*

6. S. spinosum, folis perennantibus oblongis glabris.

Sideroxylon spinosum. *Sp. pl.* 279. (excluso synonymo Rhedi.)

Lycio similis frutex indicus spinosus, Buxi folio. *Commel. amstel.* 1. *p.* 161. *t.* 83.

Thorny Iron-wood, or Argan.

*Nat.* of Morocco.

                              *Cult.*

*Cult.* 1711, by the Dutchefs of Beaufort. *Br. Muf.*
*H. S.* 141. *fol.* 39.
*Fl.* July. S. ♄.


R H A M N U S. *Gen. pl.* 265.

*Cal.* tubulofus: fquamis ftamina munientibus. *Cor.*
nulla. *Bacca.*

* *Spinofi.*

1. R. fpinis terminalibus, floribus quadrifidis dioicis, fo-          *catharti-*
    liis ovatis, caule erecto. *Syft. veget.* 232.                      *cus.*
   Purging Buckthorn.
   *Nat.* of England.
   *Fl.* May and June.                                           H. ♄.


2. R. fpinis terminalibus, floribus quadrifidis dioicis, cau-        *infectori-*
    libus procumbentibus. *Linn. mant.* 49.                          *us.*
   Dwarf, or Yellow-berried Buckthorn.
   *Nat.* of the South of Europe.
   *Cult.* 1683, by Mr. James Sutherland. *Sutherl. hort.*
    *edin.* 291. *n.* 3.
   *Fl.* June and July.                                          H. ♄.


3. R. fpinis terminalibus, foliis oblongis integerrimis.        *Oleoides.*
    *Sp. pl.* 279.
   Olive-leav'd Rhamnus.
   *Nat.* of Spain.
   *Introd.* 1786, by Cafimir Gomez Ortega, M. D.
   *Fl.*                                                    G. H. ♄.


4. R. ramulis fpinefcentibus, floribus quadrifidis trifidifve    *crenula-*
    dioicis, foliis oblongis obtufe ferratis femperviren-            *tus.*
    tibus.
   Teneriffe Rhamnus.

S 4                                *Nat.*

*Nat.* of the Ifland of Teneriffe.  Mr. *Francis Maffon.*
*Introd.* 1778.
*Fl.* March.                                    G. H. ♄.

*faxatilis.*   5. R. fpinis terminalibus, floribus quadrifidis dioicis, fo-
liis oblongis ferratis glabris annuis.
Rhamnus faxatilis.  *Sp. pl.* 1671.  *Jacqu. auftr.* 1.
*p.* 33. *t.* 53.
Rock Buckthorn.
*Nat.* of Switzerland, Auftria, and Italy.
*Introd.* 1775, by the Doctors Pitcairn and Fothergill.
*Fl.* May and June.                             H. ♄.


** *Inermes.*

*colubri-*   6. R. inermis, floribus monogynis hermaphroditis erec-
*nus.*   tis, capfulis tricoccis, petiolis ferrugineo-tomentofis.
*Syft. veget.* 233. *Jacqu. hort.* 3. *p.* 28. *t.* 50.
Pubefcent Rhamnus, or Red-wood.
*Nat.* of the Bahama Iflands.
*Introd.* 1726, by Mr. Mark Catefby.  *Mill. dict. edit.*
8.  Ceanothus arborefcens.
*Fl.* June.                                      S. ♄.

*volubilis.*   7. R. inermis, floribus monogynis hermaphroditis, foliis
oblongo-ovatis nervofis fubundulatis, caule volubili.
Rhamnus volubilis.  *Linn. fuppl.* 152.
Twining Rhamnus.
*Nat.* of Carolina.
*Cult.* 1714, by the Dutchefs of Beaufort.  *Br. Muf.*
*H. S.* 133. *fol.* 13.
*Fl.* June and July.                            H. ♄.

*Frangu-*   8. R. inermis, floribus monogynis hermaphroditis umbel-
*la.*   latis axillaribus, calycibus glabris, foliis ovatis integerrimis lineatis.
Rhamnus Frangula.  *Sp. pl.* 280.

Berry-

Berry-bearing Alder.
*Nat.* of England.
*Fl.* April and May. H. ♄.

9. R. inermis, floribus monogynis hermaphroditis, caly-    *latifoli-*
   cibus villosis, foliis ellipticis integerrimis. *L'Herit.*   *us.*
   *sert. angl. n.* 4. *tab.* 8.
   Azorian Rhamnus.
   *Nat.* of the Azores. Mr. *Francis Masson.*
   *Introd.* 1778.
   *Fl.* July. H. ♄.

10. R. inermis, floribus hermaphroditis racemosis, foliis   *glandulo-*
    ovatis obtuse serratis glabris basi glandulosis.          *sus.*
    Madeira Rhamnus.
    *Nat.* of Madeira and of the Canary Islands. Mr.
       *Francis Masson.*
    *Introd.* 1785.
    *Fl.* G. H. ♄.

11. R. inermis, floribus hermaphroditis subtrigynis axil-   *ellipticus.*
    laribus subumbellatis, foliis ellipticis acutis inte-
    gerrimis subtus villosiusculis.
    Rhamnus arborescens minor, foliis ovatis venosis, pe-
       dunculis umbellulatis alaribus, fructibus sphæricis.
       *Brown. jam.* 172. *t.* 29. *f.* 2.
    Oval-leav'd Rhamnus.
    *Nat.* of Jamaica.
    *Cult.* 1758, by Mr. Philip Miller.
    *Fl.* August. S. ♄.

12. R. inermis, floribus polygamis, stylo subtriplici, foliis   *Prinoi-*
    ovatis serratis. *L'Herit. sert. angl. n.* 5. *tab.* 9.         *des.*
    Celtis foliis subrotundis dentatis, flore viridi, fructu
       luteo. *Burm. afr.* 242. *tab.* 88.

                                                    Prinos-

Prinos-leav'd Rhamnus.
*Nat.* of the Cape of Good Hope.
*Cult.* before 1779, by Rob. Edw. Lord Petre.
*Fl.* Auguſt and September.                                    G. H. ♄.

*myſtaci-* 13. R. inermis, floribus hermaphroditis, ſtigmate triplici,
*nus.*           foliis cordatis, ramis cirrhiſeris.
Wiry Rhamnus.
*Nat.* of Africa.    *James Bruce,* Eſq.
*Introd.* 1775.
*Fl.* November.                                              S. ♄.
DESCR. *Caulis* fruticoſus, teres, decempedalis, debi-
    lis, cirrhis ſcandens. *Folia* alterna, breviſſime pe-
    tiolata, cordata, obtuſa cum acumine minuto, in-
    tegerrima, ſupra glabra, ſubtus piloſiuſcula, uncia-
    lia. *Cirrhi* ex axillis ſuperioribus, ſolitarii, ſimpli-
    ces. *Stipulæ* ad baſin ramulorum, ſubulatæ, cadu-
    cæ, trilineares. *Flores* umbellati, axillares. *Calyx*
    monophyllus, ex albido vireſcens, extus leviter pu-
    beſcens. *Tubus* turbinatus, diametro bilineari.
    *Limbus* quinquepartitus : laciniæ e lata baſi ovatæ,
    acutæ, patentiſſimæ, lineam longæ. *Petala* quin-
    que, margini receptaculi inter lacinias calycis in-
    ſerta, albida, cymbiformia, lateribus convoluta, ſta-
    mina includentia, divergentia, calyce breviora. *Fi-
    lamenta* quinque, intra baſin petalorum inſerta, fili-
    formia. *Antheræ* ſubrotundæ, parvæ. *Germen*
    immerſum in *Receptaculo* niveo, tubum calycis re-
    plente. *Stylus* triqueter, viridis, brevis. *Stigma*
    trifidum.

*alnifolius.* 14. R. inermis, floribus hermaphroditis, foliis ovalibus
           acuminatis ſerratis ſubtus reticulatis. *L'Herit. ſert.
           angl. n.* 2.
Alder-leav'd Rhamnus.
                                                            *Nat.*

*Nat.*
*Introd.* 1778, by Meffrs. Lee and Kennedy.
*Fl.* May.                                                  H. ♄.

15. R. inermis, floribus dioicis, ftigmate triplici, foliis     *Alater-*
    ferratis. *Sp. pl.* 281.                                   *nus.*
 α foliis ovato-oblongis æqualiter ferratis.                    latifolius.
    Common Alaternus.
 β Alaternus foliis lanceolatis profunde ferratis glabris.      angufti-
    *Mill. dict.*                                               folius.
    Jagged-leav'd Alaternus.
    *Nat.* of the South of Europe.
    *Cult.* 1629. *Park. parad.* 603.
    *Fl.* April——June.                                         H. ♄.

             * * * *Aculeati.*
16. R. aculeis geminatis: inferiore reflexo, floribus tri-     *Paliurus.*
    gynis. *Sp. pl.* 281.
    Common Chrift's-thorn.
    *Nat.* of the South of Europe.
    *Cult.* 1596, by Mr. John Gerard. *Hort. Ger.*
    *Fl.* June and July.                                       H. ♄.

17. R. aculeis folitariis recurvis, pedunculis aggregatis,     *Jujuba.*
    floribus femidigynis, foliis retufis fubtus tomento-
    fis. *Sp. pl.* 282.
    Blunt-leav'd Rhamnus.
    *Nat.* of the Eaft Indies.
    *Cult.* 1731, by Mr. Philip Miller. *Mill. dict. edit.* I.
    Ziziphus 3.
    *Fl.*                                                      S. ♄.

18. R. aculeis geminatis: altero recurvo, floribus digy-       *Zizyphus.*
    nis, foliis ovato-oblongis. *Syft. veget.* 235.
    Shining-leav'd Rhamnus.
                                                  *Nat.*

*Nat.* of the South of Europe.

*Cult.* 1640. *Park. theat.* 251.

*Fl.* Auguſt and September. H. ♄.


PHYLICA. *Gen. pl.* 266.

*Perianth.* 5-partitum, turbinatum. *Petala* o. *Squamæ*
5 ſtamina munientes. *Capſ.* tricocca, infera.

*ericoides.* 1. P. foliis linearibus verticillatis. *Sp. pl.* 283.
Heath-leav'd Phylica.
*Nat.* of the Cape of Good Hope.
*Cult.* 1731, by Mr. Philip Miller. *Mill. dict. edit.* 1.
Alaternoides 2.
*Fl.* September——March. G. H. ♄.

*pubeſcens.* 2. P. foliis linearibus acutis piloſis ſubtus tomentoſis, flo-
ribus ſpicatis, bracteis villoſis foliiformibus.
Downy Phylica.
*Nat.* of the Cape of Good Hope. Mr. *Francis Maſſon.*
*Introd.* 1774.
*Fl.* February —— April. G. H. ♄.

*eriopho-*
*ros.* 3. P. foliis linearibus piloſiuſculis ſubtus tomentoſis
margine revolutis, capitulis terminalibus, floribus
tomentoſis.
Phylica eriophoros. *Berg. cap.* 52.
Elichryſum æthiopicum fruteſcens, Coridis foliis in-
canis, capitulis parvis glomeratis, inter ramulos
diſperſis. *Pluk. amalth.* 72. *t.* 445. *f.* 1.
Pale-flower'd Phylica.
*Nat.* of the Cape of Good Hope.
*Introd.* 1774, by Mr. Francis Maſſon.
*Fl.* November. G. H. ♄.

4. P.

4. P. foliis lineari-subulatis: summis hirsutis. *Sp. pl.* *plumosa,*
283.
Woolly-leav'd Phylica.
*Nat.* of the Cape of Good Hope.
*Cult.* 1759, by Mr. Ph. Miller. *Mill. dict. edit.* 7. *n.* 2.
*Fl.* March——May. G. H. ♄.

5. P. foliis oblongis cordatis acuminatis pilosis subtus *callosa.*
tomentosis, floribus subcapitatis.
Phylica callosa. *Linn. suppl.* 153.
Heart-leav'd Phylica.
*Nat.* of the Cape of Good Hope. Mr. *Francis Masson.*
*Introd.* 1774.
*Fl.* March and April. G. H. ♄.

6. P. foliis oblongis cordatis acuminatis subtus tomento- *spicata.*
sis, spicis cylindraceis, floribus longitudine bractea-
rum.
Phylica spicata. *Linn. suppl.* 153.
Spiked Phylica.
*Nat.* of the Cape of Good Hope. Mr. *Francis Masson.*
*Introd.* 1774.
*Fl.* November and December. G. H. ♄.

7. P. foliis ovato oblongis subtus tomentosis, floribus *buxifolia.*
subglomeratis.
Phylica buxifolia. *Sp. pl.* 283.
Box-leav'd Phylica.
*Nat.* of the Cape of Good Hope.
*Cult.* 1759, by Mr. Ph. Miller. *Mill. dict. edit.* 7. *n.* 3.
*Fl.* most part of the Year. G. H. ♄.

CEANO-

## CEANOTHUS. *Gen. pl.* 267.

*Petala* 5, faccata, fornicata. *Bacca* ficca, 3-locularis,
3-fperma.

america-
nus.

1. C. foliis trinerviis. *Sp. pl.* 284.
American Ceanothus, or New Jerfey Tea.
*Nat.* of Carolina and Virginia.
*Cult.* before 1713, by Bifhop Compton. *Mill. ic.* 57.
*t.* 86.
*Fl.* July——Oƈtober.        H. ♄.

afiaticus.

2. C. foliis ovatis enerviis. *Sp. pl.* 284.
Afian Ceanothus.
*Nat.* of Ceylon.
*Introd.* 1781.
*Fl.*                 S. ♄.

africanus.

3. C. foliis lanceolatis enerviis, ftipulis fubrotundis. *Sp.*
*pl.* 284.
African ever-green Ceanothus.
*Nat.* of the Cape of Good Hope.
*Cult.* 1712. *Philofoph. tranf. n.* 333. *p.* 421. *n.* 77.
*Fl.* March and April.       G. H. ♄.

## ARDUINA. *Linn. mant.* 7.

*Cor.* 1-petala. *Stigma* 2-fidum. *Bacca* 2-locularis.
*Sem.* folitaria.

bifpinofa.

1. ARDUINA. *Linn. mant.* 52.
Two-fpin'd Arduina.
*Nat.* of the Cape of Good Hope.
*Cult.* 1760, by Mr. Ph. Miller. *Mill. ic.* 200. *t.* 300.
*Fl.* March——Auguft.       G. H. ♄.

MYRSINE.

## M Y R S I N E. *Gen. pl.* 269.

*Cor.* femi 5-fida, connivens. *Germen* corollam replens.
*Bacca* 1-fperma: *nucleo* 5-loculari.

1. M. foliis ellipticis acutis.                              *africana.*
 Myrfine africana. *Sp. pl.* 285.
 African Myrfine.
 *Nat.* of the Cape of Good Hope.
 *Cult.* 1691, in the Royal Garden at Hampton-court.
  *Pluk. phyt. t.* 80. *f.* 5.
 *Fl.* March——May.                              G. H. ♄.

2. M. foliis obovatis obtufis apice emarginato-denticu-  *retufa.*
  latis.
 Round-leav'd Myrfine, or Tamaja.
 *Nat.* of the Azores. Mr. *Francis Maffon.*
 *Introd.* 1778.
 *Fl.* June.                              G. H. ♄.

## C E L A S T R U S. *Gen. pl.* 270.

*Cor.* 5-petala, patens. *Capf.* 3-angularis, 3-locularis.
            *Sem.* calyptrata.

1. C. inermis, caule volubili, foliis ferrulatis. *Sp. pl.* 285.  *fcandens.*
 Climbing Staff tree.
 *Nat.* of North America.
 *Introd.* 1736, by Peter Collinfon, Efq. *Coll. mfs.*
 *Fl.* May and June.                              H. ♄.

2. C. inermis, foliis ovatis utrinque acutis laxe dentatis  *Caffinoi-*
  perennantibus, floribus axillaribus. *L'Herit. fert.*  *des.*
  *angl. n.* 2. *tab.* 10.
 Crenated Staff-tree.
 *Nat.* of the Canary Iflands. Mr. *Francis Maffon.*
                              *Introd.*

*Introd.* 1779.
*Fl.* Auguſt and September.                                   G. H. ♄.

*octogonus.*   3. **C.** inermis, foliis ellipticis angulatis ſubenervibus
                  perennantibus, capſulis bivalvibus monoſpermis.
                  *L'Herit. ſert. angl. n.* 4.
                  Angular-leav'd Staff-tree.
                  *Nat.* of Peru.
                  *Introd.* 1786, by Monſ. Thouin.
                  *Fl.* October.                              S. ♄.

*undula-*     4. **C.** inermis, foliis ſuboppoſitis lanceolatis undulatis,
*tus.*            capſulis bivalvibus polyſpermis. *L'Herit. ſert. ang.*
                  *n.* 5.
                  Wáved-leav'd Staff-tree.
                  *Nat.* of the Iſland of Bourbon.
                  *Introd.* 1785, by Meſſrs. Lee and Kennedy.
                  *Fl.*                                       S. ♄.

*buxifoli-*   5. **C.** ſpinis folioſis, ramis angulatis, foliis obtuſis. *Sp.*
*us.*             *pl.* 285.
                  Box-leav'd Staff-tree.
                  *Nat.* of the Cape of Good Hope.
                  *Cult.* 1759, by Mr. Ph. Miller. *Mill. dict. edit.* 7. *n.* 4.
                  *Fl.* May and June.                         G. H. ♄.

*pyracan-*    6. **C.** ſpinis nudis, ramis teretibus, foliis acutis. *Sp. pl.*
*thus.*           285.
                  Pyracantha-leav'd Staff-tree.
                  *Nat.* of the Cape of Good Hope.
                  *Cult.* 1752, by Mr. Ph. Miller. *Mill. dict. edit.* 6. *n.* 2.
                  *Fl.* moſt part of the Summer.              G. H. ♄.

*lucidus.*    7. **C.** foliis ovalibus nitidis integerrimis marginatis.
                  *Linn. mant.* 49.   *L'Herit. ſtirp. nov. p.* 49. *t.* 25.
                                                              Shining

Shining Staff-tree, or Small Hottentot Cherry.
*Nat.* of the Cape of Good Hope.
*Introd.* 1722, by James Sherard, M.D. *Knowlton mfcr.*
*Fl.* April——September. G. H. ♄.

EVONYMUS. *Gen. pl.* 271.

*Cor.* 5-petala. *Capf.* 5-gona, 5-locularis, 5-valvis,
colorata. *Sem.* calyptrata.

1. E. floribus plerifque tetrandris, pedunculis com-   *europæus.*
preffis multifloris, ftigmatibus fubulatis, foliis gla-
bris.
Evonymus europæus, α. tenuifolius. *Sp. pl.* 286.
Evonymus vulgaris. *Scop. carn.* 1. *p.* 166. *Du Roi*
*hort. harbecc.* 222.
Common Spindle-tree.
*Nat.* of Britain.
*Fl.* May and June. H. ♄.

2. E. floribus plerifque pentandris, cortice lævi, pedun-   *latifolius.*
culis filiformibus teretibus multifloris.
Evonymus europæus, β. latifolius. *Sp. pl.* 286.
Evonymus latifolius. *Scop. carn.* 1. *p.* 165. *Jacqu.*
*auftr.* 3. *p.* 48. *t.* 289. *Du Roi hort. harbecc.* 2ı6.
Broad-leav'd Spindle-tree.
*Nat.* of Auftria and Hungary.
*Cult.* 1730. *Hort. angl. n.* 2.
*Fl.* June and July. H. ♄.

3. E. floribus tetrandris, cortice verrucofo, pedunculis   *verruco-*
filiformibus teretibus fubtrifloris.                         *fus.*
Evonymus europæus, γ. leprofus. *Linn. fuppl.* 154.
Evonymus verrucofus. *Scopoli carniol.* 1. *p.* 166.
*Jacqu. auftr.* 1. *p.* 30. *t.* 49. *Du Roi hort. harbecc.*
225.

T                    Evonymus

Evonymus 11.  *Cuſ. hiſt.* 1. *p.* 57.
Warted Spindle-tree.
*Nat.* of Auſtria.
*Cult.* 1763, by Mr. James Gordon.
*Fl.* May and June.                              H. ♄.

*atropur-*      4. E. floribus tetrandris, pedunculis compreſſis mul-
*pureus.*            tifloris, ſtigmatibus tetragonis truncatis.
Evonymus atropurpureus. *Jacqu. hort.* 2. *p.* 55. *t.* 120.
Purple-flower'd Spindle-tree.
*Nat.* of North America.
*Introd.* 1756, by Meſſrs. Kennedy and Lee.
*Fl.* July.                                        H. ♄.

*america-*      5. E. floribus omnibus quinquefidis, foliis ſeſſilibus. *Syſt.*
*nus.*              *veget.* 238.
Ever-green Spindle-tree.
*Nat.* of North America.
*Cult.* 1713, by Biſhop Compton. *Philoſoph. tranſ.*
    *n.* 337. *p.* 64. *n.* 107.
*Fl.* June and July.                              H. ♄.


D I O S M A.  *Gen. pl.* 272.

*Cor.* 5-petala.  *Nectaria* 5, ſupra germen.  *Capſ.* 3. ſ.
    5, coalitæ.  *Sem.* calyptrata.

*oppoſitifo-*   1. D. foliis ſubulatis acutis oppoſitis.  *Sp. pl.* 286.
*lia.*             Oppoſite-leav'd Dioſma.
*Nat.* of the Cape of Good Hope.
*Introd.* 1774, by Mr. Francis Maſſon.
*Fl.* March——July.                          G. H. ♄.

*hirſuta.*      2. D. foliis linearibus hirſutis.  *Sp. pl.* 286.
Hairy-leav'd Dioſma.

                                                    *Nat.*

*Nat.* of the **Cape of Good Hope.**
*Cult.* 1731, by Mr. Philip Miller. *Mill. dict. edit.* 1.
Spiræa 4.
*Fl.* March——September.                    G. H. ♄.

3. D. foliis lineari-lanceolatis fubtus convexis bifariam    *ericoides.*
imbricatis. *Sp. pl.* 287.
Sweet-fcented Diofma.
*Nat.* of the Cape of Good Hope.
*Cult.* 1756, by Mr. Ph. Miller. *Mill. ic.* 84. *t.* 124.
*f.* 2.
*Fl.* March——September.                    G. H. ♄.

4. D. foliis ovatis mucronatis imbricatis ciliatis. *Syft.*    *imbrica-*
*veget.* 239.                                                   *ta.*
Imbricated Diofma.
*Nat.* of the Cape of Good Hope.
*Introd.* 1774, by Mr. Francis Maffon.
*Fl.* April——June.                         G. H. ♄.

5. D. foliis lanceolatis ciliatis rugofis. *Syft. veget.* 239.    *ciliata.*
Ciliated Diofma.
*Nat.* of the Cape of Good Hope.
*Introd.* 1774, by Mr. Francis Maffon.
*Fl.* April and May.                        G. H. ♄.

6. D. foliis lanceolato-ovalibus oppofitis glandulofo-    *crenata.*
crenatis, floribus folitariis. *Syft. veget.* 239.
Crenated Diofma.
*Nat.* of the Cape of Good Hope.
*Introd.* 1774, by Mr. Francis Maffon.
*Fl.*                                       G. H. ♄.

*uniflora.* 7. D. foliis ovato-oblongis, floribus folitariis terminali-
bus. *Sp. pl.* 287.
One-flower'd Diofma.
*Nat.* of the Cape of Good Hope.
*Introd.* 1775, by Mr. Francis Maffon.
*Fl.* G. H. ♄ .

*pulchella.* 8. D. foliis ovatis obtufis glandulofo-crenatis, floribus
geminis axillaribus. *Sp. pl.* 288.
Oval-leav'd Diofma.
*Nat.* of the Cape of Good Hope.
*Introd.* 1787, by Mr. Francis Maffon.
*Fl.* G. H. ♄ .

B R U N I A. *Gen. pl.* 274.

*Flores* aggregati. *Filam.* unguibus petalorum inferta.
*Stigma* 2-fidum. *Sem.* folitaria, bilocularia.

*nodiflora.* 1. B. foliis imbricatis triquetris acutis. *Syft. veget.* 240.
Imbricated Brunia.
*Nat.* of the Cape of Good Hope.
*Introd.* 1786, by Mr. Francis Maffon.
*Fl.* G. H. ♄ .

*lanugino-* 2. B. foliis linearibus patulis apice callofis. *Sp. pl.* 288.
*fa.* Heath-leav'd Brunia.
*Nat.* of the Cape of Good Hope.
*Introd.* 1774, by Mr. Francis Maffon.
*Fl.* moft part of the Summer. G. H. ♄ .

*abrota-* 3. B. foliis lineari-lanceolatis patentibus triquetris apice
*noides.* callofis. *Sp. pl.* 288.
Thyme-leav'd Brunia.
*Nat.* of the Cape of Good Hope.
*Introd.* 1787, by Mr. Francis Maffon.
*Fl.* G. H. ♄ .

4. B.

4. B. foliis linearibus triquetris, calyce radiante: foliolis   *radiata.*
    intimis coloratis. *Syſt. veget.* 240.
    Phylica radiata. *Sp. pl.* 283.
    Radiated Brunia.
    *Nat.* of the Cape of Good Hope.
    *Introd.* 1787, by Mr. Francis Maſſon.
    *Fl.*                                          G. H. ♄.


### I T E A. *Gen. pl.* 275.

*Petala* longa, calyci inſerta. *Capſ.* 1-locularis, 2-valvis.

1. I. foliis ſerratis. *L'Herit. ſtirp. nov.*              *virginica.*
    Itea virginica. *Sp. pl.* 289.
    Virginian Itea.
    *Nat.* of North America.
    *Cult.* 1744, by Archibald Duke of Argyle.
    *Fl.* June——Auguſt.                       H. ♄.

2. I. foliis integerrimis. *L'Herit. ſtirp. nov. tab.* 66.   *Cyrilla.*
    Cyrilla racemiflora. *Linn. mant.* 50. *Jacqu. ic. coll.* 1.
    *p.* 162.
    Entire-leav'd Itea.
    *Nat.* of Carolina.
    *Introd.* 1765, by Mr. John Cree.
    *Fl.* July and Auguſt.                       G. H. ♄.


### C E D R E L A. *Gen. pl.* 277.

*Cal.* marceſcens.   *Cor.* 5-petala, infundibuliformis,
    baſi ad ⅓ receptaculo adnata. *Capſ.* lignoſa, 5-locu-
    laris, 5-valvis. *Sem.* deorſum imbricata, ala mem-
    branacea.

1. C. floribus paniculatis. *Sp. pl.* 289.              *odorata.*
    Barbadoes Baſtard-cedar.
    *Nat.* of the Weſt Indies.

T 3                                 *Cult.*

*Cult.* 1739, by Mr. Philip Miller. *Rand. chel. p.* 214.
*n.* 23.
*Fl.*                                                          S. ♄.


### ELÆODENDRUM. *Jacquin.*
*Cor.* 5-petala. *Drupa* ovata, *nuce* 2-loculari.

*orientale.*    1. ELÆODENDRUM. *Jacqu. ic. Syſt. veget.* 241.
Oriental Elæodendrum.
*Nat.*
*Introd.* about 1771, by Mr. John Buſh.
*Fl.*                                                          S. ♄.


### MANGIFERA. *Gen. pl.* 278.
*Cor.* 5-petala. *Drupa* reniformis.

*indica.*    1. MANGIFERA. *Sp. pl.* 290.
Mango-tree.
*Nat.* of the Eaſt Indies.
*Cult.* 1690, in the Royal Garden at Hampton-court.
*Catal. mſcr.*
*Fl.*                                                          S. ♄.


### HIRTELLA. *Gen. pl.* 280.
*Petala* 5. *Filamenta* longiſſima, perſiſtentia, ſpiralia.
*Bacca* 1-ſpērma. *Stylus* lateralis.

*america-*    1. HIRTELLA. *Sp. pl.* 290.
*na.*        American Hirtella.
*Nat.* of the Weſt Indies.
*Introd.* 1782, by Mr. Alexander Anderſon.
*Fl.*                                                          S. ♄.


RIBES.

## RIBES. *Gen. pl.* 281.

*Petala* 5 et *flamina* calyci inferta. *Stylus* 2-fidus.
*Bacca* polyfperma, infera.

\* Ribefia *inermia.*

1. R. inerme, racemis glabris pendulis, floribus planiuf-    *rubrum.*
   culis. *Sp. pl.* 290.
α baccis rubris.                                              rutilum.
   Red Currant.
β baccis albis.                                              album.
   White Currant.
   *Nat.* of Britain.
   *Fl.* April.                                    H. ♄ .

2. R. inerme, floribus racemofis planiufculis, pedicellis   *glandulo-*
   calycibufque pilofis: pilis glandulofis.                  *fum.*
   Ribes glandulofum. *Weber dec. plant. min. cogn. p.* 2.
   Ribes proftratum. *L'Heritier ftirp. nov. p.* 3. *tab.* 2.
   Glandulous Currant.
   *Nat.* of North America.
   *Introd.* 1777, by John Fothergill, M. D.
   *Fl.* April and May.                             H. ♄ .

3. R. inerme, racemis erectis, bracteis flore longioribus.  *alpinum.*
   *Sp. pl.* 291. *Jacqu. auftr.* 1. *p.* 29. *t.* 47.
   Alpine Currant.
   *Nat.* of Britain.
   *Fl.* May.                                       H. ♄ .

4. R. inerme, racemis pilofis, floribus oblongis. *Sp. pl.*  *nigrum.*
   291.
   Common black Currant.
   *Nat.* of Britain.
   *Fl.* April and May.                             H. ♄ .

                    T 4                            5. R.

*floridum.* 5. R. inerme, racemis pendulis, floribus cylindricis,
bracteis flore vix brevioribus. *L'Herit. stirp. nov.* 4.

Ribes americanum. *Mill. dict.*

Ribes americanum nigrum. *Moench hort. weissenst.*
104. *tab.* 7.

Ribes nigrum β. *Sp. pl.* 291.

Ribesium nigrum pensilvanicum, floribus oblongis.
*Dill. elth.* 324. *t.* 244. *f.* 315.

American black Currant.

*Cult.* 1732, by James Sherard, M. D. *Dill. elth.*
*loc. cit.*

*Fl.* April and May.                                H. ♄.


** Grossulariæ *aculeatæ.*

*diacan-*   6. R. foliis incisis, aculeis geminis ad gemmas. *Linn.*
*tha.*            *suppl.* 157.

Two-spin'd Gooseberry.

*Nat.* of Siberia.

*Introd.* 1781, by Mr. John Bush.

*Fl.*                                              H. ♄.


*Grossula-*  7. R. ramis aculeatis, petiolorum ciliis pilosis, baccis
*ria.*            hirsutis. *Sp. pl.* 291.

Common Gooseberry.

*Nat.* of Europe.

*Cult.* 1629. *Park. parad.* 561.

*Fl.* March and April.                            H. ♄.


*Uva cris-*  8. R. ramis aculeatis, baccis glabris, pedicellis bractea
*pa.*             monophylla. *Sp. pl.* 292,

Smooth-fruited Gooseberry.

*Nat.* of Europe.

*Cult.* 1597. *Ger. herb.* 1143.

*Fl.* March and April.                            H. ♄.

9. R.

9. R. ramis undique aculeatis. *Sp. pl.* 291.       *oxyacan-*
Hawthorn-leav'd Currant.                             *thoides.*
*Nat.* of Canada.
*Cult.* 1705, by Mr. Reynardson. *Pluk. amalth.* 212.
*Fl.* April and May.                                H. ♄.

10. R. aculeis subaxillaribus, baccis aculeatis racemosis.  *cynosbati.*
*Sp. pl.* 292. *Jacqu. hort.* 2. *p.* 56. *t.* 123.
Prickly-fruited Currant.
*Nat.* of Canada.
*Cult.* 1759, by Mr. Ph. Miller. *Mill. dict. edit.* 7. *n.* 5.
*Fl.* April.                                        H. ♄.

## G R O N O V I A. *Gen. pl.* 282.

*Petala* 5 et *stamina* calyci campanulato inserta. *Bacca*
sicca, monosperma, infera.

1. GRONOVIA. *Sp. pl.* 292.                         *scandens.*
Climbing Gronovia.
*Nat.* of Jamaica.
*Introd.* 1731, by William Houstoun, M. D. *Mart.*
*dec.* 4. *p.* 40.
*Fl.*                                               S. ♃.

## H E D E R A. *Gen. pl.* 283.

*Petala* 5, oblonga. *Bacca* 5-sperma, calyce cincta.

1. H. foliis ovatis lobatisque. *Sp. pl.* 292. *Curtis lond.*  *Helix.*
Common Ivy.
*Nat.* of Britain.
*Fl.* September.                                    H. ♄.

2. H. foliis quinatis ovatis serratis. *Sp. pl.* 292.  *quinque-*
Five-leav'd Ivy.                                    *folia.*
*Nat.* of North America.
                                            *Cult.*

*Cult.* 1629.   *Park. parad.* 609. *f.* 7.
*Fl.* June and July.                                      H. ♄.

V I T I S.   *Gen. pl.* 284.

*Petala* apice cohærentia, emarcida.   *Bacca* 5-fperma,
fupera.

*vinifera.*   1. V. foliis lobatis finuatis nudis.   *Sp. pl.* 293.   *Jacqu.*
    *ic. coll.* 1. *p.* 160.
    Common Vine.
    *Nat.* of moft of the temperate parts of the World.
    *Fl.* June and July.                               H. ♄.

*Labruf-*   2. V. foliis cordatis fubtrilobis dentatis fubtus tomento-
*ca.*        fis.   *Sp. pl.* 293.
    Downy-leav'd Vine.
    *Nat.* of North America.
    *Cult.* 1656, by Mr. John Tradefcant, Jun.   *Trad.*
    *muf.* 177.
    *Fl.*                                              H. ♄.

*vulpina.*   3. V. foliis cordatis dentato-ferratis utrinque nudis.   *Sp.*
    *pl.* 293.
    Fox Grape, or Vine.
    *Nat.* of Virginia.
    *Cult.* 1656, by Mr. John Tradefcant, Jun.   *Trad.*
    *muf.* 177.
    *Fl.*                                              H. ♄.

*laciniofa.*   4. V. foliis quinatis : foliolis multifidis.   *Sp. pl.* 293.
    Parfley-leav'd Vine.
    *Nat.*
    *Cult.* 1656, by Mr. John Tradefcant, Jun.   *Trad.*
    *muf.* 177.
    *Fl.* June and July.                               H. ♄.

5. V.

5. V. foliis fupradecompofitis: foliolis lateralibus pin-   *arborea.*
   natis. *Sp. pl.* 294.
   Pepper Vine.
   *Nat.* of North America.
   *Cult.* before 1700, by Samuel Reynardfon, Efq. *Plnk.*
   *mant.* 85. *t.* 412. *f.* 2.
   *Fl.* July——September.                          H. ♄.


LEEA. *Linn. mant.* 17. (Aquilicia. *Linn. mant.* 146.)

*Cor.* 1-pet. *Nect.* 1-phyll. tubo corollæ impofitum, 5-
   fidum, erectum. *Bacca* 5-fperma, infera.

1. L. caule tereti pubefcente. *Linn. mant.* 124.   *æquata.*
   Shrubby Leea.
   *Nat.* of the Eaft Indies.
   *Introd.* before 1777, by Mr. James Lee.
   *Fl.*                                            S. ♄.

2. L. caule angulato fimbriato. *Linn. mant.* 124.   *crijpa.*
   Fringed-ftalk'd Leea.
   *Nat.* of the Cape of Good Hope.
   *Introd.* 1767, by Mr. William Malcolm.
   *Fl.* October.                                   S. ♃.


LAGOECIA. *Gen. pl.* 285.

*Involucr.* univerfale et partiale. *Petala* 2-fida. *Sem.*
               folitaria, infera.

1. LAGOECIA. *Sp. pl.* 294.                         *Cuminoi-*
   Wild Cumin.                                      *des.*
   *Nat.* of the Levant.
   *Cult.* 1708, by the Dutchefs of Beaufort. *Br. Muf.*
   H. S. 137. *fol.* 6.
   *Fl.* June and July.                             H. ☉.
                                          C L A Y-

## CLAYTONIA. *Gen. pl.* 287.

*Cal.* 2-valvis. *Cor.* 5-petala. *Stigma* 3-fidum. *Capſ.*
3-valvis, 1-locularis, 3-ſperma.

*virginica.* 1. C. foliis lineari-lanceolatis, petalis integris.

α foliis linearibus, calycibus obtuſis.

Claytonia virginica. *Sp. pl.* 294.

Virginian Claytonia.

β foliis lanceolatis, calycibus acutiuſculis.

Spear-leav'd Claytonia.

*Nat.* of North America.

*Introd.* before 1759, by Mr. John Clayton. *Mill.
dict. edit.* 7.

*Fl.* March——May. H. ♃.


## HELICONIA. *Linn. mant.* 147.

*Spathæ.* *Cal.* 0. *Cor.* 3-petala. *Nectarium* 2-phyl-
lum. *Peric.* tricoccum. *Sem.* ſolitaria.

*Bihai.* 1. H. foliis oblongis, ſpathis navicularibus patentibus
amplexicaulibus.

Muſa Bihai. *Sp. pl.* 1477.

Baſtard, or Wild Plantain.

*Nat.* of the Weſt Indies.

*Introd.* 1786, by Mr. Alexander Anderſon.

*Fl.* S. ♃.

Obs. Linnæus filius in *ſupplem. plant. p.* 157. ſub no-
mine Heliconæ Bihai tres plantas diverſas confundit, hanc Americanam et binas Africanas, Strelitz-
iam nimirum Reginæ, et aliam ejuſdem generis
ſpeciem floribus albis et foliis reticulatis.


STRE-

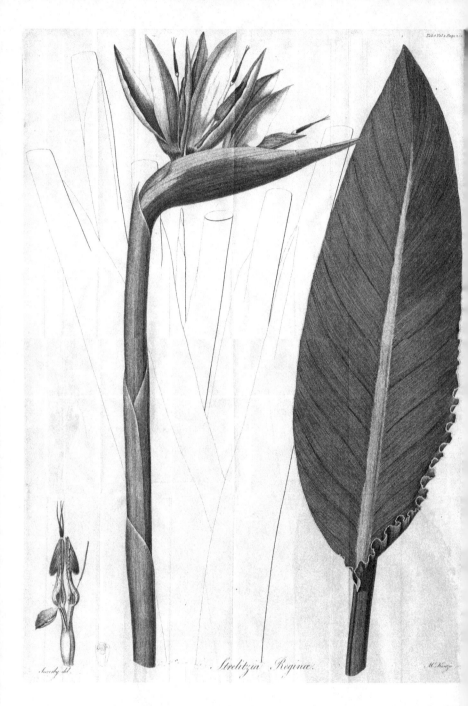

*Strelitzia Reginæ.*

## STRELITZIA.

*Spathæ.* *Cal.* o. *Cor.* 3-petala. *Nectar.* 3-phyllum, genitalia involvens. *Peric.* 3-loculare, polyfper- mum. .

1. STRELITZIA. TAB. 2. *Reginæ.*
Heliconia Bihai. *J. Mill. ic. tab.* 5, 6.
Canna-leav'd Strelitzia.
*Nat.* of the Cape of Good Hope.
*Introd.* 1773, by Sir Jofeph Banks, Bart.
*Fl.* April and May. S. ♄.
DESCR. *Folia* omnia radicalia, petiolata, oblonga, integerrima, margine inferne undulato crifpo, gla- berrima, fubtus glaucefeentia, coriacea, pedalia, perfiftentia. *Petioli* fubcompreffi, tripedales et ul- tra, craffitie pollicis, vaginantes, erecti, glabri. *Scapus* longitudine et craffitie petiolorum, erectus, teres, tectus *vaginis* alternis, remotis, acuminatis, viridibus margine purpurafcente. *Spatha* univer- falis fpithamæa, extus viridis, margine purpuraf- cens; fpathæ partiales albidæ. *Petala* lutea, qua- driuncialia. *Nectarium* cæruleum.
OBS. Differentia fpecifica Heliconiæ albæ in *Linn. fuppl.* 157. hujus eft plantæ, fed nomen triviale ad aliam pertinet fpeciem Africanam, in hortis Euro- pæis nondum obviam.

## ACHYRANTHES. *Gen. pl.* 288.

*Cal.* 5-phyllus. *Cor.* o. *Stigma* 2-fidum. *Semina* folitaria.

1. A. caule fruticofo erecto, calycibus reflexis fpicæ ad- *afpera.* preffis. *Syft. veget.* 246.

*a* Amaran-

ficula.     *α* Amaranthus Siculus fpicatus.   *Bocc. fic.* 16. *t.* 9.
         Upright Achyranthes.

indica.     *β* Amaranthus fpicatus zeylanicus, foliis obtufis, amaran-
         tho ficulo Boccone fimilis. *Burm. zeyl.* 16. *t.* 5. *f.* 3.
         Dwarf Achyranthes.
         *Nat. α.* of Sicily ; *β.* of both Indies.
         *Cult.* 1713.   *Philofoph. tranf. n.* 337. *p.* 181. *n.* 17.
         *Fl.* moft part of the Year.                S. ♄.

*lappacea.*    2. A. caule fruticofo diffufo proftrato, fpicis interruptis :
         flofculis lateralibus utrinque fafciculo fetarum unci-
         nato. *Syft. veget.* 246.
         Spreading Achyranthes.
         *Nat.* of the Eaft Indies.
         *Cult.* 1759, by Mr. Ph. Miller. *Mill. dict. edit.* 7. *n.* 3.
         *Fl.* Auguft——October.              S. ♄.

*muricata.*    3. A. caule fruticofo patulo, foliis alternis, floribus fpi-
         catis remotis ovatis, calycibus fquarrofis.   *Syft.*
         *veget.* 246.
         Prickly Achyranthes.
         *Nat.* of India.
         *Introd.* 1777, by Monf. Thouin.
         *Fl.* Auguft——November.            S. ♃.

*nivea.*     4. A. foliis verticillatis ovatis tomentofis, corymbis com-
         pactis dichotomis, floribus corollatis.
         White Achyranthes.
         *Nat.* of the Canary Iflands.   Mr. *Francis Maffon.*
         *Introd.* 1780.
         *Fl.* May——July.               G. H. ♄.

C E L O.

## CELOSIA. *Gen. pl.* 289.

*Cal.* 3-phyllus, corollæ 5-petalæ facie. *Stam.* bafi nectario plicato conjuncta. *Capf.* horizontaliter dehifcens.

1. C. foliis lanceolatis, ftipulis fubfalcatis, pedunculis *argentea.* angulatis, fpicis fcariofis. *Sp. pl.* 296.
Silvery-fpiked Celofia.
*Nat.* of China.
*Cult.* 1714, by the Dutchefs of Beaufort. *Br. Muf.*
H. S. 131. *fol.* 70.
*Fl.* June——September. S. ⊙.

2. C. foliis oblongo-ovatis, pedunculis teretibus fub- *criftata.* ftriatis, fpicis oblongis. *Syft. veget.* 247.
α Amaranthus panicula conglomerata. *Bauh. pin.* 121.
Common Celofia, or Cock's-comb.
β Amaranthus major paniculis furrectis flavefcentibus.
*Herm. lugdb.* 30.
Buff-colour'd Celofia, or Cock's-comb.
*Nat.* of Afia.
*Cult.* 1570. *Lobel. adv.* 95.
*Fl.* July——September. S. ⊙.

3. C. foliis ovato-oblongis, caule affurgente panicula- *panicula-* to, fpicis alternis terminalibus remotis. *Syft. veget.* *ta.* 247.
Panicled Celofia.
*Nat.* of Jamaica.
*Introd.* about 1732, by William Houftoun, M. D.
*Mill. dict. edit.* 8.
*Fl.* moft part of the Summer. S. ♃.

4. C. foliis ovatis ftrictis inauriculatis, caule fulcato, *coccinea.* fpicis multiplicibus criftatis. *Sp. pl.* 297.
Scarlet

Scarlet Celosia, or Chinese Cock's-comb.
*Nat.* of China.
*Cult.* 1597.   *Ger. herb.* 254. *f.* 2.
*Fl.* July——September.                                  S. ☉.

*castrensis.*  5. C. foliis lanceolato-ovatis lineatis acuminatissimis, stipulis falcatis, spicis cristatis.   *Sp. pl.* 297.
Branched Celosia, or Cock's-comb.
*Nat.* of the East Indies.
*Cult.* 1739, by Mr. Philip Miller.   *Rand. chel.* Amaranthus 8.
*Fl.* July——September.                                  S. ☉.

*Monso-*  6. C. spicis compactis cylindraceis, ramis brachiatis, fo-
*niæ.*        liis subulatis.
Celosia Monsonia.   *Retzii obs. bot.* 2. *p.* 13. *n.* 26.
Illecebrum Monsoniæ.   *Linn. suppl.* 161.
Amaranthoides spicatum indicum ramosissimum Spergulæ foliis, spica alopecuroide candida.   *Pluk. mant.* 11. *t.* 334. *f.* 4.
Downy Celosia.
*Nat.* of the East Indies.
*Introd.* 1778, by Sir Joseph Banks, Bart.
*Fl.* August and September.                             S. ☉.

*trigyna.*  7. C. foliis ovato-oblongis, racemo laxo, pistillo trifido.
*Linn. mant.* 212.   *Jacqu. hort.* 3. *p.* 12. *t.* 15.
Oval-leav'd Celosia.
*Nat.* of Senegal.
*Introd.* 1777, by Monf. Thouin.
*Fl.* August and October.                               S. ☉.

*nodiflora.*  8. C. foliis cuneiformibus acutiusculis, spicis globosis
lateralibus.   *Sp. pl.* 298.
Knotted Celosia.

*Nat.*

*Nat.* of the Eaft Indies.
*Introd.* 1780.
*Fl.* July and Auguft.      S. ⊙.

9. C. caulibus decumbentibus, pedunculis longiffimis *procum-*
aphyllis, fpicis ovatis approximatis, capfulis com- *bens.*
preffis criftato-alatis. *Syft. veget.* 247.
Gomphrena interrupta. *L'Heritier ftirp. nov. p.* 5. *t.* 3.
Procumbent Celofia.
*Nat.* of S. Domingo.
*Introd.* 1784, by Monf. Thouin.
*Fl.* July.      S. ⊙.

## ILLECEBRUM. *Gen. pl.* 290.

*Cal.* 5-phyllus, cartilagineus. *Cor.* nulla. *Stigm.* fimplex. *Capf.* 5-valvis, monofperma.

1. I. foliis ovatis pilofiufculis, fpicis lateralibus, calycibus *lanatum.*
lanatis.
Illecebrum lanatum. *Syft. veget.* 248.
α fpicis fubaggregatis folio brevioribus, ramis longis
virgatis.
Achyranthes lanata. *Sp. pl.* 296.
Small woolly Illecebrum.
β fpicis folitariis ramulorum patentium.
Great woolly Illecebrum.
*Nat.* of the Eaft Indies.
*Cult.* 1691, in the Royal Garden at Hampton-court.
*Pluk. phyt. t.* 75. *f.* 8.
*Fl.* moft part of the Year.      S. ♂.

2. I. foliis lanceolatis tomentofis, fpicis cylindraceis nu- *javani-*
merofis terminalibus. *cum.*
Illecebrum javanicum. *Syft. veget.* 248.
       U          Celofia

Celofia lanata. *Sp. pl.* 298. *Syft. veget.* 247.
Spear-leav'd Illecebrum.
*Nat.* of the Eaft Indies.
*Cult.* 1768, by Mr. Philip Miller. *Mill. dict. edit.*
8. Celofia lanata.
*Fl.* moft part of the Summer.                    S. ♃.

*verticil-*
*latum.*  3. I. floribus verticillatis nudis, caulibus procumbenti-
bus. *Sp. pl.* 298.
Verticil'd Illecebrum, or Knot-grafs.
*Nat.* of England.
*Fl.* July.                                      H. ♃.

*fuffruti-*
*cofum.*  4. I. floribus lateralibus folitariis, caulibus fuffruticofis.
*Sp. pl.* 298.
Shrubby Illecebrum, or Knot-grafs.
*Nat.* of the South of Europe.
*Cult.* 1739, by Mr. Philip Miller. *Mill. dict. vol.* 2.
Paronychia 4.
*Fl.* May——Auguft.                             G. H. ♄.

*arifta-*
*tum.*  5. I. floribus fubfafciculatis, foliis lanceolatis fericeis
ariftatis.
Bearded Illecebrum.
*Nat.* of the Canary Iflands. Mr. *Francis Maffon.*
*Introd.* 1780.
*Fl.* June and July.                            G. H. ♂.

*Parony-*
*chia.*  6. I. floribus bracteis nitidis obvallatis, caulibus procum-
bentibus, foliis lævibus. *Syft. veget.* 248.
Mountain Illecebrum, or Knot-grafs.
*Nat.* of the South of Europe.
*Cult.* 1640. *Park. theat.* 445. *n.* 1.
*Fl.* May——Auguft.                             G. H. ♃.

7. I.

7. I. floribus bracteatis subfasciculatis, pedunculis dicho- *divarica-*
tomis paniculatis, foliis ovato-oblongis petiolatis. *tum.*
Forked Illecebrum.
*Nat.* of the Canary Islands. Mr. *Francis Masson.*
*Introd.* 1779.
*Fl.* July and August. G. H. ☉.

8. I. caulibus repentibus pilosis, foliis ovatis mucronatis : *Achyran-*
opposito minore, capitulis subglobosis subspinosis. *tha.*
*Sp. pl.* 299.
Creeping Illecebrum.
*Nat.* of Buenos Ayres.
*Cult.* 1732, by James Sherard, M. D. *Dill. elth.* 8.
*t. 7. f. 7.*
*Fl.* June——August. S. ♃.

9. I. caulibus repentibus bifarie tomentosis, foliis lanceo- *sessile.*
latis subsessilibus, capitulis oblongis glabris. *Syst.*
*veget.* 249.
Sessile-flower'd Illecebrum.
*Nat.* of the East Indies.
*Introd.* 1778, by Mons. Thouin.
*Fl.* July——October. S. ☉.

G L A U X. *Gen. pl.* 291.

*Cal.* 1-phyllus. *Cor.* nulla. *Capf.* 1-locularis, 5-valvis,
5-sperma.

1. GLAUX. *Sp. pl.* 301. *mariti-*
Sea Milk-wort. *ma.*
*Nat.* of Britain.
*Fl.* May and June. H. ♃.

# PLOCAMA.

*Cor.* 5-fida.   *Cal.* 5-dentatus, fuperus.   *Bacca* 3-locu-
laris.   *Sem.* folitaria.

*pendula.*    1. PLOCAMA.
      Pendulous Plocama.
      *Nat.* of the Canary Iflands.   Mr. *Francis Maffon.*
      *Introd.* 1779.
      *Fl.*                              G. H. ♄.

# THESIUM.   *Gen. pl.* 292.

*Cal.* 1-phyllus, cui ftamina inferta.   *Sem.* 1, inferum.

*Linophyl-*    1. T. panicula foliacea, foliis linearibus. *Syft. veget.* 249.
*lum.*        Common Thefium, or Baftard Toad-flax.
      *Nat.* of England.
      *Fl.* June and July.                        H. ♃.

*umbella-*    2. T. floribus umbellatis, foliis oblongis. *Sp. pl.* 302.
*tum.*        Umbel'd Thefium.
      *Nat.* of North America.
      *Introd.* 1782, by John Hope, M. D.
      *Fl.* June.                            H. ♃.

*amplexi-*    3. T. racemis terminalibus, foliis cordatis feffilibus. *Linn.*
*caule.*          *mant.* 213.
      Heart-leav'd Thefium.
      *Nat.* of the Cape of Good Hope.
      *Introd.* 1787, by Mr. Francis Maffon.
      *Fl.*                             G. H. ♄.

# RAUVOLFIA.   *Gen. pl.* 293.

Contorta.   *Bacca* fucculenta, difperma.

*nitida.*    1. R. glaberrima nitidiffima.   *Sp. pl.* 303.
      Shining Rauvolfia.

                                           *Nat.*

*Nat.* of South America.

*Cult.* 1739, by Mr. Ph. Miller. *Mill. dict. vol.* 2. *n.* 1.

*Fl.* June——September. S. ♄.


C E R B E R A. *Gen. pl.* 294.

Contorta. *Drupa* monofperma.

1. C. foliis ovatis. *Sp. pl.* 303. *Ahouaj.*
Oval-leav'd Cerbera.
*Nat.* of Brazil.
*Cult.* 1739, by Mr. Philip Miller. *Mill. dict. vol.* 2.
Ahouai 1.
*Fl.* June and July. S. ♄.

2. C. foliis linearibus longiffimis confertis, *Sp. pl.* 304. *Thevetia.*
Linear-leav'd Cerbera.
*Nat.* of South America.
*Introd.* 1735, by Mr. Robert Millar. *Mill. dict.*
*vol.* 2. Ahouai 2.
*Fl.* S. ♄.


G A R D E N I A. *Gen. pl.* 296. *Linn. fuppl.* 23.

Contorta. *Bacca* infera 2-f. 4-locularis, polyfperma.
*Stigma* lobatum.

\* *Inermes.*

1. G. inermis, foliis ellipticis, corollis hypocrateriformibus, *florida.*
calycinis laciniis verticalibus lanceolato-fubulatis.
Gardenia florida. *Sp. pl.* 305. *Thunb. gardenia, n.* 2.
α flore fimplici.
Single-flower'd Gardenia, or Cape Jafmine.
β flore pleno.
Double-flower'd Gardenia, or Cape Jafmine.
*Nat.* of Cochinchina, China, Japan, and the South-
Sea Iflands.

*Introd.*

*Introd.* about 1754, by Capt. Hutchinfon. *Phil. tranfact. vol.* 51. *pag.* 933.

*Fl.* July——October.                              S. ♄.

*Thunber-*    2. G. inermis, foliis ellipticis, corollis hypocrateriformi-
*gia.*              bus, calycibus latere rumpentibus : laciniis apice
                   dilatatis.

Gardenia Thunbergia. *Thunb. gardenia, n.* 3. *Linn. fuppl.* 162.

Thunbergia. *Montin act. ftockh.* 1773. *p.* 289. *t.* 11.

Bergkias. *Sonnerat iter nov. guin. p.* 48. *t.* 17, 18.

Starry Gardenia.

*Nat.* of the Cape of Good Hope.

*Introd.* 1773, by Sir James Cockburn, Bart.

*Fl.*                                             S. ♄.

*latifolia.*   3. G. inermis, foliis obovato-fubrotundis, corollis hypo-
                   crateriformibus, laciniis calycinis fubulatis obtufe
                   carinatis.

Broad-leaved Gardenia.

*Nat.* of the Eaft Indies.

*Introd.* 1787, by Richard Anthony Salifbury, Efq.

*Fl.*                                             S. ♄.

*Roth-*        4. G. inermis, foliis oblongis, corollis infundibuliformi-
*mannia.*           bus, laciniis calyciais fubulatis.

Gardenia Rothmannia. *Thunb. gardenia, n.* 6. *Linn. fuppl.* 165.

Rothmannia capenfis. *Thunberg act. ftockh.* 1776. *p.* 65. *t.* 2.

Spotted-flower'd Gardenia.

*Nat.* of the Cape of Good Hope. Mr. *Francis Maffon.*

*Introd.* 1774.

*Fl.*                                             S. ♄.

5. G.

** *Spinofæ.*

5. G. fpinis oppofitis foliis longioribus, germinibus gla- *dumeto-*
bris. *rum.*

Gardenia dumetorum. *Retzii obf. bot.* 2. *p.* 14. *n.* 31.

Gardenia fpinofa. *Linn fuppl.* 164. *(diftincta a Gar-*
*denia fpinofa Thunbergii e China.)*

Spiney Gardenia.

*Nat.* of the Eaft Indies. *John Gerard Koenig*, M.D.

*Introd.* 1777, by Sir Jofeph Banks, Bart.

*Fl.* S. ♄.

6. G. fpinis oppofitis floribufque foliis brevioribus, ra- *aculeata.*
mis glabris.

Randia aculeata. *Sp. pl.* 214.

Round-leav'd thorney Gardenia.

*Nat.* of the Weft Indies.

*Introd.* before 1733, by William Houftoun, M.D.
*Mill. dict. edit.* 8.

*Fl.* S. ♄.

## VINCA. *Gen. pl.* 295.

Contorta. *Folliculi* 2, erecti. *Semina* nuda.

1. V. caulibus procumbentibus, foliis lanceolato-ovatis, *minor.*
floribus pedunculatis. *Sp. pl.* 304. *Curtis lond.*

Small Periwinkle.

*Nat.* of Britain.

*Fl.* March——September. H. ♄.

2. V. caulibus erectis, foliis ovatis, floribus peduncula- *major.*
tis. *Sp. pl.* 304. *Curtis lond.*

Greater Periwinkle.

*Nat.* of England.

*Fl.* March——September. H. ♄.

U 4 3. V.

*rosea.*　3. V. caule frutescente erecto, floribus geminis sessilibus, foliis ovato-oblongis, petiolis basi bidentatis.
*Syst. veget.* 252.

α flore carneo.
Madagascar red Periwinkle.

β flore albo, umbone carneo.
Madagascar white Periwinkle.

*Nat.* of the East Indies.

*Cult.* 1757, by Mr. Philip Miller.　*Mill. ic.* 124.
*t.* 186.

*Fl.* most part of the Year.　　　　　　　　　S. ♃.

*parviflo-*　4. V. caule herbaceo erecto, foliis lanceolatis acutis.
*ra.*　　　Vinca parviflora.　*Retzii obs. bot.* 2. *p.* 14. *n.* 33.
Vinca pusilla.　*Linn. suppl.* 166.
Cupa-veela.　*Rheed. mal.* 9. *p.* 61. *t.* 33.
Small-flower'd Periwinkle.

*Nat.* of the East Indies.

*Introd.* 1778, by Sir Joseph Banks, Bart.
*Fl.* August.　　　　　　　　　　　　　　S. ☉.

## N E R I U M.　*Gen. pl.* 297.

Contorta.　*Folliculi* 2, erecti.　*Sem.* plumosa.　*Cor.*
tubus terminatus corona lacera.

*Oleander.*　1. N. foliis lineari-lanceolatis ternis, foliolis calycinis
squarrosis, nectariis planis tricuspidatis.
Nerium Oleander.　*Sp. pl.* 305. (excluso synonymo
Rheedi.)

α flore roseo simplici.
Common Rosebay, or Oleander.

β flore albo.
White Rosebay, or Oleander.

γ flore pleno.

Double

Double Rofebay, or Oleander.
*Nat.* of Spain, Portugal, and the Levant.
*Cult.* 1596, by Mr. John Gerard. *Hort. Ger.*
*Fl.* June —— October. G. H. ♄ .

2. N. foliis lineari-lanceolatis ternis, foliolis calycinis *odorum.*
 erectis, nectariis multipartitis : laciniis filiformibus.
Tsjovanna-areli. *Rheed. mal.* 9. *p.* 1. *t.* 1.
Belutta-areli. *Rheed. mal.* 9. *p.* 3. *t.* 2.
Sweet-fcented Rofebay, or Oleander.
*Nat.* of the Eaft Indies.
*Cult.* 1758, by Mr. Philip Miller.
*Fl.* June —— Auguft. S. ♄ .

3. N. foliis oblongo-ovatis, paniculis terminalibus. *antidy-*
Nerium antidyfentericum. *Sp. pl.* 306. *fenteri-*
Oval-leav'd Rofebay. *cum.*
*Nat.* of the Eaft Indies.
*Introd.* 1778, by Patrick Ruffell, M.D.
*Fl.* S. ♄ .

4. N. foliis ellipticis, pedunculis ex dichotomia ramorum *coronari-*
geminis, bifloris. *um.*
Nerium coronarium. *Jacqu. ic. collect.* 1. *pag.* 138.
Jafminum zeylanicum folio oblongo, flore albo pleno
odoratiffimo. *Burm. zeyl.* 129. *t.* 59.
Flos manilhanus. *Rumph. amb.* 4. *p.* 87. *t.* 39.
Nandi-Ervatam major et minor. *Rheed. mal.* 2.
*p.* 105. *t.* 54, 55.
Broad-leav'd Rofebay.
*Nat.* of the Eaft Indies.
*Cult.* 1770, by Mr. James Gordon.
*Fl.* moft part of the Summer. S. ♄ .

ECHITES.

## ECHITES. *Gen. pl.* 299.

Contorta. *Folliculi* 2, longi, recti. *Sem.* pappofa.
*Cor.* infundibuliformis, fauce nuda.

*fuberecta.*   1. E. pedunculis racemofis, foliis fubovatis obtufis mu-
          cronatis. *Sp. pl.* 307.
          Oval-leav'd Echites, or Savanna-flower.
          *Nat.* of Jamaica.
          *Cult.* 1759, by Mr. Philip Miller. *Mill. dict. edit.* 7.
          Apocynum 4.
          *Fl.*                                            S. ♄ .

*torulofa.*   2. E. pedunculis fubracemofis, foliis lanceolatis acumi-
          natis. *Sp. pl.* 307.
          Climbing Echites.
          *Nat.* of Jamaica.
          *Introd.* 1778, by William Wright, M.D.
          *Fl.*                                            S. ♄ .

*umbella-*   3. E. pedunculis umbellatis, foliis ovatis obtufis mucro-
*ta.*          natis, caule volubili. *Sp. pl.* 307.
          Umbel'd Echites.
          *Nat.* of Jamaica.
          *Introd.* before 1733, by William Houftoun, M.D.
          *Mill. dict. edit.* 8. Apocynum obliquum.
          *Fl.*                                            S. ♄ .


## PLUMERIA. *Gen. pl.* 298.

Contorta. *Folliculi* 2, reflexi. *Semina* membranæ
propriæ inferto.

*rubra.*   1. P. foliis ovato-oblongis, petiolis biglandulofis. *Sp.*
          *pl.* 306.
          Red Plumeria.
                                                          *Nat.*

*Nat.* of Jamaica.
*Cult.* 1690, in the Royal Garden at Hampton-court.
  *Catal. mſcr.*
*Fl.* July and Auguſt.                    S. ♄.

2. P. foliis lanceolatis revolutis, pedunculis superne tu-  *alba.*
   beroſis. *Sp. pl.* 306.
White Plumeria.
*Nat.* of Jamaica.
*Introd.* before 1733, by William Houſtoun, M.D.
  *Mill. dict. edit.* 8.
*Fl.* July and Auguſt.                    S. ♄.

3. P. foliis lanceolatis petiolatis obtuſis. *Sp. pl.* 307.  *obtuſa.*
Blunt-leav'd Plumeria.
*Nat.* of the Weſt Indies.
*Introd.* 1783, by John Greg, Eſq.
*Fl.* July.                               S. ♄.


TABERNÆMONTANA. *Gen. pl.* 301.

Contorta.  *Folliculi* 2, horizontales.  *Sem.* pulpæ
             immerſa.

1. T. foliis oppoſitis ovatis, floribus lateralibus glomerato-  *citrifolia.*
   umbellatis. *Sp. pl.* 308.
Citron-leav'd Tabernæmontana.
*Nat.* of Jamaica.
*Cult.* 1739, by Mr. Philip Miller. *Rand. chel.*
*Fl.*                                     S. ♄.

2. T. foliis oppoſitis ovalibus obtuſiuſculis. *Sp. pl.* 308.  *laurifolia.*
Laurel-leav'd Tabernæmontana.
*Nat.* of the Weſt Indies.
*Cult.* 1768, by Mr. Philip Miller. *Mill. dict. edit.* 8.
*Fl.*                                     S. ♄.

                                          3. T.

*Amſonia.* 3. T. foliis alternis ovato-lanceolatis, caulibus herbaceis
  glaberrimis.
  Tabernæmontana Amſonia. *Sp. pl.* 308.
  Alternate-leav'd Tabernæmontana.
  *Nat.* of North America.
  *Cult.* 1759, by Mr. Philip Miller.
  *Fl.* May and June. H. ♃.

*anguſtifo-* 4. T. foliis linearibus ſparſis, caule piloſo herbaceo.
*lia.*   Narrow-leav'd Tabernæmontana.
  *Nat.* of North America.
  *Introd.* 1774, by Mr. James Gordon.
  *Fl.* May and June. H. ♃.


### C E R O P E G I A. *Gen. pl.* 302.

Contorta. *Folliculi* 2, erecti. *Semina* plumoſa. *Corollæ* limbus connivens.

*ſagittata.* 1. C. umbellis ſubſeſſilibus, foliis ſagittatis. *Linn. mant.*
  215.
  Arrow-leav'd Ceropegia.
  *Nat.* of the Cape of Good Hope.
  *Introd.* 1775, by Mr. Francis Maſſon.
  *Fl.* G. H. ♄.


# D I G Y N I A.

### P E R I P L O C A. *Gen. pl.* 303.

Contorta. *Nectarium* ambiens genitalia, filamenta
  5 exſerens.

*græca.* 1. P. floribus interne hirſutis terminalibus. *Syſt. veget.*
  256.
  Common

Common Virginian Silk, or Periploca.
*Nat.* of Syria.
*Cult.* 1597, by Mr. John Gerard. *Ger. herb.* 754.
*Fl.* July and August. H. ♄.

2. P. floribus interne hirsutis paniculatis, foliis lanceolato-   *Secamone.*
    ellipticis. *Syst. veget.* 256.
Green Periploca.
*Nat.*
*Cult.* 1775, by John Fothergill, M.D.
*Fl.* July. G. H. ♄.

3. P. corollis glabris, cymis trichotomis, foliis oblongo-   *lævigata.*
    lanceolatis lævibus, caule glabro.
Smooth Periploca.
*Nat.* of the Canary Islands. Mr. *Francis Masson.*
*Introd.* 1779.
*Fl.* G. H. ♄.

4. P. caule hirsuto. *Sp. pl.* 309.   *africana.*
African Periploca.
*Nat.* of the Cape of Good Hope.
*Cult.* 1726, by Mr. Philip Miller. *R. S. no.* 209.
*Fl.* June——September. G. H. ♄.

## CYNANCHUM. *Gen. pl.* 304.

Contorta. *Nectarium* cylindricum, 5-dentatum.

1. C. caule volubili perenni aphyllo. *Syst. veget.* 257.   *viminale.*
Naked Cynanchum.
*Nat.* of the Cape of Good Hope.
*Cult.* 1759, by Mr. Philip Miller. *Mill. dict. edit.* 7.
    Euphorbia. 15.
*Fl.* D. S. ♄.

2. C.

*acutum.*   2. C. caule volubili herbaceo, foliis cordato-oblongis gla-
bris. *Sp. pl.* 310.
Acute-leav'd Cynanchum.
*Nat.* of Spain and Sicily.
*Cult.* 1596, by Mr. John Gerard. *Hort. Ger.*
*Fl.* July.                                              H. ♃.

*fuberofum.*   3. C. caule volubili inferne fuberofo fiſſo, foliis cordatis
acuminatis. *Sp. pl.* 310.
Cork-bark'd Cynanchum.
*Nat.* of America.
*Cult.* 1732, by James Sherard, M.D. *Dill. elth.* 308.
*t.* 229. *f.* 296.
*Fl.* July——September.                         S. ♄.

*hirtum.*   4. C. caule volubili fruticofo inferne fuberofo fiſſo, foliis
ovato-cordatis. *Sp. pl.* 310.
Hairy Cynanchum.
*Nat.* of America.
*Introd.* before 1733, by William Houſtoun, M.D.
*Mill. dict. edit.* 8.
*Fl.*                                                    S. ♄.

*crifpiflo-rum.*   5. C. caule volubili, foliis fubtus villoſis oblongis corda-
tis : ſinu clauſo, petalis apice crifpis.
Periploca florum divifuris circinatis et crifpis. *Plum.
ic.* 210. *tab.* 216. *fig.* 1.
Curl'd-flower'd Cynanchum.
*Nat.* of South America.
*Cult.* before 1741, by Robert James Lord Petre.
*Fl.* July.                                              S. ♄.

*monfpe-liacum.*   6. C. caule volubili herbaceo, foliis reniformi-cordatis
acutis. *Sp. pl.* 311.
Montpelier Cynanchum.

                                                         *Nat.*

*Nat.* of the South of Europe.

*Cult.* 1597, by Mr. John Gerard.   *Ger. herb.* 718.

*Fl.* Auguft and September.                               H. 4.

7. C. caule volubili frutefcente, foliis cordatis acutis, pe-    *extenfum.*
   dunculis elongatis : pedicellis filiformibus, corollis
   margine hirfutis, folliculis ramentaceis.

Cynanchum extenfum.   *Jacqu. ic. mifcell.* 2. *p.* 353.

Cynanchum cordifolium. *Retzii obf. bot.* 2. *p.* 15. *n.* 37.

Hairy-flower'd Cynanchum.

*Nat.* of the Eaft Indies.

*Introd.* 1777, by Patrick Ruffell, M. D.

*Fl.* July and Auguft.                                    S. ♄.

8. C. caule erecto divaricato, foliis cordatis glabris. *Sp.*    *erectum.*
   *pl.* 311.  *Jacqu. hort.* 1. *p.* 14. *t.* 38.

Upright Cynanchum.

*Nat.* of Syria.

*Cult.* 1640.   *Park. theat.* 385. *f.* 1.

*Fl.* July and Auguft.                                    S. ♄.

## A P O C Y N U M.   *Gen. pl.* 305.

*Cor.* campanulata. *Filamenta* 5, cum ftaminibus alterna.

1. A. caule rectiufculo herbaceo, foliis ovatis utrinque    *androfæ-*
   glabris, cymis terminalibus.   *Sp. pl.* 311.             *mifolium.*

Tutfan-leav'd Dog's-bane.

*Nat.* of North America.

*Cult.* 1731, by Mr. Ph. Miller. *Mill. dict. edit.* 1. *n.* 4.

*Fl.* July——September.                                    H. 4.

2. A. caule rectiufculo herbaceo, foliis oblongis, cymis    *cannabi-*
   lateralibus folio longioribus.                            *num.*

Apocynum cannabinum.   *Sp. pl.* 311.

                                                    Hemp

Hemp Dog's-bane.

*Nat.* of North America.

*Cult.* 1699, by the Dutchefs of Beaufort. *Br. Muf.*
    *Sloan. mff.* 525 and 3329.

*Fl.* July and September.        H. ♃.

*hyperici-*
*folium.*    3. A. caule rectiufculo herbaceo, foliis oblongis cordatis
       glabris, cymis folio brevioribus.

       Apoeynum fibiricum.  *Jacqu. hort.* 3. *p.* 37. *t.* 66.
          *Syft. veget.* 258.

       St. John's-wort-leav'd Dog's-bane.

       *Nat.* of North America.

       *Cult.* 1758, by Mr. Philip Miller.

       *Fl.* June and July.        H. ♃.

*venetum.*    4. A. caule rectiufculo herbaceo, foliis ovato-lanceolatis.
       *Sp. pl.* 311.

       Spear-leav'd Dog's-bane.

       *Nat.* of the Iflands in the Adriatic Sea.

       *Cult.* 1690, in the Royal Garden at Hampton-court.
          *Catal. mfcr.*

       *Fl.* July and Auguft.        H. ♃.

## ASCLEPIAS. *Gen. pl.* 306.

Contorta.  *Nectaria* 5, ovata, concava, corniculum
exferentia.

\* *Foliis oppofitis planis.*

*undulata.*    1. A. foliis feffilibus oblongis lanceolatis undulatis gla-
       bris, petalis ciliatis.  *Syft. veget.* 258.

       Waved-leav'd Swallow-wort.

       *Nat.* of the Cape of Good Hope.

       *Introd.* 1783, by Mr. John Græfer.

       *Fl.* July.        G. H. ♄.

                                          2. A.

2. A. foliis cordato-lanceolatis undulatis fcabris oppofitis, *crifpa.*
umbella terminali. *Linn. fuppl.* 170.
Curl'd-leav'd Swallow-wort.
*Nat.* of the Cape of Good Hope.
*Introd.* 1774, by Mr. Francis Maffon.
*Fl.* G. H. ♄.

3. A. foliis obovato-oblongis, petiolis breviffimis, corol- *procera.*
lis fubcampanulatis.
Afclepias gigantea. *Jacqu. obf.* 3. *p.* 17. *t.* 69. *Houtt.*
*nat. hift.* 7. *p.* 749. *t.* 44.
Zja-raek. *Le Brun it. perf. p.* 315. *t.* 184.
Bell-flower'd gigantic Swallow-wort.
*Nat.* of Perfia.
*Cult.* 1714, by the Dutchefs of Beaufort. *Br. Muf.*
*H. S.* 135. *fol.* 18.
*Fl.* July——September. S. ♄.

4. A. foliis obovato-oblongis, petiolis breviffimis, laci- *gigantea.*
niis corollæ reflexis involutis.
Afclepias gigantea. *Sp. pl.* 312. (exclufis fynonymis
Plukenetii et Alpini.)
Curl'd-flower'd gigantic Swallow-wort.
*Nat.* of the Eaft Indies.
*Cult.* 1690, in the Royal Garden at Hampton-court.
*Catal. mfcr.*
*Fl.* July——September. S. ♄.

5. A. foliis ovalibus fubtus tomentofis, caule fimpliciffi- *fyriaca.*
mo, umbellis nutantibus. *Sp. pl.* 313.
Syrian Swallow-wort.
*Nat.* of Virginia.
*Cult.* 1629, by Mr. John Parkinfon. *Park. parad.*
443. *f.* 2.
*Fl.* July and Auguft. H. ♃.

X 6. A.

*amæna.*　6. A. foliis ovatis fubtus pilofiufculis, caule fimplici, umbellis nectariifque erectis. *Sp. pl.* 313.
Oval-leav'd Swallow-wort.
*Nat.* of North America.
*Cult.* 1732, by James Sherard, M.D. *Dill. elth.* 31. *t.* 27. *f.* 30.
*Fl.* July and Auguft.　　　　　　　　H. ♃.

*purpu-*
*rafcens.*　7. A. foliis ovatis fubtus villofis, caule fimplici, umbellis erectis, nectariis refupinatis. *Sp. pl.* 313.
Purple Virginian Swallow-wort.
*Nat.* of North America.
*Cult.* 1732, by James Sherard, M.D. *Dill. elth.* 32. *t.* 28. *f.* 31.
*Fl.* Auguft and September.　　　　　　H. ♃.

*variega-*
*ta.*　8. A. foliis ovatis rugofis nudis, caule fimplici, umbellis fubfeffilibus : pedicellis tomentofis. *Sp. pl.* 312.
Variegated Swallow-wort.
*Nat.* of North America.
*Cult.* 1696. *Pluk. alm.* 34. *t.* 77. *f.* 1.
*Fl.* July.　　　　　　　　　　　H. ♃.

*curaffa-*
*vica.*　9. A. foliis lanceolatis glabris nitidis, caule fimplici, umbellis erectis folitariis lateralibus. *Syft. veget.* 259.
Curaffavian Swallow-wort.
*Nat.* of South America.
*Cult.* 1692, in the Royal Garden at Hampton-court. *Catal. mfcr.*
*Fl.* June——September.　　　　　　S. ♄.

*nivea.*　10. A. foliis ovato-lanceolatis glabriufculis, caule fimplici, umbellis erectis lateralibus folitariis. *Syft. veg.* 259.
Almond-leav'd Swallow-wort.

　　　　　　　　　　　　　　　　*Nat.*

*Nat.* of North America.
*Cult.* 1732, by James Sherard, M. D. *Dill. elth.* 33.
t. 29. *f.* 32.
*Fl.* July——September. H. ♃.

11. A. foliis lanceolatis acuminatis glabris oppofitis bafi
attenuatis, caule fuffruticofo erecto, umbellis late-
ralibus folitariis.
Small-flower'd Swallow-wort.
*Nat.* of Carolina and Eaft Florida.
*Introd.* 1774, by John Fothergill, M. D.
*Fl.* July——October. G. H. ♃.

*parviflo-*
*ra.*

12. A. foliis lanceolatis, caule fuperne divifo, umbellis
erectis geminis. *Sp. pl.* 314. *Jacqu. hort.* 2.
p. 49. t. 107.
Flefh-colour'd Swallow-wort.
*Nat.* of North America.
*Cult.* 1731. *Mill. dict. edit.* 1. Apocynum 2.
*Fl.* July and Auguft. H. ♃.

*incarna-*
*ta.*

13. A. foliis ovatis bafi barbatis, caule erecto, umbellis
proliferis. *Sp. pl.* 314.
α Afelepias albo flore. *Bauh. pin.* 303.
White officinal Swallow-wort.
β Afelepias foliis ovatis acutis, caule infirmo, umbellis
fimplicibus. *Mill. dict.*
Yellow officinal Swallow-wort.
*Nat.* of Europe.
*Cult.* 1640. *Park. theat.* 388. *f.* 1.
*Fl.* May——Auguft. H. ♃.

*Vincetox-*
*icum.*

lutea.

14. A. foliis ovatis bafi barbatis, caule fuperne fubvo-
lubili. *Sp. pl.* 315.
Black Swallow-wort.

*nigra.*

X 2 *Nat.*

*Nat.* of France and Spain.

*Cult.* 1596, by Mr. John Gerard. *Hort. Ger.*

*Fl.* June——Auguſt.                    H. ♃.

\*\* *Foliis lateribus revolutis.*

arboreſ-
cens.    15. A. foliis revolutis ovatis, caule fruticoſo ſubvilloſo.
            *Linn. mant.* 216.
         Tree Swallow-wort.
         *Nat.* of the Cape of Good Hope.
         *Cult.* 1714, by the Dutcheſs of Beaufort. *Br. Muſ.*
            H. S. 132. *fol.* 22.
         *Fl.* December.                    G. H. ♄.

fruticoſa.  16. A. foliis revolutis lineari-lanceolatis, caule fruticoſo.
            *Syſt. veget.* 260.
         Willow-leav'd Swallow-wort.
         *Nat.* of the Cape of Good Hope.
         *Cult.* 1714, by the Dutcheſs of Beaufort. *Br. Muſ.*
            H. S. 133. *fol.* 67.
         *Fl.* June——September.                G. H. ♄.

ſibirica.  17. A. foliis revolutis lineari-lanceolatis oppoſitis ter-
            niſque, caule decumbente. *Sp. pl.* 315.
         Siberian Swallow-wort.
         *Nat.* of Siberia.
         *Cult.* 1775, by Mr. James Gordon.
         *Fl.* July and Auguſt.                    H. ♃.

verticil-
lata.    18. A. foliis revolutis linearibus verticillatis, caule erecto.
            *Sp. pl.* 315.
         Whorl'd-leav'd Swallow-wort.
         *Nat.* of North America.
         *Cult.* 1759, by Mr. Philip Miller. *Mill. dict. edit.* 7.
            n. 4.
         *Fl.*                                    H. ♃.
                                        \*\*\* *Foliis*

☉ ☉ ☉ *Foliis alternis.*

19. A. foliis alternis lanceolatis, caule divaricato pilofo. *tuberofa.*
Sp. pl. 316.
Tuberous-rooted Swallow-wort, or Orange Apocy-
num.
*Nat.* of North America.
*Cult.* 1690, in the Royal Garden at Hampton-court.
*Catal. mfcr.*
*Fl.* July——September.                H. ♃.

M E L O D I N U S.   *Linn. fuppl. 23.*
*Bacca* bilocularis, polyfperma.   *Faux corollæ* coronata.

1. MELODINUS.   *Forft. gen. 19.   Linn. fuppl. 167.* *fcandens.*
*Forft. fl. auftr. 20.*
Climbing Melodinus.
*Nat.* of New Caledonia.
*Introd.* 1775, by John Reinhold Forfter, LL.D.
*Fl.*                        S. ♄.

S T A P E L I A.   *Gen. pl. 307.*
Contorta. *Nectarium* duplici ftellula tegente genitalia.

1. S. denticulis ramorum patentibus, floribus peduncu-   *variega-*
latis, corollis glabris fupra rugulofis: laciniis ovatis   *ta.*
acuminatis planis.
Stapelia variegata.   *Sp. pl. 316.   Jacqu. mifcell. 1.*
*p. 27. t. 4.   Curtis magaz. 26.*
Variegated Stapelia.
*Nat.* of the Cape of Good Hope.
*Introd.* 1690, by Mr, Bentick.   *Br. Muf. Sloan. mff.*
3370.
*Fl.* June and July.                D. S. ♄.

2. S. denticulis ramorum erectis, floribus pedunculatis,   *hirfuta.*
corollis fupra rugulofis centro hirfutis margine ciliatis: laciniis ovato-oblongis acutis planis.
                X 3                Stapelia

Stapelia hirſuta. *Sp. pl.* 316. *Jacqu. miſcell.* I. *p.* 28. *t.* 3.

Hairy Stapelia.

*Nat.* of the Cape of Good Hope.

*Introd.* 1714, by Profeſſor Richard Bradley. *Bradl. ſucc.* 3. *p.* 5. *t.* 23.

*Fl.* June and July.    D. S. ♄.

*pulla.* 3. S. ſubhexagona erecta, aculeis patentiſſimis, floribus ſeſſilibus aggregatis: corollarum laciniis lanceolatis ſupra holoſericeis replicatis.

Black-flower'd Stapelia.

*Nat.* of the Cape of Good Hope. Mr. *Francis Maſſon.*

*Introd.* 1774.

*Fl.* Auguſt and September.    D. S. ♄.

*articula-ta.* 4. S. articulis ramorum oblongis teretibus reticulatim obſolete verrucoſis: ſpinulis minutis, floribus ſubſeſſilibus, corollis ſupra papilloſis: laciniis triangularibus.

Jointed Stapelia.

*Nat.* of the Cape of Good Hope. Mr. *Francis Maſſon.*

*Introd.* 1774.

*Fl.* Auguſt and September.    D. S. ♄.

*mammil-laris.* 5. S. denticulis ramorum obtuſis mucronatis. *Linn. mant.* 216.

Prickly Stapelia.

*Nat.* of the Cape of Good Hope.

*Introd.* 1774, by Mr. *Francis Maſſon.*

*Fl.* June and July.    D. S. ♄.

## HERNIARIA. *Gen. pl.* 308.

*Cal.* 5-partitus. *Cor.* o. *Stam.* 5 ſterilia. *Capſ.* 1-ſperma.

*glabra.* 1. H. glabra herbacea. *Sp. pl.* 317.

Smooth

*Fl.* July.                                          H. ⊙.


2. H. hirfuta herbacea.  *Sp. pl.* 317.                      *hirfuta.*
Hairy Rupture-wort.
*Nat.* of England.
*Fl.* July and Auguft.                               H. ⊙.


## CHENOPODIUM. *Gen. pl.* 309.

*Cal.* 5-phyllus, 5-gonus.  *Cor.* 0.  *Sem.* 1, lenticulare,
fuperum.

\* *Foliis angulofis.*

1. C. foliis triangulari-fagittatis integerrimis, fpicis com-      *Bonus*
pofitis aphyllis axillaribus. *Syft. veg.* 261. *Curtis lond.*   *Henri-*
Angular-leav'd Goofefoot, or Englifh Mercury.                 *cus.*
*Nat.* of Britain.
*Fl.* May——Auguft.                                  H. ♃.


2. C. foliis triangularibus fubdentatis, racemis confertis      *urbicum.*
ftrictiffimis cauli approximatis longiffimis.  *Sp. pl.*
318.
Upright Goofefoot.
*Nat.* of Britain.
*Fl.* Auguft.                                        H. ⊙.


3. C. foliis rhombeo-ovatis lanceolatifque: inferioribus       *Atripli-*
finuato-dentatis, paniculis axillaribus ramofis, caule       *cis.*
erecto.
Chenopodium Atriplicis.  *Linn. fuppl.* 171.
Chenopodium purpurafcens.  *Jacqu. hort.* 3. *p.* 43.
*t.* 80.
Purple Goofefoot.
*Nat.* of China.

X 4                          *Introd.*

*Introd.* 1780, by Monſ. Thouin.

*Fl.* Auguſt and September.                                    S. ⊙.

rubrum.     4. C. foliis cordato-triangularibus obtuſiuſculis dentatis,
              racemis erectis compoſitis ſubfolioſis caule brevio-
              ribus. *Syſt. veget.* 261.

            Red Gooſefoot.

            *Nat.* of Britain.

            *Fl.* Auguſt and September.                       H. ⊙.

murale.     5. C. foliis ovatis nitidis acutis dentatis, racemis ramoſis
              nudis. *Sp. pl.* 318.

            Wall Gooſefoot, or Sowbane.

            *Nat.* of Britain.

            *Fl.* Auguſt.                                      H. ⊙.

ſerotinum.  6. C. foliis deltoideis ſinuato-dentatis rugoſis glabris
              uniformibus, racemis terminalibus. *Sp. pl.* 319.

            Fig-leav'd Gooſefoot.

            *Nat.* of England.

            *Fl.* Auguſt and September.                        H. ⊙.

album.      7. C. foliis rhomboideo-triangularibus eroſis poſtice in-
              tegris: ſummis oblongis, racemis erectis. *Sp. pl.*
              319. *Curtis lond.*

            Common Gooſefoot.

            *Nat.* of Britain.

            *Fl.* July.                                        H. ⊙.

viride.     8. C. foliis rhomboideis dentato-ſinuatis, racemis ramoſis
              ſubfoliatis. *Sp. pl.* 319. *Curtis lond.*

            Green Gooſefoot.

            *Nat.* of Britain.

            *Fl.* Auguſt.                                      H. ⊙.

hybridum.   9. C. foliis cordatis angulato-acuminatis, racemis ramo-
              ſis nudis. *Sp. pl.* 319. *Curtis lond.*

                                                              Baſtard

Baſtard Gooſefoot.
*Nat.* of Britain.
*Fl.* Auguſt——October. H. ⊙.

10. C. foliis oblongis finuatis, racemis nudis multifidis. *Botrys.*
Sp. pl. 320.
Cut-leav'd Gooſefoot.
*Nat.* of the South of Europe.
*Cult.* 1551. *Turn. herb. ſign. G 1.*
*Fl.* June——September. H. ⊙.

11. C. foliis lanceolatis dentatis, racemis foliatis ſimpli-   *ambroſi-*
cibus. *Sp. pl.* 320.                                          *oides.*
Mexican Gooſefoot.
*Nat.* of Mexico.
*Cult.* 1640. *Park. theat.* 89. *n.* 2.
*Fl.* June——October. H. ⊙.

12. C. foliis ovato-oblongis dentatis, racemis aphyllis.   *anthel-*
*Sp. pl.* 320.                                             *minticum.*
Shrubby Gooſefoot.
*Nat.* of America.
*Cult.* 1732, by James Sherard, M.D. *Dill. eltb.*
77. *t.* 66. *f.* 76.
*Fl.* July and Auguſt. G. H. ♄.

13. C. foliis ovato-oblongis repandis, racemis nudis ſim-   *glaucum.*
plicibus glomeratis. *Sp. pl.* 320.
Oak-leav'd Gooſefoot.
*Nat.* of England.
*Fl.* July and Auguſt. H. ⊙.

** *Foliis integris.*

14. C. foliis caulinis lanceolatis obtuſis; ramorum ob-   *laterale.*
longis, peduncu is lateralibus ſolitariis unifloris.
Branching oblong-leav'd Gooſefoot.
*Nat.*

*Nat.*
*Introd.* 1781, by P. M. A. Brouſſonet, M.D.
*Fl.* Auguſt and September.　　　　　　　S. ⊙.

*Vulvaria.*　15. C. foliis integerrimis rhombeo-ovatis, floribus con-
glomeratis axillaribus. *Sp. pl.* 321.
Chenopodium olidum. *Curtis lond.*
Stinking Gooſefoot.
*Nat.* of Britain.
*Fl.* Auguſt.　　　　　　　　　　　　　H. ⊙.

*polyſper-*　16. C. foliis integerrimis ovatis, caule decumbente, cy-
*mum.*　　　mis dichotomis aphyllis axillaribus. *Sp. pl.* 321.
*Curtis lond.*
Round-leav'd Gooſefoot, or Alſeed.
*Nat.* of England.
*Fl.* July and Auguſt.　　　　　　　　H. ⊙.

*Scoparia.*　17. C. foliis lineari-lanceolatis planis integerrimis. *Sp.*
*pl.* 321.
Linear-leav'd Gooſefoot, or Summer Cypreſs.
*Nat.* of Greece.
*Cult.* 1633. *Ger. emac.* 554. *f.* 11.
*Fl.* June——September.　　　　　　　H. ⊙.

*mariti-*　18. C. foliis ſubulatis ſemicylindricis. *Sp. pl.* 321.
*mum.*　　　Sea Gooſefoot.
*Nat.* of Britain.
*Fl.* Auguſt.　　　　　　　　　　　　　H. ⊙.

*ariſtatum.*　19. C. foliis lanceolatis ſubcarnoſis integerrimis, corym-
bis dichotomis ariſtatis axillaribus. *Sp. pl.* 321.
Bearded Gooſefoot.
*Nat.* of Virginia and Siberia.
*Introd.* 1771, by Monſ. Richard.
*Fl.* Auguſt.　　　　　　　　　　　　H. ⊙.

BETA.

B E T A. *Gen. pl.* 310.

*Cal.* 5-phyllus. *Cor.* o. *Sem.* reniforme, intra fub-
ftantiam bafeos calycis.

1. B. floribus geminis. *Syft. veget.* 262.                    *mariti-*
Sea Beet.                                                       *ma.*
*Nat.* of Britain.
*Fl.* Auguft.                                        H. ♂ .

2. B. floribus ternis. *Syft. veget.* 262.              *Cicla.*
White Beet.
*Nat.* of Portugal.
*Cult.* 1656, by Mr. John Tradefcant, Jun. *Trad.*
*muf.* 90.
*Fl.* Auguft.                                        H. ♂ .

3. B. floribus congeftis, foliis inferioribus ovatis.   *vulgaris.*
Beta vulgaris. *Sp. pl.* 322.
α Beta rubra vulgaris. *Bauh. pin.* 118.
Red Beet.
β Beta pallide virens major. *Bauh. pin.* 118.
Green Beet.
*Nat.*
*Cult.* 1596, by Mr. John Gerard. *Hort. Ger.*
*Fl.* Auguft.                                        H. ♂ .

4. B. floribus congeftis, foliis omnibus lineari-lanceola-   *patula.*
tis, ramis divaricatis.
Spreading Beet.
*Nat.* of Madeira.  Mr. *Francis Maffon.*
*Introd.* 1778.
*Fl.* Auguft.                                    G. H. ♂ .
OBS. *Caulis* brevis, vix pedalis, ramofiffimus : *rami*
longi divaricati. *Calycis* foliola bafi dilatata, non
vero dentata.

S A L S O-

## SALSOLA. *Gen. pl.* 311.

*Cal.* 5-phyllus.   *Cor.* o.   *Capf.* 1-fperma.   *Sem.* coch-
leatum.

*Kali.*     1. S. herbacea decumbens, foliis fubulatis fpinofis fcabris,
calycibus marginatis axillaribus. *Syft. veg.* 263.
Prickly Salt-wort.
*Nat.* of Britain.
*Fl.* July and Auguft.       H. ⊙.

*Soda.*     2. S. herbacea patula, foliis inermibus. *Sp. pl.* 323.
*Jacqu. hort.* 1. *p.* 28. *t.* 68.
Long flefhy-leav'd Salt-wort.
*Nat.* of the South of Europe.
*Cult.* 1759, by Mr. Ph. Miller. *Mill. dict. edit.* 7. *n.* 3.
*Fl.* July and Auguft.       H. ⊙.

*fativa.*     3. S. herbacea diffufa, foliis teretibus glabris, floribus
conglomeratis. *Sp. pl.* 323.
Cultivated Salt-wort.
*Nat.* of Spain.
*Introd.* 1783, by Abbé Pourret.
*Fl.* Auguft.       H. ♃.

*altiffima.*     4. S. herbacea erecta ramofiffima, foliis filiformibus acu-
tiufculis bafi pedunculiferis. *Syft. veget.* 263.
Grafs-leav'd Salt-wort.
*Nat.* of Italy.
*Introd.* 1775, by John Earl of Bute.
*Fl.* July and Auguft.       H. ⊙.

*falfa.*     5. S. herbacea erectiufcula, foliis linearibus fubcarnofis
muticis, calycibus fucculentis diaphanis. *Syft. veget.*
263. *Jacqu. hort.* 3. *p.* 44. *t.* 83.

Strip'd-

Strip'd-ſtalk'd Salt-wort.
*Nat.* of Aſtracan.
*Introd.* 1782, by P. M. A. Brouſſonet, M.D.
*Fl.* Auguſt and September.　　　　　　H. ⊙.

6. S. frutefcens, ramis diffuſis, foliis lanceolatis fericeis, *ſericea.*
　　calycibus muticis.
　Chenolea diffuſa. *Thunberg nov. gen.* 1. *p.* 10. *Syſt.*
　　*veget.* 247.
　Silky Salt-wort.
　*Nat.* of the Cape of Good Hope.
　*Cult.* 1758, by Mr. Philip Miller.
　*Fl.* Auguſt and September.　　　　　G. H. ♄.

7. S. frutefcens proſtrata, foliis linearibus piloſis inermi- *proſtrata.*
　　bus. *Sp. pl.* 323. *Jacqu. auſtr.* 3. *p.* 52. *t.* 294.
　Trailing Salt-wort.
　*Nat.* of the South of Europe.
　*Introd.* 1780, by Peter Simon Pallas, M.D.
　*Fl.*　　　　　　　　　　　　　　　G. H. ♄.

8. S. fruticoſa erecta, foliis filiformibus obtuſiuſculis. *fruticoſa.*
　　*Sp. pl.* 324.
　Shrubby Salt-wort, or Stone-crop.
　*Nat.* of England.
　*Fl.* Auguſt.　　　　　　　　　　　H. ♄.

9. S. fruticoſa patula, ramulis hirſutis, calycibus ſpino- *muricata.*
　　ſis. *Syſt. veget.* 263.
　Baſſia. *Allioni in miſcell. taurin.* 3. *p.* 177. *t.* 4. *f.* 2.
　Hairy Salt-wort.
　*Nat.* of Egypt.
　*Introd.* 1773, by John Earl of Bute.
　*Fl.* July and Auguſt.　　　　　　　H. ⊙.

GOMPHRE-

## G O M P H R E N A. *Gen. pl.* 314.

*Cal.* coloratus : exterior 3-phyllus : foliolis 2 con-
niveantibus carinatis. *Petala* 5 rudia villosa. *Nec-*
*tarium* cylindricum, 5-dentatum. *Capf.* 1-fperma.
*Stylus* femibifidus.

*globosa.* 1. G. caule ereĉto, foliis ovato-lanceolatis, capitulis fo-
litariis, pedunculis diphyllis. *Sp. pl.* 326.
Annual Globe Amaranth.
*Nat.* of India.
*Cult.* 1714, by the Dutchefs of Beaufort. *Br. Muf.*
H. *S.* 133. *fol.* 12.
*Fl.* May——Oĉtober. S. ☉.

*perennis.* 2. G. foliis lanceolatis, capitulis diphyllis, flofculis pe-
rianthio proprio diftinĉtis. *Sp. pl.* 326.
Perennial Globe Amaranth.
*Nat.* of South America.
*Cult.* 1732, by James Sherard, M.D. *Dill. elth.* 24.
*t.* 20. *f.* 22.
*Fl.* July——Oĉtober. S. ♃.

## B O S E A. *Gen. pl.* 315.

*Cal.* 5-phyllus. *Cor.* o. *Bacca* 1-fperma.

*Yervamo-* 1. BOSEA. *Sp. pl.* 326.
*ra.* Golden-rod Tree.
*Nat.* of the Canary Iflands.
*Cult.* before 1728, by Mr. Philip Miller. *Mill. diĉt.*
*edit.* 8.
*Fl.* G. H. ♄.

ULMUS.

## U L M U S. *Gen. pl.* 313.

*Cal.* 5-fidus. *Cor.* 0. *Bacca* exſucca compreſſo-membranacea.

1. U. fol. duplicato-ſerratis : baſi inæqualibus. *Sp. pl.* 327.     *campeſ-tris.*

α Ulmus vulgatiſſima folio lato ſcabro. *Ger. em.* 1480.    *vulgaris.*
*Raj. ſyn.* 468. *Hudſ. angl.* 109. α.
Ulmus campeſtris. *Mill. dict. Du Roi hort. harb.* 495.
Common Elm.

β Ulmus minor folio anguſto ſcabro. *Ger. em.* 1480.    *ſtricta.*
*Raj. ſyn.* 469. *Hudſ. angl.* 109. β.
Ulmus ſativa. *Mill. dict. Du Roi hort. harb.* 502.
Corniſh Elm.

γ Ulmus folio latiſſimo ſcabro. *Ger. em.* 1481. *Raj.*    *latifolia.*
*ſyn.* 469. *Hudſ. angl.* 109. γ.
Ulmus ſcabra. *Mill. dict. Du Roi hort. harb.* ad-ditam.
Witch Elm, or Hazel.

δ Ulmus folio glabro. *Ger. em.* 1481. *Raj. ſyn.* 469.    *glabra.*
*Hudſ. angl.* 109. δ.
Ulmus glabris. *Mill. dict.*
Smooth Elm.

ε Ulmus *Hollandica*, foliis ovatis acuminatis rugoſis in-    *fungoſa.*
æqualiter ſerratis, cortice fungoſo. *Mill. dict. Du Roi hort. harb.* 505.
Dutch Elm.
*Nat.* of Britain.
*Fl.* April and May.          H. ♄.

2. U. foliis æqualiter ſerratis : baſi inæqualibus. *Sp. pl.* 327.     *america-na.*

α foliis ovatis rugoſis ſcabris, ramis rubris.       *rubra.*

Ulmus

Ulmus altitudinis et craffitiei minoris, foliis latioribus
rugofis. *Clayton in Gronov. virg.* 39.

American red Elm.

alba. β foliis oblongis fcabris, ramis albicantibus.

Ulmus procerior foliis anguftioribus, trunco per inter-
valla viminibus denfe congeftis infra ramos obfito.
*Clayton in Gron. virg.* 39.

American white Elm.

pendula. γ foliis oblongis glabriufculis, ramis pendulis.

Drooping American Elm.

*Nat.* of North America.

*Cult.* 1752, by Mr. James Gordon.

*Fl.*                                                    H. ♄.

nemora-    3. U. foliis oblongis glabriufculis æqualiter ferratis bafi
lis.            fubæqualibus, floribus feffilibus.

Hornbeam-leav'd Elm.

*Nat.* of North America.

*Cult.* 1760, by Mr. *James Gordon.*

*Fl.*                                                    H. ♄.

pumila. 4. U. foliis æqualiter ferratis : bafi æqualibus. *Sp. pl.*
327.

Dwarf Elm.

*Nat.* of Siberia.

*Introd.* 1771, by Monf. Richard.

*Fl.*                                                    H. ♄.

## HEUCHERA. *Gen. pl.* 320.

*Petala* 5.    *Capf.* 2-roftris, 2-locularis.

america-    1. Heuchera. *Sp. pl.* 328.
na.            American Heuchera, or Sanicle.

*Nat.* of Virginia.

*Cult.* 1704.  *Raj. hift.* 3. *p.* 509.

*Fl.* May——July.                                         H. ♃.

VELEZIA.

## V E L E Z I A.  *Gen. pl.* 447.

*Cor.* 5-petala, parva. *Cal.* filiformis, 5-dentatus. *Capf.*
1-locularis. *Sem.* plurima, ferie fimplici.

1. VELEZIA. *Sp. pl.* 474.  *rigida.*
Rigid Velezia.
*Introd.* 1785, by Cafimir Gomez Ortega, M.D.
*Nat.* of Spain.
*Fl.* July.  H. ☉.

## S W E R T I A.  *Gen. pl.* 321.

*Cor.* rotata. *Nectariferi* pori ad bafin laciniarum corollæ. *Capf.* 1-locularis, 2-valvis.

1. S. corollis quinquefidis, foliis radicalibus ovalibus.  *perennis.*
*Sp. pl.* 328. *Jacqu. auftr.* 3. *p.* 25. *t.* 243.
Marfh Swertia.
*Nat.* of England.
*Fl.* July.  H. ♃.

## G E N T I A N A.  *Gen. pl.* 322.

*Cor.* monopetala. *Capf.* 2-valvis, 1-locularis : *Recep-*
*taculis* 2, longitudinalibus.

  * *Corollis quinquefidis fubcampaniformibus.*
1. G. corollis quinquefidis monogvnis, paniculis tricho-  *vifcofa.*
tomis, bracteis perfoliatis, foliis oblongis trinerviis.
Clammy Gentian.
*Nat.* of the Canary Iflands.  Mr. *Francis Maffon.*
*Introd.* 1781.
*Fl.*  G. H. ♃ :

2. G. corollis fubquinquefidis rotatis verticillatis, caly-  *lutea.*
cibus fpathaceis. *Syft. veget.* 267.
Y  Yellow

Yellow Gentian.
*Nat.* of the Alps in Europe.
*Cult.* 1596, by Mr. John Gerard. *Hort. Ger.*
*Fl.* June and July.                                    H. ♃.

*purpurea.*  3. G. corollis fubquinquefidis campanulatis verticillatis,
calycibus truncatis. *Syft. veget.* 267.
Purple Gentian.
*Nat.* of the Alps in Europe.
*Introd.* 1768, by Profeffor de Sauffure.
*Fl.*                                                   H. ♃.

*punctata.*  4. G. corollis fubquinquefidis campanulatis punctatis,
calycibus quinquedentatis. *Syft. veget.* 267. *Jacqu.*
*auftr.* 5. *p.* 42. *tab. app.* 28.
Spotted-flower'd Gentian.
*Nat.* of the Alps in Europe.
*Introd.* 1775, by the Doctors Pitcairn and Fothergill.
*Fl.* July.                                             H. ♃.

*afclepia-*  5. G. corollis quinquefidis campanulatis oppofitis feffili-
*dea.*          bus, foliis amplexicaulibus. *Sp. pl.* 329. *Jacqu.*
*auftr.* 4. *p.* 15. *t.* 328.
Swallow-wort-leav'd Gentian.
*Nat.* of Auftria and Switzerland.
*Cult.* 1629. *Parkinf. parad. p.* 350. *n.* 2.
*Fl.* July and Auguft.                                  H. ♃.

*Pneumo-*  6. G. corollis quinquefidis campanulatis oppofitis pe-
*nanthe.*      dunculatis, foliis linearibus. *Sp. pl.* 330.
Marfh Gentian, or Calathian Violet.
*Nat.* of England.
*Fl.* Auguft and September.                             H. ♃.

7. G.

7. G. corollis quinquefidis campanulatis ventricosis ver- *Sapona-*
  ticillatis, foliis trinerviis. *Sp. pl.* 330. *ria.*
  Soap-wort-leav'd Gentian.
  *Nat.* of North America.
  *Introd.* 1776, by Mr. William Young.
  *Fl.* Auguſt and September. H. ♃.

8. G. corolla quinquefida campanulata caulem excedente. *acaulis.*
  *Sp. pl.* 330. *Jacqu. auſtr.* 2. *p.* 21. *t.* 135. *Curtis*
  *magaz.* 52.
  Dwarf Gentian, or Gentianella.
  *Nat.* of the Alps in Europe.
  *Cult.* 1629. *Park. parad.* 351. *f.* 3.
  *Fl.* April and May. H. ♃.

  ** *Corollis quinquefidis infundibuliformibus.*
9. G. corolla quinquefida infundibuliformi ſerrata, foliis *bavarica.*
  ovatis obtuſis. *Syſt. veget.* 268.
  Bavarian Gentian.
  *Nat.* of Switzerland and Germany.
  *Introd.* 1775, by the Doctors Pitcairn and Fothergill.
  *Fl.* H. ♃.

10. G. corollis quinquefidis infundibuliformibus, caule *Centauri-*
  dichotomo, piſtillo ſimplici. *Syſt. veget.* 268. *um.*
  Chironia Centaurium. *Curtis lond.*
  Centory Gentian, or Leſſer Centory.
  *Nat.* of Britain.
  *Fl.* June——Auguſt. H. ☉.

11. G. corollis quinquefidis infundibuliformibus, ſtylis *maritima.*
  geminis, caule dichotomo paucifloro. *Linn. mant.*
  55.
  Procumbent Sea Gentian.

Y 2 *Nat.*

*Nat.* of the South of Europe, and the Azores.
*Introd.* 1777, by Mr. Francis Maſſon.
*Fl.* July and Auguſt. G. H. ♃.

*Amarella.* 12. G. corollis quinquefidis hypocrateriformibus fauce
barbatis. *Sp. pl.* 334.
Autumnal Gentian.
*Nat.* of Britain.
*Fl.* Auguſt. H. ☉.

\*\*\* *Corollis quadrifidis.*

*campeſ-* 13. G. corollis quadrifidis fauce barbatis. *Sp. pl.* 334.
*tris.* Field Gentian.
*Nat.* of England.
*Fl.* Auguſt. H. ☉.

*Cruciata.* 14. G. corollis quadrifidis imberbibus, floribus verticil-
latis feſſilibus. *Sp. pl.* 334. *Jacqu. auſtr.* 4.
*p.* 37. *t.* 372.
Croſs-wort Gentian.
*Nat.* of Auſtria and Switzerland.
*Cult.* 1596, by Mr. John Gerard. *Hort. Ger.*
*Fl.* June and July. H. ♃.

P H Y L L I S. *Gen. pl.* 323.
*Stigmata* hiſpida. *Fructif.* ſparſæ.

*Nobla.* 1. P. ſtipulis dentatis. *Sp. pl.* 335.
Baſtard Hare's-ear.
*Nat.* of the Canary Iſlands.
*Cult.* 1699, by the Dutcheſs of Beaufort. *Br. Muſ.*
*Sloan. mſſ.* 3343.
*Fl.* June and July. G. H. ♄.

F A L K I A.

FALKIA. *Linn. suppl.* 30.

*Cal.* campanulatus, 5-fidus. *Cor.* campanulata. *Stigmata* orbicularia, peltata. *Sem.* 4, arillata.

1. FALKIA. *Thunb. nov. gen.* 1. *p.* 17. *Linn. suppl.* 211.    *repens.*
Creeping Falkia.
*Nat.* of the Cape of Good Hope. Mr. *Francis Maſſon.*
*Introd.* 1774.
*Fl.* May——Auguſt.              G. H. ♃.

ERYNGIUM. *Gen. pl.* 324.

*Flores* capitati. *Receptaculum* paleaceum.

1. E. foliis radicalibus lanceolatis ſerratis; floralibus mul-   *fœtidum.*
tifidis, caule dichotomo. *Syſt. veget.* 270.
Stinking Eryngo.
*Nat.* of the Weſt Indies.
*Cult.* 1714, by the Dutcheſs of Beaufort. *Br. Muſ.*
H. S. 139. *fol.* 3.
*Fl.* Auguſt——Octuober.            S. ♃.

2. E. foliis gladiatis ſerrato-ſpinoſis: floralibus indiviſis,   *aquati-*
caule ſimplici. *Sp. pl.* 336.                  *cum.*
Marſh Eryngo.
*Nat.* of North America.
*Introd.* before 1699, by the Rev. John Baniſter. *Moriſ.*
*hiſt.* 3. *p.* 167. *n.* 21.
*Fl.* July——September.             H. ♃.

3. E. foliis radicalibus ovalibus planis crenatis, capitulis   *planum.*
pedunculatis. *Sp. pl.* 336. *Jacqu. auſtr.* 4. *p.* 48.
*t.* 391.
Flat-leav'd Eryngo.

                *Nat.*

*Nat.* of Europe.
*Cult.* 1596, by Mr. John Gerard.   *Hort. Ger.*
*Fl.* July——September.                    H. ♃

*pusillum.*   4. E. foliis radicalibus oblongis incisis, caule dichotomo,
            capitulis sessilibus.   *Sp. pl.* 337.
            Dwarf Eryngo.
            *Nat.* of Spain and the Levant.
            *Cult.* 1759, by Mr. Ph. Miller.  *Mill. dict. edit.* 7. *n.* 8.
            *Fl.* June——August.                    H. ♃.

*mariti-*   5. E. foliis radicalibus subrotundis plicatis spinosis, ca-
*mum.*        pitulis pedunculatis, paleis tricuspidatis. *Syst. veget.*
            271.
            Sea Eryngo, or Holly.
            *Nat.* of Britain.
            *Fl.* July——October.                    H. ♃.

*campef-*   6. E. foliis radicalibus amplexicaulibus pinnato-lanceo-
*tre.*        latis. *Syst. veget.* 271. *Jacqu. austr.* 2. *p.* 35.
            *t.* 155.
            Common Eryngo.
            *Nat.* of England.
            *Fl.* July and August.                    H. ♃.

*amethyf-*  7. E. foliis radicalibus trifidis basi subpinnatis.  *Syst.*
*tinum.*      *veget.* 271.
            Amethystian Eryngo.
            *Nat.* of Styria.
            *Cult.* 1664.   *Evelyn's kalend. hort.* 73.
            *Fl.* July and August.                    H. ♃.

*Bourgati.* 8. E. foliis omnibus digitatis laciniatis suborbiculatis,
            capitulis subrotundis, paleis subulatis integris.
            Eryngium Bourgati.  *Gouan illustr. p.* 7. *t.* 3.
                                                Eryngium

Eryngium pallefcente. *Mill. dict.*
Cut-leav'd Eryngo.
*Nat.* of the South of France.
*Cult.* 1731, by Mr. Ph. Miller. *Mill. dict. edit.* 1. *n.* 6.
*Fl.* June——Auguft. H. ♃.

9. E. foliis radicalibus cordatis indivifis; caulinis digi- *alpinum.*
    tato-laciniatis, capitulis fubcylindricis, involucro
    pinnatifido frondofo, paleis trifidis.
Eryngium alpinum. *Sp. pl.* 337. *Jacqu. ic.*
Alpine Eryngo.
*Nat.* of the Alps of Switzerland and Italy.
*Cult.* 1752, by Mr. Philip Miller. *Mill. dict. edit.* 6.
    *n.* 11.
*Fl.* July and Auguft. H. ♃.


HYDROCOTYLE. *Gen. pl.* 325.

*Umbella* fimplex : *Involucro* 4-phyllo. *Petala* integra.
      *Sem.* femiorbiculato-compreffa.

1. H. foliis peltatis, umbellis quinquefloris. *Sp. pl.* 338. *vulgaris.*
    Common Marfh Penny-wort.
    *Nat.* of Britain.
    *Fl.* May and June. H. ♃.

2. H. foliis reniformibus dentato-crenatis. *Syft. veget.* *afiatica.*
    272.
    African Penny-wort.
    *Nat.* of the Eaft Indies, and of the Cape of Good Hope.
    *Introd.* 1774, by Mr. Francis Maffon.
    *Fl.* July and Auguft. G. H. ♃.

## SANICULA. *Gen. pl.* 326.

*Umbellæ* confertæ, fubcapitatæ. *Fructus* fcaber. *Flores* difci abortientes.

*europæa.*    1. S. foliis radicalibus fimplicibus, flofculis omnibus feffilibus. *Sp. pl.* 339.
Common Sanicle.
*Nat.* of Britain.
*Fl.* June and July.            H. ♃.

*marilan-*    2. S. flofculis mafculis pedunculatis; hermaphroditis fef-
*dica.*           filibus. *Sp. pl.* 339.
Maryland Sanicle.
*Nat.* of Virginia and Maryland.
*Introd.* 1765, by Mr. John Cree.
*Fl.* June and July.            H. ♃.

## ASTRANTIA. *Gen. pl.* 327.

*Involucra partialia* lanceolata, patentia, æqualia, longiora, colorata. *Flores* plurimi abortientes.

*major.*    1. A. foliis quinquelobis: lobis trifidis. *Sp. pl.* 339.
Great Black Mafter-wort.
*Nat.* of the Alps in Europe.
*Cult.* 1596, by Mr. John Gerard. *Hort. Ger.*
*Fl.* May——September.          H. ♃.

*minor.*    2. A. foliis digitatis ferratis. *Sp. pl.* 340.
Small Black Mafter-wort.
*Nat.* of the Alps of Switzerland.
*Cult.* 1759, by Mr. Ph. Miller. *Mill. dict. edit.* 7. *n.* 3.
*Fl.* May and June.            H. ♃.

BUPLEU-

## BUPLEURUM. *Gen. pl.* 328.

*Involucra* umbellulæ majora, 5-phylla. *Pet.* involuta.
*Fructus* fubrotundus, compreffus, ftriatus.

### * *Herbacea.*

1. B. involucris univerfalibus nullis, foliis perfoliatis.    *rotundi-*
   *Sp. pl.* 340.                                       *folium.*
   Round-leav'd Hare's-ear, or Thorow-wax.
   *Nat.* of England.
   *Fl.* June and July.                        H. ☉.

2. B. involucellis coadunatis; univerfali triphyllo. *Sp.*    *ftellatum.*
   *pl.* 340.
   Starry Hare's-ear.
   *Nat.* of the Alps of Switzerland.
   *Introd.* 1775, by the Doctors Pitcairn and Fothergill.
   *Fl.* May——July.                       H. ♃.

3. B. involucellis coadunatis; univerfali pentaphyllo.    *petræum.*
   *Sp. pl.* 340. *Jacqu. ic. collect.* 1. *p.* 209.
   Rock Hare's-ear.
   *Nat.* of the Alps of Switzerland.
   *Introd.* 1768, by Profeffor de Sauffure.
   *Fl.* May——July.                       H. ♃.

4. B. involucellis pentaphyllis orbiculatis; univerfali tri-    *angula-*
   phyllo ovato, foliis amplexicaulibus cordato-lanceo-    *fum.*
   latis. *Sp. pl.* 341.
   Angular-leav'd Hare's-ear.
   *Nat.* of Switzerland.
   *Cult.* 1759, by Mr. Ph. Miller. *Mill. dict. edit.* 7. *n.* 2.
   *Fl.* May——July.                       H. ♃.

5. B.

*longifoli-*      5. B. involucellis pentaphyllis ovatis; univerſali ſubpen-
*um.*                taphyllo, foliis amplexicaulibus.  *Sp. pl.* 341.
             Long-leav'd Hare's-ear.
             *Nat.* of Germany and Switzerland.
             *Cult.* 1713.  *Philoſoph. tranſ. n.* 337. *p.* 190. *n.* 47.
             *Fl.* May——July.                              H. ♃.

*falcatum.*   6. B. involucellis pentaphyllis acutis; univerſali ſubpen-
                 taphyllo, foliis lanceolatis, caule flexuoſo.  *Sp. pl.*
                 341. *Jacqu. auſtr.* 2. *p.* 38. *t.* 158.
             Twiſted-ſtalk'd Hare's-ear.
             *Nat.* of Germany.
             *Cult.* 1739, by Mr. Ph. Miller.  *Mill. dict. vol.* 2. *n.* 2.
             *Fl.* May——September.                          H. ♃.

*odontites.*  7. B. involucellis pentaphyllis acutis ; univerſali tri-
                 phyllo, floſculo centrali altiore, ramis divaricatis.
                 *Sp. pl.* 342. *Jacqu. hort.* 3. *p.* 47. *t.* 91.
             Narrow-leav'd Hare's-ear.
             *Nat.* of Switzerland and Italy.
             *Cult.* 1759, by Mr. Ph. Miller.  *Mill. dict. edit.* 7. *n.* 3.
             *Fl.* June——Auguſt.                           H. ☉.

*ſemicom-*    8. B. umbellis compoſitis ſimulque ſimplicibus.  *Sp. pl.*
*poſitum.*        342.
             Dwarf Hare's-ear.
             *Nat.* of Spain.
             *Introd.* 1778, by Monſ. Thouin.
             *Fl.* July and Auguſt.                          H. ☉.

*tenuiſſi-*   9. B. umbellis ſimplicibus alternis pentaphyllis ſubtri-
*mum.*           floris.  *Sp. pl.* 343.
             Leaſt Hare's-ear.
             *Nat.* of England.
             *Fl.* July and Auguſt.                          H. ☉.
                                                           10. B.

10. B. caule erecto paniculato, foliis linearibus, involu-   *junceum.*
cris triphyllis, involucellis pentaphyllis. *Sp. pl.*
343.
Linear-leav'd Hare's-ear.
*Nat.* of France and Italy.
*Cult.* 1722, in Chelsea Garden. *R. S. n.* 3.
*Fl.* July and August.                          H. ⊙.

11. B. caule ramoso aphyllo, foliis radicalibus decompo-   *nudum.*
sitis planis incisis, involucris involucellisque lan-
ceolato-oblongis.
Naked-stalk'd Hare's-ear.
*Nat.* of the Cape of Good Hope.
*Introd.* 1778, by Patrick Russell, M.D.
*Fl.* October.                                   G. H. ♃.

** *Frutescentia.*

12. B. frutescens, foliis obovatis integerrimis. *Sp. pl.*   *frutico-*
343.                                                         *sum.*
Common shrubby Hare's-ear.
*Nat.* of the South of France, and the Levant.
*Cult.* 1596, by Mr. John Gerard. *Hort. Ger.*
*Fl.* July and August.                           H. ♄.

13. B. frutescens, foliis lanceolatis coriaceis obliquis.    *coriace-*
*L'Herit. stirp. nov. tab.* 67.                              *um.*
Thick-leav'd shrubby Hare's-ear.
*Nat.* of Gibraltar. Mr. *Francis Masson.*
*Introd.* 1784.
*Fl.*                                            G. H. ♄.

14. B. frutescens, foliis linearibus, involucro universali   *frutices-*
partialique. *Sp. pl.* 344.                                  *cens.*
Grass-leav'd shrubby Hare's-ear.
*Nat.* of Spain.

*Cult.*

*Cult.* 1739, by Mr. Ph. Miller. *Mill. dict. vol.* 2. *n.* 14.
*Fl.* Auguſt and September. G. H. ♄.

*difforme.* 15. B. frutefcens, foliis vernalibus decompofitis planis incifis ; æſtivalibus filiformibus angulatis trifidis. *Sp. pl.* 344.
Various-leav'd Hare's-ear.
*Nat.* of the Cape of Good Hope.
*Cult.* 1752, by Mr. Philip Miller. *Mill. dict. edit.* 6. *n.* 19ᵢ
*Fl.* June——Auguſt. G. H. ♄.

## HASSELQUISTIA. *Gen. pl.* 341.

*Cor.* radiatæ : difci maſculæ. *Sem. ambitus* gemi-nata, margine crenata ; *difci* folitaria, urceolata, hemifphærica.

*ægyptia-ca.* 1. H. foliis pinnatis : foliolis pinnatifidis. *Syſt. veget.* 275. *Jacqu. hort.* 1. *p.* 37. *t.* 87.
Egyptian Haſſelquiſtia.
*Nat.* of Egypt.
*Cult.* 1768, by Mr. Philip Miller. *Mill. dict. edit.* 8.
*Fl.* July. G. H. ☉.

*cordata.* 2. H. foliis cordatis. *Linn. ſuppl.* 179. *Jacqu. hort.* 2. *p.* 91. *t.* 193.
Heart-leav'd Haſſelquiſtia.
*Nat.*
*Introd.* 1787, by Mr. Zier.
*Fl.* July. H. ☉.

TORDY-

TORDYLIUM. *Gen. pl.* 330.

*Cor.* radiatæ, omnes hermaphroditæ. *Fruɛus* fuborbiculatus, margine crenatus. *Involucra* longa, indivifa.

1. T. involucris umbella longioribus.    *Sp. pl.* 345.    *fyriacum.*
*Jacqu. hort.* 1. *p.* 21. *t.* 54.
Syrian Hart-wort.
*Nat.* of Syria.
*Cult.* 1597, by Mr. John Gerard. *Ger. herb.* 885. *f.* 1.
*Fl.* July.                                    H. ⊙.

2. T. involucris partialibus longitudine florum, foliolis    *officinale.*
ovatis laciniatis. *Sp. pl.* 345.
Officinal Hart-wort.
*Nat.* of England.
*Fl.* July.                                    H. ⊙.

3. T. umbellulis remotis, foliis pinnatis: pinnis fubro-    *apulum.*
tundis laciniatis. *Sp. pl.* 345.
Small Hart-wort.
*Nat.* of Italy.
*Cult.* 1739, by Mr. Philip Miller. *Rand. chel. n.* 4.
*Fl.* July.                                    H. ⊙.

4. T. umbellis confertis radiatis, foliolis lanceolatis in-    *maxi-*
cifo-ferratis. *Sp. pl.* 345. *Jacqu. auftr.* 2. *p.* 26.    *mum.*
*t.* 142.
Great Hart-wort.
*Nat.* of Italy.
*Cult.* 1683, by Mr. James Sutherland. *Sutherl. hort.*
*edin.* 336. *n.* 6.
*Fl.* July.                                    H. ⊙.

## CAUCALIS. *Gen. pl.* 331.

*Cor.* radiatæ: difci mafculæ. *Pet.* inflexo-emarginata.
*Fructus* fetis hifpidus. *Involucra* integra.

*grandi-*
*flora.*
1. C. involucris fingulis pentaphyllis: foliolo unico duplo majore. *Syft. veget.* 275. *Jacqu. auftr.* 1. *p.* 33.
*t.* 54.
Great-flower'd Caucalis.
*Nat.* of the South of Europe.
*Introd.* 1775, by Jofeph Nicholas de Jacquin, M.D.
*Fl.* July and Auguft.                                    H. ⊙.

*leptophyl-*
*la.*
2. C. involucro univerfali fubnullo, umbella bifida, invo-
lucellis pentaphyllis. *Syft. veget.* 276.
Caucalis pumila. *Jacqu. hort.* 2. *p.* 92. *t.* 195.
Fine-leav'd Caucalis.
*Nat.* of England.
*Fl.* July.                                               H. ⊙.

*latifolia.*
3. C. involucris involucellifque membranaceis, umbella
univerfali fubquadriradiata, feminum fetis confertis
hifpidis, foliis pinnatis incifis pilofis.
Caucalis latifolia. *Syft. veget.* 276. *Jacqu. hort.* 2.
*p.* 59. *t.* 128.
Tordylium latifolium. *Sp. pl.* 345.
Broad-leav'd Caucalis.
*Nat.* of England.
*Fl.* July and Auguft.                                    H. ⊙.

*arvenfis.*
4. C. involucro univerfali fubnullo, feminibus ovatis,
ftylis reflexis, foliis decompofitis: foliolo extimo
lineari-lanceolato, caule ramofiffimo.
Caucalis arvenfis. *Hudf. angl.* 113. (exclufo fynonymo
Scandicis infeftæ.)

Caucalis

Caucalis helvetica. *Jacqu. hort.* 3. *p.* 12. *t.* 16.
Corn Caucalis.
*Nat.* of Britain.
*Fl.* Auguſt.

5. C. involucris polyphyllis, feminibus ovatis, ſtylis re-    *Anthriſ-*
flexis, foliis decompoſitis: foliolo extimo lineari-    *cus.*
lanceolato.
Caucalis Anthriſcus.  *Hudſ. angl.* 114.
Tordylium Anthriſcus.  *Sp. pl.* 346. *Jacqu. auſtr.* 3.
*p.* 34. *t.* 261.
Hedge Caucalis.
*Nat.* of Britain.
*Fl.* Auguſt.                                    H. ♂ .

6. C. umbellis ſimplicibus ſubſeſſilibus, foliis ſuprade-    *nodoſa.*
compoſitis.
Caucalis nodoſa.  *Hudſ. angl.* 114.
Tordylium nodoſum.  *Sp. pl.* 346.
Knotted Caucalis.
*Nat.* of Britain.
*Fl.* May——Auguſt.                             H. ☉ .

### A R T E D I A.   *Gen. pl.* 332.

*Involucra* pinnatifida.  *Floſculi* diſci maſculi.  *Fructus*
ſquamis hiſpidus.

1. ARTEDIA.  *Sp. pl.* 347.                         *ſquamata.*
Fennel-leav'd Artedia.
*Nat.* of the Levant.
*Introd.* 1788, by Monſ. Thouin.
*Fl.* July.                                        G. H. ☉ .

### D A U C U S.

## DAUCUS. *Gen. pl.* 333.

*Cor.* fubradiatæ, omnes hermaphroditæ. *Fructus* pilis
hifpidus.

*Carota.*   1. D. feminibus hifpidis, petiolis fubtus nervofis. *Sp. pl.*
348.

α Paftinaca tenuifolia fativa radice lutea. *Bauh. pin.*
151.
Yellow Garden Carrot.

β Paftinaca tenuifolia fativa radice atrorubente. *Bauh.*
*pin.* 151.
Red Garden Carrot.
*Nat.* of Britain.
*Fl.* June and July.                                                H. ♂.

*mauri-*   2. D. feminibus hifpidis, flofculo centrali fterili carnofo,
*tanicus.*        receptaculo communi hemifphærico. *Sp. pl.* 348.
Fine-leav'd Carrot.
*Nat.* of Spain.
*Cult.* 1768, by Mr. Philip Miller. *Mill. dict. edit.* 8.
*Fl.* June and July.                                                H. ♂.

*Vifnaga.*   3. D. feminibus lævibus, umbella univerfali bafi coalita.
*Syft. veget.* 277. *Jacqu. hort.* 3. *p.* 17. *t.* 26.
Spanifh Carrot, or Pick-tooth.
*Nat.* of the South of Europe.
*Cult.* 1597, by Mr. John Gerard. *Ger. herb.* 885. *f.* 2.
*Fl.* June——Auguft.                                                H. ☉.

*Gingidi-*   4. D. radiis involucri planis : laciniis recurvis. *Sp. pl.* 348.
*um.*     Shining-leav'd Carrot.
*Nat.* of the South of France.

*Cult.*

*Cult.* 1722, in Chelſea Garden. *R. S. n.* 16.
*Fl.* June and July. H. ⊙.

5. D. ſeminibus triglochidi-aculeatis. *Syſt. veget.* 277. *murica-*
Prickly-ſeeded Carrot. *tus.*
*Nat.* of Barbary.
*Cult.* 1699. *Moriſ. hiſt.* 3. *p.* 308. *n.* 4.
*Fl.* July. H. ♂.

### A M M I. *Gen. pl.* 334.

*Involucra* pinnatifida. *Corollæ* radiatæ, omnes hermaphroditæ. *Fructus* lævis.

1. A. foliis inferioribus pinnatis lanceolatis ferratis ; ſu- *majus.*
perioribus multifidis linearibus. *Sp. pl.* 349.
Biſhop's-weed.
*Nat.* of the South of Europe.
*Cult.* 1597, by Mr. John Gerard. *Ger. herb.* 881. *f.* 1.
*Fl.* June and July. H. ⊙.

2. A. foliis ſupradecompoſitis linearibus, ſeminibus mu- *copticum.*
ricatis. *Linn. mant.* 56. *Jacqu. hort.* 2. *p.* 92.
*t.* 196.
Prickly-ſeeded Biſhop's-weed.
*Nat.* of Egypt.
*Introd.* 1773, by John Earl of Bute.
*Fl.* July. H. ⊙.

### B U N I U M. *Gen. pl.* 335.

*Corolla* uniformis. *Umbella* conferta. *Fructus* ovati.

1. BUNIUM. *Sp. pl.* 349. *Curt. lond.* *Bulbocaſ-*
Earth-nut. *tanum.*
*Nat.* of Britain.
*Fl.* May and June. H. ♃.

Z                    CONIUM.

## C O N I U M. ' *Gen. pl.* 336.

*Involucella* dimidiata, fubtriphylla. *Fruƈus* fubglobo-
fus, 5-ftriatus, utrinque crenatus.

*macula-*
*tum.*
1. C. feminibus ftriatis.　*Sp. pl.* 349.　*Curtis lond.*
　　*Jacqu. auftr.* 2. *p.* 36. *t.* 156.
Common Hemlock.
*Nat.* of Britain.
*Fl.* June and July.　　　　　　　　　　　H. ♂.

*africa-*
*num.*
2. C. feminibus muricatis, petiolis pedunculifque lævi-
　　bus. *Syft. veget.* 278. *Jacqu. hort.* 2. *p.* 91. *t.* 194.
Rue-leav'd Hemlock.
*Nat.* of the Cape of Good Hope.
*Cult.* 1759, by Mr. Ph. Miller. *Mill. diƈ. edit.* 7. *n.* 3.
*Fl.* June——September.　　　　　　　　　S. ☉.

## S E L I N U M.　*Gen. pl.* 337.

*Fruƈus* ovali-oblongus, compreffo-planus, in medio ftriatus.　*Involucr.* reflexum.　*Petala* cordata, æqualia.

*paluftre.*
1. S. fublaƈefcens, radice unica.　*Sp. pl.* 350.
Marfh Selinum, or Milk-parfley.
*Nat.* of England.
*Fl.* July and Auguft.　　　　　　　　　H. ♃.

*Carvifo-*
*lia.*
2. S. caule fulcato acutangulo, involucro univerfali eva-
　　nido, ftylis ereƈis, petalis conniventibus.　*Syft.*
　　*veget.* 278. *Jacqu. auftr.* 1. *p.* 13. *t.* 16.
Caraway-leav'd Selinum.
*Nat.* of Auftria and Siberia.
*Introd.* 1774, by Monf. Richard.
*Fl.* July and Auguft.　　　　　　　　　H. ♃.

3. S.

3. S. involucro univerſali nullo, piſtillis divaricatiſſimis. *Seguieri.*
   *Syſt. veget.* 279. *Jacqu. hort.* 1. *p.* 24. *t.* 61.
   Fennel-leav'd Selinum.
   *Nat.* of Italy.
   *Introd.* 1774, by Joſeph Nicholas de Jacquin, M. D.
   *Fl.* July.                                         H. ♃.

4. S. umbellis confertis, involucro univerſali reflexo, ſe- *Monnie-*
   minum coſtis quinque membranaceis. *Sp. pl.* 351. *ri.*
   *Jacqu. hort.* 1. *p.* 25. *t.* 62.
   Annual Selinum.
   *Nat.* of the South of France.
   *Introd.* 1771, by Monſ. Richard.
   *Fl.* July and Auguſt.                              H. ☉.


ATHAMANTA. *Gen. pl.* 338.

*Fruĉtus* ovato-oblongus, ſtriatus. *Pet.* inflexa, emar-
                                   ginata.

1. A. foliis bipinnatis planis, umbella hemiſphærica, ſe- *Libano-*
   minibus hirſutis. *Sp. pl.* 351. *Jacqu. auſtr.* 4. *tis.*
   *p.* 48. *t.* 392.
   α Libanotis minor, apii folio. *Bauh. pin.* 157.
   Mountain Spignel.
   β Athamanta pyrenaica. *Jacqu. hort.* 2. *p.* 93. *t.* 197.
   Pyrenean mountain Spignel.
   *Nat.* of England.
   *Fl.* June and July.                                H. ♃.

2. A. foliolis pinnatis decuſſatis inciſo-angulatis, ſemini- *Cerva-*
   bus nudis. *Sp. pl.* 352. *Jacqu. auſtr.* 1. *p.* 44. *t.* 69. *ria.*
   Broad-leav'd Spignel.
   *Nat.* of Europe.
   *Cult.* 1597, by Mr. John Gerard. *Ger. herb.* 858. *f.* 2.
   *Fl.* July and Auguſt.                              H. ♃.

3. A.

*fibirica.*　3. A. foliis pinnatis incifo-angulatis. *Linn. mant.* 56.
　　　　　　Siberian Spignel.
　　　　　　*Nat.* of Siberia.
　　　　　　*Introd.* 1771, by Monf. Richard.
　　　　　　*Fl.* July and Auguft.　　　　　　　　　　H. ♃.

*condenfa-*　4. A. foliis fubbipinnatis: foliolis deorfum imbricatis,
*ta.*　　　　　　umbella lentiformi. *Syft. veget.* 279.
　　　　　　Clofe-headed Spignel.
　　　　　　*Nat.* of Siberia.
　　　　　　*Introd.* 1773, by John Earl of Bute.
　　　　　　*Fl.* July——September.　　　　　　　　H. ♃.

*Oreofeli-*　5. A. foliolis divaricatis. *Sp. pl.* 352. *Jacqu. auftr.* 1.
*num.*　　　　　　*p.* 43. *t.* 68.
　　　　　　Divaricated Spignel.
　　　　　　*Nat.* of England.
　　　　　　*Fl.* July.　　　　　　　　　　　　　　H. ♃.

*ficula.*　　6. A. foliis inferioribus nitidis, umbellis primordialibus
　　　　　　fubfeffilibus, feminibus pilofis. *Sp. pl.* 352.
　　　　　　Flix-weed-leav'd Spignel.
　　　　　　*Nat.* of Sicily.
　　　　　　*Cult.* 1713. *Philofoph. tranf. n.* 337. *p.* 189. *n.* 48.
　　　　　　*Fl.* June and July.　　　　　　　　　H. ♃.

*annua.*　　7. A. foliis multipartitis: laciniis linearibus teretiufculis
　　　　　　acuminatis. *Sp. pl.* 353.
　　　　　　Annual Spignel.
　　　　　　*Nat.* of Candia.
　　　　　　*Introd.* 1770, by Monf. Richard.
　　　　　　*Fl.* July.　　　　　　　　　　　　　H. ☉.

PEUCE-

## PEUCEDANUM. *Gen. pl.* 339.

*Fructus* ovatus, utrinque ftriatus, ala cinctus. *Involucra* breviffima.

1. P. foliis quinquies tripartitis filiformibus linearibus.   *officinale.*
   *Sp. pl.* 353.
α Peucedanum germanicum.  *Bauh. pin.* 149.
   Common Sulphur-wort.
β Peucedanum majus italicum.  *Bauh. pin.* 149.
   Italian Sulphur-wort.
   *Nat. α.* of England, *β.* of Italy.
   *Fl.* May and June.          H. ♃.

2. P. foliolis pinnatifidis : laciniis oppofitis, involucro  *Silaus.*
   univerfali diphyllo.  *Sp. pl.* 354. *Jacqu. auftr.* 1.
   *p.* 12. *t.* 15.
   Meadow Sulphur-wort, or Saxifrage.
   *Nat.* of England.
   *Fl.* June——Auguft.        H. ♃.

3. P. foliolis pinnatifidis : lacinulis trifidis obtufiufculis.  *alfaticum.*
   *Sp. pl.* 354. *Jacqu. auftr.* 1. *p.* 45. *t.* 70.
   Small-headed Sulphur-wort.
   *Nat.* of Germany.
   *Introd.* 1774, by Monf. Richard.
   *Fl.* June and July.        H. ♃.

4. P. foliis tripinnatis : foliolis caulinis lineari-lanceola-  *aureum.*
   tis; radicalibus oblongis multifidis.
   Golden Sulphur-wort.
   *Nat.* of the Canary Iflands.  Mr. *Francis Maffon.*
   *Introd.* 1779.
   *Fl.* June.        G. H. ♂.

Z 3    CRITHMUM.

## CRITHMUM. *Gen. pl.* 340.

*Fructus* ovalis, compreſſus.   *Floſculi* æquales.

*mariti-*        1. C. foliolis lanceolatis carnoſis.   *Sp. pl.* 354.  *Jacqu.*
*mum.*               *hort.* 2. *p.* 88. *t.* 187.
                Sea Samphire.
                *Nat.* of Britain.
                *Fl.* July——September.                    H. ♃.


*latifoli-*      2. C. foliolis cuneiformibus fiſſis.
*um.*                Crithmum latifolium.   *Linn. ſuppl.* 180.
                Wedge-leav'd Samphire.
                *Nat.* of the Canary Iſlands.   Mr. *Francis Maſſon.*
                *Introd.* 1780.
                *Fl.* July.                                G. H. ♂.


## CACHRYS. *Gen. pl.* 342.

*Fructus* ſubovatus, angulatus, ſuberoſo-corticatus.

*Libano-*       1. C. foliis bipinnatis : foliolis linearibus acutis multifidis,
*tis.*               ſeminibus ſulcatis lævibus.   *Syſt. veget.* 280.
                Smooth-ſeeded Cachrys.
                *Nat.* of Sicily.
                *Cult.* 1597, by Mr. John Gerard.  *Ger. herb.* 858. *f.* 3.
                *Fl.* July and Auguſt.                     H. ♃.


*ſicula.*       2. C. foliis bipinnatis : foliolis linearibus acutis, ſemini-
                bus ſulcatis hiſpidis.   *Sp. pl.* 355.
                Hairy-ſeeded Cachrys.
                *Nat.* of Spain and Sicily.
                *Cult.* 1739, by Mr. Philip Miller.  *Rand. chel. n.* 2.
                *Fl.* Auguſt.                              H. ♃.


                                          FERULA.

FERULA. *Gen. pl.* 343.

*Fruilus* ovalis, compreffo-planus, ftriis utrinque 3.

1. F. foliolis linearibus longiffimis fimplicibus. *Sp. pl.*    *communis.*
    355.
Common Gigantic Fennel.
*Nat.* of the South of Europe.
*Cult.* 1597, by Mr. John Gerard. *Ger. herb.* 898.
*Fl.* June and July.            H. ♃.

2. F. foliis fupradecompofitis: foliolis lanceolato-linea-   *glauca.*
    ribus planis. *Sp. pl.* 355.
Glaucous Gigantic Fennel.
*Nat.* of Sicily and Italy.
*Cult.* 1768, by Mr. Philip Miller. *Mill. dict. edit.* 8.
*Fl.* June and July.            H. ♃.

3. F. foliolis laciniatis: lacinulis tridentatis inæqualibus  *tingitana.*
    nitidis. *Sp. pl.* 355.
Tangier Gigantic Fennel.
*Nat.* of Spain and Barbary.
*Cult.* 1683, by Mr. James Sutherland. *Sutherl. hort.*
    *edin.* 119.
*Fl.* June and July.            H. ♂.

4. F. foliorum pinnis bafi nudis: foliolis fetaceis. *Sp.*  *orientalis.*
    *pl.* 356.
Narrow-leav'd Gigantic Fennel.
*Nat.* of the Levant.
*Cult.* 1759, by Mr. Ph. Miller. *Mill. dict. edit.* 7. *n.* 5.
*Fl.*            H. ♃.

5. F. foliolis appendiculatis, umbellis fubfeffilibus. *Sp.*  *nodiflora.*
    *pl.* 356. *Jacqu. auftr.* 5. *p.* 28. *tab. app.* 5.

           Knotted

Knotted Gigantic Fennel.
*Nat.* of the South of Europe.
*Cult.* 1759, by Mr. Ph. Miller. *Mill. dict. edit.* 7. *n.* 7.
*Fl.* June and July.                                          H. ♃.


## LASERPITIUM. *Gen. pl.* 344.

*Fructus* oblongus: angulis 8 membranaceis. *Pet.*
inflexa, emarginata, patentia.

*latifoli-*
*um.*

1. L. foliolis cordatis incifo-ferratis. *Sp. pl.* 356. *Jacqu.*
     *auftr.* 2. *p.* 28. *t.* 146.
   Broad-leav'd Laffer-wort.
   *Nat.* of Europe.
   *Cult.* 1640. *Park. theat.* 952. *f.* 1.
   *Fl.* June and July.                                       H. ♃.


*trilobum.*

2. L. foliolis trilobis incifis. *Sp. pl.* 357.
   Columbine-leav'd Laffer-wort.
   *Nat.* of the Levant.
   *Cult.* 1640. *Park. theat.* 952. *n.* 4.
   *Fl.* May——July.                                          H. ♃.


*gallicum.*

3. L. foliolis cuneiformibus furcatis. *Sp. pl.* 357.
   French Laffer-wort.
   *Nat.* of the South of Europe.
   *Cult.* 1683, by Mr. James Sutherland. *Sutherl. hort.*
     *edin.* 186. *n.* 2.
   *Fl.* June and July.                                       H. ♃.


*angufti-*
*folium.*

4. L. foliolis lanceolatis integerrimis feffilibus. *Sp. pl.*
     357.
   Narrow-leav'd Laffer-wort.
   *Nat.* of the South of Europe.
   *Cult.* 1759, by Mr. Ph. Miller. *Mill. dict. edit.* 7. *n.* 5.
   *Fl.* June and July.                                       H. ♃.

                                                             5. L.

5. L. foliolis ovali-lanceolatis integerrimis petiolatis. *Siler.*
   *Sp. pl.* 357. *Jacqu. auftr.* 2. *p.* 27. *t.* 145.
Mountain Laffer-wort.
*Nat.* of Auftria, Switzerland, and France.
*Cult.* 1640. *Park. theat.* 909. *f.* 1.
*Fl.* May——July.           H. ♃.

6. L. foliis fupradecompofitis lineari-fubulatis glabris, *lucidum.*
   involucris univerfalibus pinnatis.
Laferpitium foliis triplicato pinnatis, pinnulis lan-
   ceolatis, involucris femitrifidis. *Hall. hift. n.* 796.
Shining Laffer-wort.
*Nat.* of Switzerland.
*Introd.* 1775, by the Doctors Pitcairn and Fothergill.
*Fl.* July.           H. ♂.

7. L. foliolis linearibus. *Sp. pl.* 358.      *ferulace-*
   Fennel-leav'd Laffer-wort.            *um.*
*Nat.* of the Levant.
*Cult.* 1768, by Mr. Philip Miller. *Mill. dict. edit.* 8.
*Fl.* June.           H. ♃.

HERACLEUM. *Gen. pl.* 345.

*Fructus* ellipticus, emarginatus, compreffus, ftriatus,
   marginatus. *Cor.* difformis, inflexo-emarginata.
*Involucr.* caducum.

1. H. foliolis pinnatifidis lævibus, floribus uniformibus. *Sphondy-*
   *Syft. veget.* 282.               *lium.*
Common Cow Parfnep.
*Nat.* of Britain.
*Fl.* May——July.           H. ♃.

2. H.

*angustifo-*
*lium.*   2. H. foliis cruciato-pinnatis : foliolis linearibus, corollis
          flosculosis.  *Syst. veget.* 282.

          α Heracleum angustifolium.  *Jacqu. austr.* 2. *p.* 46.
              *t.* 173.
              Narrow-leav'd Cow Parsnep.

          β Heracleum longifolium.  *Jacqu. austr.* 2. *p.* 46. *t.* 174.
              Long-leav'd Cow Parsnep.
              *Nat.* α. of England ; β. of Austria.
              *Fl.* June and July.                         H. ♃.


*sibiricum.*   3. H. foliis pinnatis : foliolis quinis : intermediis sessili-
               libus, corollulis uniformibus.  *Sp. pl.* 358.
               Siberian Cow Parsnep.
               *Nat.* of Siberia.
               *Cult.* 1768, by Mr. Philip Miller.  *Mill. dict. edit.* 8.
               *Fl.* May——July.                         H. ♂.


*Panaces.*   4. H. foliis pinnatis : foliolis quinis : intermediis sessili-
             bus, floribus radiatis.  *Sp. pl.* 358.
             Palmated Cow Parsnep.
             *Nat.* of Italy and Siberia.
             *Cult.* 1597, by Mr. John Gerard.  *Ger. herb.* 850. *f.* 1.
             *Fl.* July and August.                         H. ♂.


*austria-*
*cum.*   5. H. foliis pinnatis utrinque rugosis scabris, floribus sub-
         radiatis.  *Syst. veget.* 282.  *Jacqu. austr.* 1. *p.* 38.
         *t.* 61.
         Austrian Cow Parsnep.
         *Nat.* of Austria.
         *Cult.* 1748, by Mr. Philip Miller.  *Mill. dict. edit.* 5.
         Sphondylium 7.
         *Fl.* June and July.                         H. ♃.


*alpinum.*   6. H. foliis simplicibus, floribus radiatis.  *Sp. pl.* 359.
             Alpine Cow Parsnep.

                                                          *Nat.*

*Nat.* of the **Alps** of Switzerland.

*Cult.* 1739, by Mr. Philip Miller. *Rand. chel.*
Sphondylium 3.

*Fl.* June and July.	H. ♃.

## LIGUSTICUM. *Gen. pl.* 346.

*Fructus* oblongus, 5-fulcatus utrinque. *Cor.* æquales:
Petalis involutis, integris.

1. L. foliis multiplicibus : foliolis fuperne incifis. *Sp. pl.*	*Levifti-*
359.	*cum.*
Common Lovage.
*Nat.* of the Alps of Italy.
*Cult.* 1596, by Mr. John Gerard. *Hort. Ger.*
*Fl.* June and July.	H. ♃.

2. L. foliis biternatis. *Sp. pl.* 359.	*fcoticum.*
Scotch Lovage.
*Nat.* of Britain.
*Fl.* June——Auguft.	H. ♃.

3. L. foliis multiplicato-pinnatis: foliolis pinnatim in-	*pelopo-*
cifis. *Sp. pl.* 360. *Jacqu. auftr.* 5. *p.* 33. *t. app.* 13.	*nenfe.*
Hemlock-leav'd Lovage.
*Nat.* of the Alps of Switzerland.
*Cult.* 1596, by Mr. John Gerard. *Hort. Ger.*
*Fl.* May——July.	H. ♃.

4. L. foliis bipinnatis : foliolis confluentibus incifis inte-	*auftria-*
gerrimis. *Sp. pl.* 360. *Jacqu. auftr.* 2. *p.* 32.	*cum.*
*t.* 151.
Auftrian Lovage.
*Nat.* of Auftria.
*Cult.* 1759, by Mr. Ph. Miller. *Mill. dict. edit.* 7. *n.* 3.
*Fl.* June——Auguft.	H. ♃.

5. L.

348 PENTANDRIA DIGYNIA. Ligusticum.

peregri-
num.
5. L. umbellæ primariæ involucro subnullo; lateralium
basi membranacea: radiis subramosis. *Syst. veget.*
283. *Jacqu. hort.* 3. *p.* 13. *t.* 18.
Parsley-leav'd Lovage.
*Nat.* of Portugal.
*Introd.* 1776, by Monf. Thouin.
*Fl.* June and July.                                    H. ♂.

candi-
cans.
6. L. foliis supradecompositis : foliolis cuneiformibus incisis glabris, involucro universali diphyllo subfolia-
ceo, costis seminum membranaceis glabris.
Pale Lovage.
*Nat.*
*Introd.* about 1780.
*Fl.* July and August.                                  H. ♃.


A N G E L I C A.    *Gen. pl.* 347.
*Fructus* subrotundus, angulatus, solidus, *stylis* reflexis.
*Corollæ* æquales: petalis incurvis.

Archan-
gelica.
1. A. foliorum impari lobato.    *Sp. pl.* 360.
Garden Angelica.
*Nat.* of Lapland.
*Cult.* 1568.    *Turn. herb. part* 3. *p.* 5.
*Fl.* June——August.                                     H. ♂.

sylvestris.
2. A. foliolis æqualibus ovato-lanceolatis serratis. *Sp. pl.*
361.
Wild Angelica.
*Nat.* of Britain.
*Fl.* June and July.                                    H. ♃.

verticil-
laris.
3. A. foliis divaricatissimis : foliolis ovatis serratis, caule
pedunculis verticillato.    *Linn. mant.* 217.    *Jacqu.*
*hort.* 2. *p.* 60. *t.* 130.
Whorl'd-

Whorl'd-flower'd Angelica.
*Nat.* of Italy?
*Introd.* 1774, by Monf. Richard.
*Fl.* July. H. ♃.

4. A. extimo foliorum pari coadunato: foliolo termi-   *atropur-*
nali petiolato. *Sp. pl.* 361.   *purea.*
Dark-purple Angelica.
*Nat.* of Canada.
*Cult.* 1759, by Mr. Ph. Miller. *Mill. dict. edit.* 7. *n.* 4.
*Fl.* July and Auguft. H. ♃.

5. A. foliolis æqualibus ovatis incifo-ferratis. *Sp. pl.*   *lucida.*
361. *Jacqu. hort.* 3. *p.* 16. *t.* 24.
Shining-leav'd Angelica.
*Nat.* of Canada.
*Cult.* 1640, by Mr. John Parkinfon. *Park. theat.*
949. *f.* 3.
*Fl.* July and Auguft. H. ♂.

S I U M. *Gen. pl.* 348.

*Fructus* fubovatus, ftriatus. *Involucrum* polyphyllum.
*Petala* cordata.

1. S. foliis pinnatis, umbellis terminalibus. *Sp. pl.* 361.   *latifoli-*
*Jacqu. auftr.* 1. *p.* 42. *t.* 66.   *um.*
Great Water Parfnep.
*Nat.* of England.
*Fl.* July and Auguft. H. ♃.

2. S. foliis pinnatis, umbellis axillaribus pedunculatis,   *anguftifo-*
involucro univerfali pinnatifido. *Syft. veget.* 284.   *lium.*
*Jacqu. auftr.* 1. *p.* 42. *t.* 67.
Narrow-leav'd Water Parfnep.
*Nat.* of Britain.
*Fl.* July. H. ♃.

3. S.

*nodiflo-*      3. S. foliis pinnatis, umbellis axillaribus feffilibus. *Sp.*
*rum.*              *pl.* 361.
              Creeping Water Parfnep.
              *Nat.* of Britain.
              *Fl.* July and Auguft.                    H. ♃.

*Sifarum.*      4. S. foliis pinnatis: floralibus ternatis.  *Sp. pl.* 361.
              Skirret.
              *Nat.*
              *Cult.* 1597.   *Ger. herb.* 871.
              *Fl.* July and Auguft.                    H. ♃.

*rigidius.*      5. S. foliis pinnatis: foliolis lanceolatis fubintegerrimis.
                  *Sp. pl.* 362.
              Virginian Water Parfnep.
              *Nat.* of Virginia.
              *Introd.* 1774, by Mr. William Young.
              *Fl.* July and Auguft.                    H. ♃.

*Falcaria.*      6. S. foliolis linearibus decurrentibus connatis.  *Sp. pl.*
                  362.  *Jacqu. auftr.* 3. *p.* 32. *t.* 257.
              Decurrent Water Parfnep.
              *Nat.* of Europe.
              *Cult.* 1759, by Mr. Ph. Miller. *Mill. dict. edit.* 7. *n.* 5.
              *Fl.* July and Auguft.                    H. ♃.

                    S I S O N.   *Gen. pl.* 349.

              *Fructus* ovatus, ftriatus.   *Involucra* fub 4-phylla.

*Amomum.*      1. S. foliis pinnatis, umbellis erectis.  *Sp. pl.* 362.
                  *Jacqu. hort.* 3. *p.* 13. *t.* 17.
              Field Hone-wort.
              *Nat.* of England.
              *Fl.* July and Auguft.                    H. ♂.

                                                      2. S.

2. S. foliis pinnatis, umbellis cernuis. *Sp. pl.* 362. *Jacqu.* *fegetum.*
   *hort.* 2. *p.* 63. *t.* 134.
   Corn Hone-wort.
   *Nat.* of England.
   *Fl.* July and Auguft.                    H. ♂.

3. S. foliis ternatis. *Sp. pl.* 363.                    *canadenfe.*
   Three-leav'd Hone-wort.
   *Nat.* of North America.
   *Introd.* before 1699, by William Sherard, Efq. *Morif.*
   *hift.* 3. *p.* 301. *n.* 4.
   *Fl.* July and Auguft.                    H. ♃.

4. S. repens, umbellis bifidis. *Sp. pl.* 363.             *inunda-*
   Water Hone-wort.                                        *tum.*
   *Nat.* of Britain.
   *Fl.* May and June.                    H. ☉.

5. S. foliolis verticillatis capillaribus. *Sp. pl.* 363.    *verticil-*
   Whorl'd-leav'd Hone-wort.                                *latum.*
   *Nat.* of Britain.
   *Fl.* Auguft.                    H. ♃.


          B U B O N.  *Gen. pl.* 350.

       *Fructus* ovatus, ftriatus, villofus.

1. B. foliolis rhombeo-ovatis incifo-dentatis: dentibus   *macedoni-*
   acuminatis, umbellis numerofiffimis, feminibus hir-     *cum.*
   tis.
   Bubon macedonicum. *Sp. pl.* 364.
   Macedonian Bubon, or Parfley.
   *Nat.* of Barbary and Greece.
   *Cult.* 1596, by Mr. John Gerard. *Hort. Ger.*
   *Fl.* June——Auguft.                    G. H. ♂.

                                        2. B.

*Galba-*
*num.*

2. B. foliolis ovato-cuneiformibus acutis argute ferratis, umbellis paucis, feminibus glabris, caule frutefcente glauco.

Bubon Galbanum. *Sp. pl.* 364. *Jacqu. hort.* 3. *p.* 21. *t.* 36.

Lovage-leav'd Bubon.

*Nat.* of the Cape of Good Hope.

*Cult.* 1596, by Mr. John Gerard. *Hort. Ger.*

*Fl.* July and Auguſt.      G. H. ♄.

*læviga-*
*tum.*

3. B. foliolis lanceolatis obtufiffime obfoletiffimeque crenatis, feminibus glabris, caule frutefcente.

Smooth Bubon.

*Nat.* of the Cape of Good Hope. Mr. *Francis Maſſon.* *Introd.* 1774.

*Fl.* December——February.      G. H. ♄.

*gummife-*
*rum.*

4. B. foliolis incifis acuminatis : inferioribus latioribus, feminibus glabris, caule frutefcente.

Bubon gummiferum. *Sp. pl.* 364.

Gum-bearing Bubon.

*Nat.* of the Cape of Good Hope.

*Cult.* 1731. *Mill. dict. edit.* 1. Ferula 7.

*Fl.* July.      G. H. ♄.

## C U M I N U M.   *Gen. pl.* 351.

*Fructus* ovatus, ſtriatus. *Umbellulæ* 4. *Involucra* 4-fida.

*Cyminum.*

1. CUMINUM. *Sp. pl.* 365.

Cumin.

*Nat.* of Egypt.

*Cult.* 1594, by Sir Hugh Plat. *Plat's Garden of Eden,* part 2. *p.* 134.

*Fl.* June and July.      H. ☉.

O E N A N-

## ŒNANTHE. *Gen. pl.* 352.

*Flosculi* difformes : in difco feffiles, fteriles. *Fructus* calyce et piftillo coronatus.

1. Œ. ftolonifera, foliis caulinis pinnatis filiformibus fif- *fiftulofa.*
　　tulofis. *Sp. pl.* 365.
　　Common Water-dropwort.
　　*Nat.* of Britain.
　　*Fl.* June and July.　　　　　　　　　　H. ♃.

2. Œ. foliis omnibus multifidis obtufis fubæqualibus. *crocata.*
　　*Sp. pl.* 365. *Jacqu. hort.* 3. *p.* 32. *t.* 55.
　　Hemlock Water-dropwort.
　　*Nat.* of Britain.
　　*Fl.* June.　　　　　　　　　　　　　　H. ♃.

3. Œ. umbellularum pedunculis marginalibus longiori- *prolifera.*
　　bus ramofis mafculis. *Sp. pl.* 365. *Jacqu. hort.*
　　3. *p.* 35. *t.* 62.
　　Proliferous Water-dropwort.
　　*Nat.* of Sicily and Italy.
　　*Cult.* 1739, by Mr. Philip Miller. *Rand. chel. n.* 9.
　　*Fl.* June and July.　　　　　　　　　　H. ♃.

4. Œ. fructibus globofis. *Sp. pl.* 365. *globulofa.*
　　Globular-headed Water-dropwort.
　　*Nat.* of Portugal.
　　*Cult.* 1739, by Mr. Philip Miller. *Rand. chel. n.* 8.
　　*Fl.* June and July.　　　　　　　　　　H. ♂.

5. Œ. foliolis radicalibus cuneatis fiffis ; caulinis integris *pimpinel-*
　　lincaribus longiffimis fimplicibus. *Syft. veget.* 286. *loides.*
　　*Jacqu. auftr.* 4. *p.* 49. *t.* 394.
　　Parfley Water-dropwort.

A a　　　　　　　　　　*Nat.*

*Nat.* of England.
*Fl.* June.                                                        H. ♃.


## PHELLANDRIUM. *Gen. pl.* 353.

*Flofculi* difci minores.  *Fructus* ovatus, lævis, coronatus perianthio et piftillo.

*aquati-*    1. P. foliorum ramificationibus divaricatis. *Sp. pl.* 366.
*cum.*           Water Phellandrium.
                 *Nat.* of Britain.
                 *Fl.* June and July.                             H. ♂.

*Mutelli-*   2. P. caule fubnudo, foliis bipinnatis.  *Sp. pl.* 366.
*na.*               *Jacqu. auftr.* 1. *p.* 35. *t.* 56.
                 Alpine Phellandrium.
                 *Nat.* of the Alps of Switzerland and Auftria.
                 *Introd.* 1774, by the Doctors Pitcairn and Fothergill.
                 *Fl.* July and Auguft.                           H. ♃.


## C I C U T A.  *Gen. pl.* 354.

*Fructus* fubovatus, fulcatus.

*virofa.*    1. C. umbellis oppofitifoliis, petiolis marginatis obtufis.
                 *Sp. pl.* 366.
                 Water Hemlock.
                 *Nat.* of Britain·
                 *Fl.* July.                                      H. ♃.


## Æ T H U S A.  *Gen. pl.* 355.

*Involucella* dimidiata, 3-phylla, pendula.  *Fructus*
ftriatus.

*Cynapi-*   1. Æ. foliis conformibus. *Syft. veget.* 286.  *Curtis lond.*
*um.*           Common Fools Parfley.

                                                          *Nat.*

*Nat.* of Britain.

*Fl.* Auguft and September.                                    H. ☉.

2. *Æ.* foliis radicalibus pinnatis, caulinis multipartito-   *Bunius.*
   fetaceis. *Syft. veget.* 286.
   Coriander-leav'd Fools Parfley.
   *Nat.* of the Pyrenean Mountains.
   *Introd.* 1778, by Monf. Thouin.
   *Fl.* July.                                                 H. ♂.

3. *Æ.* foliis omnibus multipartito fetaceis: foliolis fub-   *Meum.*
   verticillatis, caule paucifolio, vaginis petiolorum
   dilatatis ventricofis, involucro univerfali mono-
   phyllo.
   Æthufa Meum. *Syft. veget.* 287.
   Athamanta Meum. *Sp. pl.* 353.
   Meum Athamanticum. *Jacqu. auftr.* 4. *p.* 2. *t.* 303.
   Common Spignel, or Bawd-money.
   *Nat.* of Britain.
   *Fl.* April —— June.                                        H. ♃.

4. *Æ.* foliis omnibus multipartito-fetaceis: foliolis fub-   *fatua.*
   verticillatis, caule multifolio, vaginis petiolorum
   anguftis, involucro univerfali polyphyllo.
   Fine-leav'd Fools Parfley.
   *Nat.*
   *Introd.* 1781, by Monf. Thouin.
   *Fl.* Auguft and September.                                 H. ♃.


CORIANDRUM. *Gen. pl.* 356.

*Cor.* radiata: *Petala* inflexo-emarginata. *Involucr.*
*univerfale* 1-phyllum: *Partialia* dimidiata. *Fruc-*
*tus* fphæricus.

1. C. fructibus globofis. *Sp. pl.* 367.                      *fativum.*

Common

Common Coriander.
*Nat.* of England.
*Fl.* June.                                                    H. ⊙.

*testicula-*
*tum.*   2. C. fructibus didymis.   *Sp. pl.* 367.
Small Coriander.
*Nat.* of the South of Europe.
*Cult.* 1739, by Mr. Philip Miller.   *Rand. chel. n.* 2.
*Fl.* June and July.                                   H. ⊙.

## S C A N D I X.   *Gen. pl.* 357.

*Cor.* radiata.   *Fructus* subulatus.   *Petala* emarginata.
*Flosculi* disci sæpe masculi.

*odorata.*   1. S. seminibus sulcatis angulatis.   *Sp. pl.* 368.   *Jacqu.*
*austr.* 5. *p.* 48. *tab. app.* 37.
Sweet-scented Cicely, or Myrrh.
*Nat.* of Britain.
*Fl.* May and June.                                    H. ♃.

*Pecten.*   2. S. seminibus rostro longissimo.   *Syst. veget.* 287.
*Jacqu. austr.* 3. *p.* 35. *t.* 263.   *Curtis lond.*
Corn Cicely, or Shepherd's-needle.
*Nat.* of Britain.
*Fl.* June and July.                                   H. ⊙.

*Cerefoli-*
*um.*   3. S. seminibus nitidis ovato-subulatis, umbellis sessilibus
lateralibus. *Sp. pl.* 368. *Jacqu. austr.* 4. *p.* 47. *t.* 390.
Garden Cicely, or Chervil.
*Nat.* of Europe.
*Cult.* 1597.   *Ger. herb.* 882. *f.* 1.
*Fl.* May and June.                                    H. ⊙.

*Anthris-*
*cus.*   4. S. seminibus ovatis hispidis, corollis uniformibus,
caule lævi. *Sp. pl.* 368. *Curtis lond. Jacqu. austr.*
2. *p.* 35. *t.* 154.

Rough

Rough Cicely, or Chervil.
*Nat.* of Britain.
*Fl.* May and June. H. ⊙.

5. S. feminibus fubulatis hifpidis, floribus radiatis, cau- *auftralis.*
libus lævibus. *Sp. pl.* 369.
Radiated Cicely, or Chervil.
*Nat.* of the South of Europe.
*Cult.* 1713. *Philofoph. tranf. n.* 337. *p.* 41. *n.* 30.
*Fl.* May and June. H. ⊙.

CHÆROPHYLLUM. *Gen. pl.* 358.

*Involucr.* reflexum, concavum. *Petala* inflexo-cor-
data. *Fruĉtus* oblongus, lævis.

1. C. caule lævi ftriato: geniculis tumidiufculis. *Syſt.* *fylveftre.*
veget. 288. *Jacqu. auftr.* 2. *p.* 31. *t.* 149. *Curtis
lond.*
Wild Chervil, or Cow-weed.
*Nat.* of Britain.
*Fl.* May and June. H. ♃.

2. C. caule lævi geniculis tumido; bafi hirto. *Syſt.* *bulbofum.*
veget. 288. *Jacqu. auftr.* 1. *p.* 40. *t.* 63.
Bulbous-rooted Chærophyllum.
*Nat.* of Europe.
*Cult.* 1739, by Mr. Ph. Miller. *Rand. chel.* Myrr-
his 5.
*Fl.* June and July. H. ♂.

3. C. caule fcabro: geniculis tumidis. *Sp. pl.* 370. *temulum.*
*Jacqu. auftr.* 1. *p.* 41. *t.* 65.
Rough Chærophyllum.

A a 3 *Nat.*

*Nat.* of Britain.

*Fl.* July and Auguſt.                                    H. ♂ .

*hirſutum.*   4. C. caule æquali, foliolis inciſis acutis, fructibus bia-
riſtatis. *Syſt. veget.* 288. *Jacqu. auſtr.* 2. *p.* 30.
*t.* 148.
Hairy-leav'd Chærophyllum.
*Nat.* of Switzerland.
*Cult.* 1768, by Mr. Philip Miller.   *Mill. dict. edit.* 8.
*Fl.* June and July.                                   H. ♃.

*aromati-*   5. C. caule æquali, foliolis ſerratis integris, fructibus
*cum.*        biariſtatis. *Syſt. veget.* 288. *Jacqu. auſtr.* 2. *p.* 32.
*t.* 150.
Aromatic Chærophyllum.
*Nat.* of Germany.
*Cult.* 1758, by Mr. Philip Miller.
*Fl.* June——Auguſt.                               H. ♃.

*aureum.*   6. C. caule æquali, foliolis inciſis, feminibus coloratis
fulcatis muticis. *Syſt. veget.* 288. *Jacqu. auſtr.* 1.
*p.* 40. *t.* 64.
Golden Chærophyllum.
*Nat.* of Germany.
*Cult.* 1570, by Mr. Penn.   *Lobel. adv.* 327.
*Fl.* July.                                               H. ♃.

IMPERATORIA.   *Gen. pl.* 359.

*Fructus* ſubrotundus, compreſſus, medio gibbus, margine cinctus.   *Petala* inflexo-emarginata.

*Oſtruthi-*   1. IMPERATORIA.   *Sp. pl.* 371.
*um.*            Common Maſter-wort.
*Nat.* of Scotland.
*Fl.* May——July.                          H. ♃.
SESELI.

SESELI. *Gen. pl.* 360.

*Umbellæ* globofæ. *Involucr.* foliolo uno alterove. *Fructus* ovatus, ftriatus.

1. S. petiolis ramiferis membranaceis oblongis integris, *monta-*
   foliis caulinis anguftiffimis. *Sp. pl.* 372. *num.*
   Sefeli multicaule. *Jacqu. hort.* 2. *p.* 59. *t.* 129.
   Long-leav'd Meadow-faxifrage.
   *Nat.* of Italy and France.
   *Cult.* 1658, in Oxford Garden. *Hort. oxon. edit.* 2.
   *p.* 109.
   *Fl.* June and July. H. ♃.

2. S. petiolis ramiferis membranaceis oblongis integris: *glaucum.*
   foliolis fingularibus binatifque canaliculatis lævibus
   petiolo longioribus. *Sp. pl.* 372. *Jacqu. auftr.* 2.
   *p.* 27. *t.* 144.
   Glaucous Meadow-faxifrage.
   *Nat.* of France.
   *Cult.* 1759, by Mr. Ph. Miller. *Mill. dict. edit.* 7. *n.* 3.
   *Fl.* July and Auguft. H. ♃.

3. S. petiolis rameis fubmembranaceis laxis integerri- *arifta-*
   mis, foliis fupradecompofitis: foliolis lanceolatis *tum.*
   ariftatis, fructibus ovatis.
   Ligufticum lucidum. *Mill. dict.*
   Bearded-leav'd Meadow-faxifrage.
   *Nat.* of the Pyrenean Mountains.
   *Cult.* 1759, by Mr. Philip Miller. *Mill. dict. edit.* 7.
   Ligufticum 4.
   *Fl.* June and July. H. ♂.

*Ammoi-*
*des.*

4. S. foliis radicalibus: foliolis imbricatis.  *Syſt. veget.*
       289.  *Jacqu. hort.* 1. *p.* 20. *t.* 52.
    Millfoil-leav'd Meadow-ſaxifrage.
    *Nat.* of the South of Europe.
    *Cult.* 1759.   *Mill. dict. edit.* 7. *n.* 6.
    *Fl.* June and July.                                    H. ☉.


*tortuo-*
*ſum.*

5. S. caule alto rigido, foliolis linearibus faſciculatis. *Sp.*
       *pl.* 373.
    Hard Meadow-ſaxifrage.
    *Nat.* of the South of Europe.
    *Cult.* 1656, by Mr. John Tradeſcant, Jun.  *Trad.*
       *muſ.* 168.
    *Fl.* October.                                         H. ♂.


*Hippo-*
*mara-*
*thrum.*

6. S. involucellis connato-monophyllis.   *Sp. pl.* 374.
       *Jacqu. auſtr.* 2. *p.* 26. *t.* 143.
    Various-leav'd Meadow-ſaxifrage.
    *Nat.* of Auſtria.
    *Cult.* 1656, by Mr. John Tradeſcant, Jun.  *Trad.*
       *muſ.* 122.
    *Fl.* July.                                            H. ♃.


# THAPSIA.  *Gen. pl.* 361.

*Fructus* oblongus, membrana cinctus.

*villoſa.*

1. T. foliolis dentatis villoſis baſi coadunatis.  *Sp. pl.*
       375.
    Deadly Carrot.
    *Nat.* of France, Spain, and Portugal.
    *Cult.* 1739, by Mr. Philip Miller.  *Rand. chel. n.* 1.
    *Fl.* June and July.                                   H. ♃.


PASTI-

## PASTINACA. *Gen. pl.* 362.

*Fru&tus* ellipticus, compreffo-planus. *Petala* involuta, integra.

1. P. foliis fimplicibus cordatis lobatis lucidis acute cre-     *lucida.*
    natis. *Linn. mant.* 58. *Gouan illuftr.* 19. *t.* 11,
    12. *Jacqu. hort.* 2. *p.* 94. *t.* 199.
    Shining-leav'd Parfnep.
    *Nat.* of the South of Europe.
    *Introd.* 1771, by Monf. Richard.
    *Fl.* June and July.                   H. ♂.

2. P. foliis fimpliciter pinnatis. *Sp. pl.* 376.     *fativa.*
    Garden Parfnep.
    *Nat.* of England.
    *Fl.* July.                           H. ♂.

3. P. foliis pinnatis : foliolis bafi antica excifis. *Syft.*     *Opopa-*
    *veget.* 290.                                       *nax.*
    Paftinaca Opopanax. *Gouan illuftr.* 19. *t.* 13, 14.
    Rough Parfnep.
    *Nat.* of the South of Europe.
    *Cult.* 1731, by Mr. Philip Miller. *Mill. dict. edit.* 1.
    *n.* 3.
    *Fl.* June and July.                   H. ♃.

## SMYRNIUM. *Gen. pl.* 363.

*Fru&tus* oblongus, ftriatus. *Petala* acuminata, carinata.

1. S. foliis caulinis fimplicibus amplexicaulibus. *Sp. pl.*     *perfolia-*
    376.                                         *tum.*
    Perfoliate Alexanders.
    *Nat.* of Candia and Italy.

                                                              *Cult.*

*Cult.* 1596, by Mr. John Gerard.    *Hort. Ger.*
*Fl.* May.                                                          H. ♂.

*Olusa-*       2. S. foliis caulinis ternatis petiolatis ferratis. *Sp. pl.* 376.
*trum.*            Common Alexanders.
                   *Nat.* of Britain.
                   *Fl.* May and June.                               H. ♂.

*aureum.*      3. S. foliis pinnatis ferratis; pofticis ternatis, flofculis
                   omnibus fertilibus.   *Sp. pl.* 377.
                   Golden Alexanders.
                   *Nat.* of North America.
                   *Cult.* before 1699, by Mr. Jacob Bobart.   *Morif.*
                   *hift.* 3. *p.* 281, *n.* 13.
                   *Fl.* May and June.                               H. ♃.

### A N E T H U M.    *Gen. pl.* 364.

*Fructus* fubovatus, compreffus, ftriatus.  *Petala* involuta, integra.

*graveo-*      1. A. fructibus compreffis.   *Sp. pl.* 377.
*lens.*            Common Dill.
                   *Nat.* of Spain and Portugal.
                   *Cult.* 1597.   *Ger. herb.* 878.
                   *Fl.* June and July.                              H. ♂.

*Fœnicu-*      2. A. fructibus ovatis.   *Sp. pl.* 377.
*lum.*             Common Fennel.
                   *Nat.* of England.
                   *Fl.* July and Auguft.                           H. ♃.

### C A R U M.    *Gen. pl.* 365.

*Fructus* ovato-oblongus, ftriatus.  *Involucr.* 1-phyl-
lum.  *Petala* carinata, inflexo-emarginata.

*Carvi.*       1. CARUM.  *Sp. pl.* 378.  *Jacqu. auftr.* 4. *p.* 49. *t.* 393.
                                                                 Common

Common Caraway.
*Nat.* of Britain.
*Fl.* May and June.　　　　　　　　H. ♂ .

PIMPINELLA, *Gen. pl.* 366.

*Fructus* ovato-oblongus. *Petala* inflexa. *Stigmata*
　　　　　ſubglobofa.

1. P. foliis pinnatis : foliolis radicalibus ſubrotundis ; *ſaxifra-*
ſummis linearibus. *Sp. pl.* 378. *Jacqu. auſtr.* 4. *ga.*
*p.* 50. *t.* 395.
Small Burnet-ſaxifrage.
*Nat.* of Britain.
*Fl.* June——Auguſt.　　　　　　　H. ♃ .

2. P. foliolis omnibus lobatis : impari trilobo. *Linn. magna.*
*mant.* 219. *Jacqu. auſtr.* 4. *p.* 50. *t.* 396.
Great Burnet-ſaxifrage.
*Nat.* of England.
*Fl.* Auguſt.　　　　　　　　H. ♃ .

3. P. foliis radicalibus pinnatis crenatis ; ſummis cunei- *peregri-*
formibus inciſis, umbellis nubilibus nutantibus. *na.*
*Syſt. veget.* 291. *Jacqu. hort.* 2. *p.* 61. *t.* 131.
Nodding Burnet-ſaxifrage.
*Nat.* of Italy.
*Cult.* 1739, by Mr. Philip Miller. *Rand. chel.* Api-
um 13.
*Fl.* June and July.　　　　　　　H: ♃ .

4. P. foliis radicalibus trifidis inciſis. *Sp. pl.* 379. *Aniſum.*
Aniſe.
*Nat.* of Egypt.
*Cult.* 1551. *Turn. herb. part* 1. *ſign. D j.*
*Fl.* July.　　　　　　　　H. ☉ .
　　　　　　　　　　　　　　　5. P.

*dioica.*   5. P. umbellis numerosiffimis compositis.simplicibusque,
       floribus dioicis.

      Pimpinella dioica. *Syst. veget.* 291. *Hudf. angl. ed.* 2.
        *p.* 128.

      Pimpinella pumila. *Jacqu. austr.* 1. *p.* 19. *t.* 28.

      Sefeli pumilum. *Sp. pl.* 373.

      Peucedanum minus. *Hudf. angl. ed.* 1. *p.* 101. *Linn.*
        *mant.* 219. *Syst. veget.* 280.

      Least Pimpinell, or Burnet-faxifrage.

      *Nat.* of England.

      *Fl.* May and June.                H. ♃.

## A P I U M.   *Gen. pl.* 367.

*Fructus* ovatus, striatus. *Involucr.* 1-phyllum. *Petala*
aequalia.

*Petrose-*   1. A. foliolis caulinis linearibus, involucellis minutis.
*linum.*     *Sp. pl.* 379.

   α Apium sativum.   *Riv. pent.* 88.
    Common Parsley.

   β Apium crispum.   *Riv. pent.* 90.
    Curled Parsley.

   γ Apium radice efculenta.   *Hort. upf.* 67.
    Large-rooted Parsley.

    *Nat.* of Sardinia.

    *Cult.* 1551.   *Turn. herb. part* 1. *fign. D iiij.*

    *Fl.* June and July.              H. ♂.

*graveo-*   2. A. foliolis caulinis cuneiformibus.   *Syst. veget.* 292.
*lens.*   α Apium palustre & Apium officinarum. *Bauh. pin.* 154.
    Smallage.

   β Apium dulce, Celeri italorum.   *Tourn. inst.* 305.
    Celery.

    *Nat.* of Britain.

    *Fl.* June and July.              H. ♂.

                             ÆGOPO-

## ÆGOPODIUM. *Gen. pl.* 368.

*Fructus* ovato-oblongus, ftriatus.

1. Æ. foliis caulinis fummis ternatis. *Sp. pl.* 379.     *Podagra-*
Gout-weed.     *ria.*
*Nat.* of Britain.
*Fl.* May——July.          H. ♃.


# TRIGYNIA.

## R H U S. *Gen. pl.* 369.

*Cal.* 5-partitus. *Petala* 5. *Bacca* 1-fperma.

1. R. foliis pinnatis obtufiufcule ferratis ovalibus fubtus    *Coriaria.*
villofis. *Sp. pl.* 379.
Elm-leav'd Sumach.
*Nat.* of the South of Europe, and the Levant.
*Cult.* 1648, in Oxford Garden. *Hort. oxon. edit.* 1.
*p.* 44.
*Fl.* July.          H. ♄.

2. R. foliis pinnatis lanceolatis argute ferratis: fubtus    *typhinum.*
tomentofis. *Sp. pl.* 380.
Virginian Sumach.
*Nat.* of Virginia and Carolina.
*Cult.* 1629. *Park. parad.* 609. *f.* 6.
*Fl.* July.          H. ♄.

3. R. foliis pinnatis lanceolatis ferratis utrinque nudis,    *glabrum.*
floribus hermaphroditis.
Rhus glabrum. *Sp. pl.* 380.
Scarlet Sumach.

                       *Nat.*

*Nat.* of North America.
*Cult.* 1726, by James Sherard, M. D. *Dill. elth.* 323.
t. 243. f. 314.
*Fl.* July and Auguft.                              H. ♄.

*elegans.* 4. R. foliis pinnatis lanceolatis ferratis utrinque nudis,
floribus dioicis.
Rhus glabrum panicula fpeciofa coccinea. *Catefb.*
*car. app.* 4. t. 4.
Carolina Sumach.
*Nat.* of South Carolina. Mr. *Mark Catefby.*
*Introd.* 1726.
*Fl.* July.                                          H. ♄.

*Vernix.* 5. R. foliis pinnatis integerrimis annuis opacis, petiolo
integro æquali. *Syft. veget.* 293.
Varnifh Sumach.
*Nat.* of North America.
*Cult.* 1713, in Chelfea Garden. *Philofoph. tranf.*
n. 337. p. 220. n. 156.
*Fl.* July.                                          H. ♄.

*fucceda-* 6. R. foliis pinnatis integerrimis perennantibus lucidis,
*neum.* petiolo integro æquali. *Linn. mant.* 221.
Red Lac Sumach.
*Nat.* of China and Japan.
*Introd.* 1773, by John Blake, Efq.
*Fl.* June.                                       G. H. ♄.

*Copalli-* 7. R. foliis pinnatis integerrimis, petiolo membranaceo
*num.* articulato. *Sp. pl.* 380.
Lentifcus-leav'd Sumach.
*Nat.* of North America.
*Cult.* 1697, by the Dutchefs of Beaufort. *Br. Muf.*
*Sloan. mff.* 3357. fol. 56.
*Fl.* Auguft and September.                       H. ♄.

8. R.

8. R. foliis ternatis: foliolis petiolatis ovatis nudis in-    *radicans.*
   tegerrimis, caule radicante.  *Sp. pl.* 381.

α Toxicodendron amplexicaule, foliis minoribus gla-    opacum.
   bris.  *Dill. elth.* 390.
   Common upright Poifon-oak, or Sumach.

β Toxicôdendron rectum, foliis minoribus glabris. *Dill.*    lucidum.
   *elth.* 389. *t.* 291.
   Small-leav'd Poifon-oak, or Sumach.
   *Nat.* of Virginia and Canada.
   *Cult.* 1727, by James Sherard, M.D.  *Dill. elth. loc.*
   *cit.*
   *Fl.* June and July.                              H. ♄.


9. R. foliis ternatis: foliolis petiolatis angulatis pubef-    *Toxico-*
   centibus, caule radicante.  *Sp. pl.* 381.    *dendron.*
   Trailing Poifon-oak, or Sumach.
   *Nat.* of North America.
   *Cult.* 1640.  *Park. theat.* 679. *f.* 5.
   *Fl.* June and July.                              H. ♄.


10. R. foliis ternatis: foliolis petiolatis ovatis fubtus to-    *Cominia.*
    mentofis remotiffime ferratis.  *Sp. pl.* 381.
    Jamaica Sumach.
    *Nat.* of Jamaica.
    *Introd.* 1778, by Thomas Clark, M.D.
    *Fl.*                                            S. ♄.


11. R. foliis ternatis: foliolis feffilibus ovato-rhombeis    *aromati-*
    incifo-ferratis pilofiufculis.    *cum.*
    Aromatic Sumach.
    *Nat.* of Carolina.  Mr. *John Bartram.*
    *Introd.* 1772.
    *Fl.* May.                                       H. ♄.


                                                  12. R.

*fuaveo-*
*lens.*

12. R. foliis ternatis : foliolis feffilibus cuneiformi-rhom-
 beis incifo-ferratis glabris.

Sweet Sumach.

*Nat.* of North America.

*Cult.* 1759, by Mr. Philip Miller.

*Fl.* May.         H. ♄.

*tomento-*
*fum.*

13. R. foliis ternatis : foliolis fubpetiolatis rhombeis angulatis fubtus tomentofis. *Sp. pl.* 382.

Woolly-leav'd Sumach.

*Nat.* of the Cape of Good Hope.

*Cult.* 1694, by Mr. Jacob Bobart. *Br. Muf. Sloan.*
 *mff.* 3343.

*Fl.*         G. H. ♄.

*villofum.*

14. R. foliis ternatis : foliolis obovatis integerrimis feffilibus utrinque pilofis.

Rhus villofum. *Linn. fuppl.* 183.

Rhus africanum trifoliatum majus, folio fubrotundo
 integro molli et incano. *Pluk. alm.* 319. *t.* 219.
 *f.* 8.

Hairy Sumach.

*Nat.* of the Cape of Good Hope.

*Cult.* 1714. *Philofoph. tranf. n.* 346. *p.* 364. *n.* 130.

*Fl.* July.         G. H. ♄.

*viminale.*

15. R. foliis ternatis : foliolis lineari-lanceolatis integerrimis glabris bafi attenuatis ; intermedio fubpe-
 tiolato.

Willow-leav'd Sumach.

*Nat.* of the Cape of Good Hope. Mr. *Francis Maffon.*
 *Introd.* 1774.

*Fl.*         G. H. ♄.

16. R.

16. R. foliis ternatis: foliolis petiolatis lineari-lanceola- *angustifo-*
tis integerrimis fubtus tomentofis. *Sp. pl.* 382. *lium.*
Narrow-leav'd Sumach.
*Nat.* of the Cape of Good Hope.
*Cult.* 1714. *Philofoph. tranf. n.* 346. *p.* 364. *n.* 129.
*Fl.* G. H. ♄.

17. R. foliis ternatis: foliolis feffilibus lanceolatis lævi- *læviga-*
bus. *Sp. pl.* 1672. *tum.*
Smooth-leav'd Sumach.
*Nat.* of the Cape of Good Hope.
*Cult.* 1758, by Mr. Philip Miller.
*Fl.* G. H. ♄.

18. R. foliis ternatis: foliolis feffilibus cuneiformibus *lucidum.*
lævibus. *Sp. pl.* 382.
α Rhus arboreum trifoliatum latifolium. *Burm. afr.*
252. *t.* 91. *f.* 2.
Great Shining-leav'd Sumach.
β Vitex trifolia minor indica rotundifolia. *Comm. hort.*
1. *p.* 181. *t.* 93.
Small Shining-leav'd Sumach.
*Nat.* of the Cape of Good Hope.
*Cult.* 1697, by the Dutchefs of Beaufort. *Br. Muf.*
*Sloan. mff.* 3343.
*Fl.* July and Auguft. G. H. ♄.

19. R. foliis fimplicibus obovatis. *Sp. pl.* 383. *Jacqu.* *Cotinus.*
*auftr.* 3. *p.* 6. *t.* 210.
Venus's Sumach.
*Nat.* of Italy and Auftria..
*Cult.* 1656, by Mr. John Tradefcant, Jun. *Trad.*
*muf.* 102.
*Fl.* June and July. H. ♄.

B b VIBUR-

# VIBURNUM. *Gen. pl.* 370.

*Cal.* 5-partitus, fuperus. *Cor.* 5-fida. *Bacca* 1-fperma.

*Tinus.*  1. V. foliis integerrimis ovatis: ramificationibus vena-
rum fubtus villofo-glandulofis. *Sp. pl.* 383. *Curtis magaz.* 38.

hirtum.  α foliis ovali-oblongis fubtus margineque hirtis.
Tinus I.  *Cluf. hiſt.* 1. *p.* 49.
Hairy Laureftine.

lucidum.  β foliis ovato-oblongis utrinque glabris lucidis.
Viburnum lucidum. *Mill. dict.*
Tinus II.  *Cluf. hiſt.* 1. *p.* 49. *fig.*
Shining Laureftine.

virga-
tum.  γ foliis lanceolato-oblongis margine venifque fubtus pilofis.
Tinus III.  *Cluf. hiſt.* 1. *p.* 49. *fig.*
Common Laureftine.

ftrictum.  δ foliis ovatis undique hirtis rigidis.
Upright Laureftine.
*Nat.* of the South of Europe.
*Cult.* 1596, by Mr. John Gerard.  *Hort. Ger.*
*Fl.* moft part of the Winter.          H. ♄ .

nudum.  2. V. foliis ovalibus fubrugofis margine revolutis obfo-
lete crenulatis.
Viburnum nudum.  *Sp. pl.* 383.  *du Roi hort. harbecc.*
2. *p.* 484.
Oval-leav'd Viburnum.
*Nat.* of North America.
*Cult.* 1758, by Mr. Ph. Miller.  *Mill. ic.* 183. *t.* 274.
*Fl.* May and June.          H. ♄ .

Caſſinoi-
des.  3. V. foliis lanceolatis lævibus margine revolutis obfolete crenulatis.

Viburnum

Viburnum Caffinoides. *Sp. pl.* 384. (exclufis fyno-
nymis).
Thick-leav'd Viburnum.
*Nat.* of North America.
*Cult.* 1761, by Mr. James Gordon.
*Fl.* June.                                               H. ♄.

4. V. foliis lineari-lanceolatis fupra nitidis obfolete fer-   *nitidum.*
ratis integrifve.
Shining-leav'd Viburnum.
*Nat.* of North America.
*Cult.* 1758, by Mr. Chriftopher Gray.
*Fl.* May and June.                                       H. ♄.

5. V. foliis lanceolatis lævibus remote ferratis bafi inte-   *læviga-*
gerrimis.                                                     *tum.*
Viburnum Caffinoides. *Mill. dict. du Roi hort. har-*
*becc.* 2. *p.* 486.
Caffine Peragua. *Linn. mant.* 220.
Caffioberry Bufh.
*Nat.* of Carolina.
*Cult.* 1724. *Furber's catal.*
*Fl.* July and Auguft.                                    H. ♄.

6. V. foliis obovato-fubrotundis ovalibufque glabris ar-   *prunifo-*
gute ferratis, petiolis marginatis.                          *lium.*
Viburnum prunifolium. *Sp. pl.* 383.
Viburnum Lentago. *du Roi hort. harbecc.* 2. *p.* 485.
*Moench hort. weiffenft.* 140. *tab.* 8.
Plum-leav'd Viburnum.
*Nat.* of North America.
*Cult.* 1731. *Mill. dict. edit.* 1.   Mefpilus 11.
*Fl.* May and June.                                       H. ♄.

7. V.

*Lentago.*    7. V. foliis lato-ovatis acuminatis argute ferratis, petio-
lis marginatis crifpis.
Viburnum Lentago.    *Sp. pl.* 384.
Pear-leav'd Viburnum.
*Nat.* of North America.
*Cult.* 1761, by Mr. James Gordon.
*Fl.* July.    H. ♄.

*dentatum.*    8. V. foliis ovatis dentato-ferratis plicatis.    *Sp. pl.* 384.
*Jacqu. hort.* 1. *p.* 13. *t.* 36.
*lucidum.*    α foliis utrinque glabris.
Shining tooth'd leav'd Viburnum.
*pubef-*    β foliis acuminatis fubtus villofis.
*cens.*    Downy tooth'd-leav'd Viburnum.
*Nat.* of North America.
*Introd.* 1736, by Peter Collinfon, Efq.  *Coll. mſſ.*
*Fl.* June and July.    H. ♄.

*Lantana.*    9. V. foliis cordatis ferratis venofis fubtus tomentofis.
*Sp. pl.* 384.  *Jacqu. auſtr.* 4. *p.* 21. *t.* 341.
*europæ-*    α foliis minoribus obfcure viridibus.
*um.*    Common way-faring Tree, or Viburnum.
*grandifo-*    β foliis majoribus læte viridibus.
*lium.*    Large-leav'd' way-faring Tree, or Viburnum.
*Nat.* α. of Britain; β. of North America.
*Fl.* June.    H. ♄.

*acerifoli-*    10. V. foliis lobatis, petiolis lævibus.    *Sp. pl.* 384.
*um.*    Maple-leav'd Viburnum.
*Nat.* of Virginia.
*Introd.* 1736, by Peter Collinfon, Efq.  *Coll. mſſ.*
*Fl.* July.    H. ♄.

*Opulus.*    11. V. foliis lobatis, petiolis glandulofis.    *Sp. pl.* 384.
*europæa.*    α ramulis viridibus opacis.
Marſh

Marſh Viburnum, or Gelder Roſe.

β ramulis rubicundis lucidis.     *america-na.*

Red-twig'd Marſh Viburnum, or Gelder Roſe.

γ Sambucus aquatica flore globoſo pleno. *Bauh. pin.* roſea.
456.

Snow-ball Viburnum, or Gelder Roſe.

*Nat. α.* of Britain ; β. of North America.

*Fl.* May and June.     H. ♄ .


### C A S S I N E. *Gen. pl.* 371.

*Cal.* 5-partitus.   *Petala* 5.   *Bacca* 3-ſperma.

1. C. foliis petiolatis ſerratis ovatis obtuſis, ramulis te- *capenſis.*
tragonis. *Linn. mant.* 220.

Cape Caſſine, or Phillyrea.

*Nat.* of the Cape of Good Hope.

*Cult.* 1726, by James Sherard, M.D. *Dill. elth.* 315.
*t.* 236. *f.* 305.

*Fl.* July and Auguſt.     G. H. ♄ .

2. C. foliis ſeſſilibus integerrimis obovatis coriaceis. *Syſt.* *Mauro-cenia.*
veget. 295.

Great Hottentot Cherry.

*Nat.* of the Cape of Good Hope.

*Introd.*1690, by Mr. Bentick. *Br. Muſ. Sloan. mſſ.*3370.

*Fl.* July and Auguſt.     G. H. ♄ .


### S A M B U C U S. *Gen. pl.* 372.

*Cal.* 5-partitus.   *Cor.* 5-fida.   *Bacca* 3-ſperma.

1. S. cymis trifidis, ſtipulis foliaceis, caule herbaceo. *Ebulus.*
*Sp. pl.* 385. *Curtis lond.*

Dwarf Elder.

*Nat.* of Britain.

*Fl.* June and July.     H. ♃ .

*canaden-*  2. S. cymis quinquepartitis, foliis fubbipinnatis, caule
*fis.*            frutefcente.  *Sp. pl.* 385.
                Canadian Elder.
                *Nat.* of North America.
                *Cult.* 1768, by Mr. Philip Miller.  *Mill. dict. edit.* 8.
                *Fl.* June——Auguft.                                    H. ♄.


*nigra.*   3. S. cymis quinquepartitis, caule arboreo.  *Sp. pl.* 385.
*pulla.*    α Sambucus fructu in umbella nigro:  *Bauh. pin.* 456.
                Common Black-berried Elder.
*viridis.*   β Sambucus fructu in umbella viridi.  *Bauh. pin.* 456.
                Green-berried Elder.
*laciniata.*  γ Sambucus laciniato folio.  *Bauh. pin.* 456.
                Sambucus laciniata.  *du Roi hort. harbecc.* 2. *p.* 413.
                   *Syft. veget.* 296.
                Parfley-leav'd Elder.
                *Nat.* of Britain.
                *Fl.* May——July.                                      H. ♄.


*racemofa.*  4. S. racemis compofitis ovatis, caule arboreo.  *Sp. pl.*
                386. *Jacqu. ic. collect.* 1. *p.* 36.
                Red-berried Elder.
                *Nat.* of the South of Europe.
                *Cult.* 1596, by Mr. John Gerard.  *Hort. Ger.*
                *Fl.* April and May.                                   H. ♄.


### S P A T H E L I A.  *Gen. pl.* 373.

*Cal.* 5-phyllus.  *Petala* 5.  *Capf.* 3-gona, 3-locularis.
*Sem.* folitaria.

*fimplex.*  1. SPATHELIA.  *Sp. pl.* 386.
                Rhus-leav'd Spathelia.
                *Nat.* of Jamaica.
                *Introd.* 1778, by William Wright, M. D.
                *Fl.*                                                  S. ♄.

STAPHY-

## STAPHYLEA. *Gen. pl.* 374.

*Cal.* 5-partitus. *Petala* 5. *Capf.* inflatæ, connatæ. *Sem.* 2, globofa cum cicatrice.

1. S. foliis pinnatis. *Sp. pl.* 368.          *pinnata.*
Five-leav'd Bladder-nut.
*Nat.* of England.
*Fl.* April——June.        H. ♄ .

2. S. foliis ternatis. *Sp. pl.* 368.         *trifolia.*
Three-leav'd Bladder-nut.
*Nat.* of Virginia.
*Introd.* before 1640, by Mr. John Tradefcant, Sen.
    *Park. theat.* 1417.
*Fl.* May and June.        H. ♄ .

## TAMARIX. *Gen. pl.* 375.

*Cal.* 5-partitus. *Petala* 5. *Capf.* 1-locularis, 3-valvis. *Sem.* pappofa.

1. T. floribus pentandris. *Sp. pl.* 386.       *gallica.*
French Tamarifk.
*Nat.* of France, Spain, and Italy.
*Cult.* 1596, by Mr. John Gerard. *Hort. Ger.*
*Fl.* May——October.        H. ♄ .

2. T. floribus decandris. *Sp. pl.* 387.     *germani-*
German Tamarifk.                      *ca.*
*Nat.* of Germany.
*Cult.* 1596, by Mr. John Gerard. *Hort. Ger.*
*Fl.* June——September.        H. ♄ .

## XYLOPHYLLA. *Linn. mant.* 147.

*Cal.* 5-partitus, coloratus. *Cor.* o. *Stigmata* lacera.
*Capf.* 3-locularis. *Sem.* gemina.

*latifolia.* 1. X. foliis rhombeis crenatis : crenis approximatis floriferis.

Xylophylla latifolia. *Syft. veget.* 296. (exclufo fynonymo fpecierum plantarum.)

Hemionitidi affinis americana epiphyllanthos, folio
fimpliciter pinnato, Hippogloffi æmulo, radice rep-
tatrice lignofa, ad foliorum crenas florida. *Pluk.
phyt. t.* 36. *f.* 7.

Broad-leav'd Sea-fide Laurel.

*Nat.* of Jamaica.

*Introd.* 1783, by Matthew Wallen, Efq.

*Fl.* Auguft——October. S. ♄.

*falcata.* 2. X. foliis lineari-lanceolatis crenatis : crenis remotis
floriferis.

Xylophylla falcata. *Swartz prodr.* 23.

Phyllanthus Epiphyllanthus. *Sp. pl.* 1392. (exclufo
fynonymo Plukenetii altero.)

Narrow-leav'd Sea-fide Laurel.

*Nat.* of the Bahama Iflands.

*Cult.* 1739, by Mr. Philip Miller. *Mill. dict. vol.* 2.
Addenda. Phyllanthos.

*Fl.* July and Auguft. S. ♄.

*ramiflo-* 3. X. foliis ellipticis, floribus axillaribus.
*ra.*
Pharnaceum fuffruticofum. *Pallas it.* 3. *p.* 716. *tab.*
E. *fig.* 2.

Siberian Xylophylla.

*Nat.* of Siberia.

*Introd.* 1783, by Mr. John Bell.

*Fl.* Auguft. H. ♄.

TUR-

## TURNERA. *Gen. pl.* 376.

*Cal.* 5-fidus, infundibulif. exterior 2-phyllus. *Petala* 5, calyci inferta. *Stigmata* multifida. *Capf.* 1-locularis, 3-valvis.

1, T. floribus feffilibus petiolaribus, foliis bafi biglandu- *ulmifolia.* lofis. *Sp. pl.* 387.
Elm-leav'd Turnera.
*Nat.* of Jamaica.
*Cult.* 1733, by Mr. Philip Miller. *R. S. n.* 597.
*Fl.* June——November.                                            S. ♂.

2. T. pedunculis axillaribus aphyllis, foliis apice ferra- *cifloides.* tis. *Syft. veget.* 297.
Betony-leav'd Turnera.
*Nat.* of Jamaica.
*Introd.* 1774, by Monf. Richard.
*Fl.* June——October.                                            S. ☉.

## TELEPHIUM. *Gen. pl.* 377.

*Cal.* 5-phyllus. *Petala* 5, receptaculo inferta. *Capf.* 1-locularis, 3-valvis.

1. T. foliis alternis. *Sp. pl.* 388.                           *imperati.*
True Orpine.
*Nat.* of the South of France and Italy.
*Cult.* 1739. *Mill. dict. vol.* 2. Telephium 1.
*Fl.* June——Auguft.                                             H. ♃.

## CORRIGIOLA. *Gen. pl.* 378.

*Cal.* 5-phyllus. *Petala* 5. *Semen* 1, triquetrum.

1. CORRIGIOLA. *Sp. pl.* 388.                                    *litoralis.*
Baftard Knot-grafs.

*Nat.*

*Nat.* of England.
*Fl.* July and Auguſt. H. ☉.


## PHARNACEUM. *Gen. pl.* 379.

*Cal.* 5-phyllus. *Cor.* o. *Capſ.* 3-locularis, polyſperma.

*Cerviana.* 1. P. pedunculis ſubumbellatis lateralibus æquantibus fo-
lia linearia. *Sp. pl.* 388.
Umbel'd Pharnaceum.
*Nat.* of Ruſſia and Spain.
*Introd.* 1771, by Monſ. Richard.
*Fl.* June. H. ☉.

*incanum.* 2. P. pedunculis communibus longiſſimis, foliis lineari-
bus, ſtipulis piloſis. *Syſt. veget.* 297.
Linear-leav'd Pharnaceum.
*Nat.* of the Cape of Good Hope.
*Introd.* 1782, by George Wench, Eſq.
*Fl.* moſt part of the Summer. G. H. ♄.

*dichoto-*
*mum.* 3. P. pedunculis axillaribus elongatis dichotomis, foliis
verticillatis linearibus. *Linn. ſuppl.* 186.
Fork'd Pharnaceum.
*Nat.* of the Cape of Good Hope.
*Introd.* 1783, by John Earl of Bute.
*Fl.* July. S. ☉.


## ALSINE. *Gen. pl.* 380.

*Cal.* 5-phyllus. *Petala* 5, æqualia. *Capſ.* 1-locularis,
3-valvis.

*media.* 1. A. petalis bipartitis, foliis ovato-cordatis. *Sp. pl.* 389.
*Curtis lond.*
Chickweed.

*Nat.*

*Nat.* of Britain.
*Fl.* April——October. H. ⊙.

2. A. petalis integris brevibus, foliis fetaceis, calycibus *mucrona-ta.*
ariftatis. *Sp. pl.* 389.
Briftly Chickweed.
*Nat.* of the South of Europe.
*Introd.* 1777, by Anthony Gouan, M.D.
*Fl.* June. H. ♂.

### D R Y P I S. *Gen. pl.* 381.

*Cal.* 5-dentatus. *Petala* 5. *Capf.* circumfciffa, monofperma.

1. DRYPIS. *Sp. pl.* 390. *Jacqu. hort.* 1. *p.* 19. *t.* 49. *fpinofa.*
Prickly Drypis.
*Nat.* of Barbary and Italy.
*Introd.* 1775, by Jofeph Nicholas de Jacquin, M.D.
*Fl.* June and July. H. ♂.

### P O R T U L A C A R I A. *Jacquin.*

*Cal.* 2-phyllus. *Petala* 5. *Sem.* 1, alato-triquetrum.

1. PORTULACARIA. *Jacqu. collect.* 1. *pag.* 160. *tab.* 22. *afra.*
Craffula Portulacaria. *Sp. pl.* 406.
Claytonia Portulacaria. *Linn. mant.* 211.
Purflane-tree.
*Nat.* of Africa.
*Cult.* 1732, by James Sherard, M.D. *Dillen. elth.*
120.
*Fl.* G. H. ♄.

BASEL-

## B A S E L L A.  *Gen. pl.* 382.

*Cal.* o.  *Cor.* 7-fida : laciniis 2 oppofitis latioribus,
tandem baccata.  *Sem.* 1.

*rubra.*  1. B. foliis planis, pedunculis fimplicibus.  *Sp. pl.* 390.
Red Malabar Nightfhade.
*Nat.* of the Eaft Indies.
*Cult.* 1739, by Mr. Philip Miller.  *Rand. chel. n.* 1.
*Fl.* July——November.                          S. ♂ .

*alba.*  2. B. foliis ovatis undatis, pedunculis fimplicibus folio
longioribus.  *Sp. pl.* 390.
White Malabar Nightfhade.
*Nat.* of China.
*Cult.* 1691, by Bifhop Compton.  *Pluk. phyt. t.* 63.
*f.* 1.
*Fl.* July——November.                          S. ♂ .

## S A R O T H R A.  *Gen. pl.* 383.

*Cal.* 5-partitus.  *Cor.* 5-petala.  *Capf.* 1-locularis,
3-valvis, colorata.

*Gentia-*  1. SAROTHRA.  *Sp. pl.* 391.
*noides.*  Baftard Gentian.
*Nat.* of Virginia and Penfylvania.
*Introd.* 1768, by Mr. John Bartram.
*Fl.* July.                                      H. ☉ .

*TETRAGY-*

# *TETRAGYNIA.*

## PARNASSIA. *Gen. pl.* 384.

*Cal.* 5-partitus. *Petala* 5. *Nectaria* 5, cordata, ciliata apicibus globofis. *Capf.* 4-valvis.

1. PARNASSIA. *Sp. pl.* 391.                    *paluſtris.*
   Grafs of Parnaffus.
   *Nat.* of Britain.
   *Fl.* July and Auguſt.                    H. ♃.

## EVOLVULUS. *Gen. pl.* 385.

*Cal.* 5-phyllus. *Cor.* 5-fida, rotata. *Capf.* 3-locularis.
                    *Sem.* folitaria.

1. E. foliis obcordatis obtufis pilofis petiolatis, caule dif-    *alſinoides.*
   fufo, pedunculis trifloris. *Sp. pl.* 392.
   Chickweed-leav'd Evolvulus.
   *Nat.* of the Eaſt Indies.
   *Introd.* 1771, by Monf. Richard.
   *Fl.* June and July.                    S. ☉.

2. E. foliis lanceolatis villofis feffilibus, caule erecto, pe-    *linifolius.*
   dunculis trifloris longis. *Sp. pl.* 392.
   Flax-leav'd Evolvulus.
   *Nat.* of Jamaica.
   *Introd.* 1782, by Mr. Francis Maffon.
   *Fl.* Auguſt and September.                    S. ☉.

# PENTAGYNIA.

### ARALIA. *Gen. pl.* 386.

*Involucr.* umbellulæ. *Cal.* 5-dentatus, superus. *Cor.*
5-petala. *Bacca* 5-sperma.

*capitata.*　1. A. arborea, foliis simplicibus ellipticis integerrimis,
　　　　paniculis terminalibus: floribus sessilibus capitatis.
　　　　*Swartz prodr.* 55.
　　　Aralia capitata. *Jacqu. hist.* 89. *t.* 61.
　　　Cluster-flower'd Aralia.
　　　*Nat.* of the West Indies.
　　　*Introd.* 1778, by Mr. William Forsyth.
　　　*Fl.* August.　　　　　　　　　　　　　　　　S. ♄.

*spinosa.*　2. A. arborescens, caule foliolisque aculeatis. *Sp. pl.* 392.
　　　Thorny Aralia, or Angelica-tree.
　　　*Nat.* of Virginia.
　　　*Cult.* 1688, by Bishop Compton. *Raj. hist.* 2. *p.* 1798.
　　　*Fl.* September.　　　　　　　　　　　　　　H. ♄.

*racemosa.*　3. A. caule folioso herbaceo lævi. *Sp. pl.* 393.
　　　Berry-bearing Aralia.
　　　*Nat.* of Canada.
　　　*Cult.* 1731, by Mr. Ph. Miller. *Mill. dict. edit.* 1. *n.* 1.
　　　*Fl.* June——September.　　　　　　　　　　H. ♃.

*nudicau-*　4. A. caule nudo, foliis binis ternatis. *Sp. pl.* 393.
*lis.*　　　Naked-stalk'd Aralia.
　　　*Nat.* of Virginia and Canada.
　　　*Cult.* 1731, by Mr. Ph. Miller. *Mill. dict. edit.* 1. *n.* 2.
　　　*Fl.* June and July.　　　　　　　　　　　　H. ♃.

STATICE.

STATICE. *Gen. pl.* 388.

*Cal.* 1-phyllus, integer, plicatus, fcariofus. *Pet.* 5.
Sem. 1, fuperum.

1. S. fcapo fimplici capitato, fquamis calycinis fubrotun-  *Armeria.*
dis obtufiffimis, foliis linearibus.
Statice Armeria. *Sp. pl.* 394.
α Caryophyllus montanus five mediterraneus. *Lob. ic.*  major.
452.
Great common Thrift.
β Caryophyllus marinus omnium minimus. *Lob. ic.* 452.  minor.
Small common Thrift.
*Nat.* of Britain.
*Fl.* May——July.                                        H. ♃.

2. S. fcapo fimplici capitato, fquamis calycinis ovatis acu-  *Cepha-*
minatis.                                                *lotes.*
Statice Armeria major. *Jacqu. hort.* 1. *p.* 16. *t.* 42.
Statice Pfeud-armeria. *Syft. veget.* 300.
Large fimple-ftalk'd Thrift.
*Nat.* of Algarbia.  Mr. *Fr. Maffon.*
*Introd.* 1775, by Jofeph Nicholas de Jacquin, M. D.
*Fl.* May——July.                                        H. ♃.

3. S. fcapo paniculato : ramis triquetris, foliis linearibus  *gramini-*
canaliculatis.                                          *folia.*
Grafs-leav'd Thrift.
*Nat.*
*Introd.* 1780, by Mr. William Malcolm.
*Fl.* June and July.                                    G. H. ♃.

4. S. fcapo paniculato tereti, foliis lævibus enerviis fub-  *Limoni-*
tus mucronatis. *Syft. veget.* 300.                     *um.*
Common Sea Thrift, or Lavender.

*Nat.*

*Nat.* of England.

*Fl.* May——July.               H. ♃.

*cordata.*   5. S. fcapo paniculato, foliis fpathulatis retufis.  *Sp. pl.*
        394.
        Blunt-leav'd Thrift.
        *Nat.* of the South of Europe.
        *Cult.* 1752, by Mr. Philip Miller.  *Mill. dict. edit.* 6.
        Limonium 9.
        *Fl.* May——July.               H. ♃.

*reticula-*   6. S. fcapo paniculato proftrato, ramis fterilibus retro-
*ta.*          flexis nudis, foliis cuneiformibus muticis. *Syst. veg.*
        301.
        Matted Thrift, or Sea Lavender.
        *Nat.* of England.
        *Fl.* July and Auguft.           H. ♃.

*echioides.*   7. S. fcapo paniculato tereti articulato, foliis fcabris.
        *Syst. veget.* 301.
        Rough-leav'd Thrift.
        *Nat.* of the South of Europe.
        *Cult.* 1752, by Mr. Philip Miller.  *Mill. dict. edit.* 6.
        Limonium 10.
        *Fl.* July and Auguft.        G. H. ♂.

*fpeciofa.*   8. S. fcapo dichotomo ancipiti, foliis ovatis mucronatis,
        floribus aggregatis.  *Syst. veget.* 301.
        Plantain-leav'd Thrift.
        *Nat.* of Ruffia.
        *Introd.* 1776, by Chevalier Murray.
        *Fl.* July and Auguft.           H. ♂.

*tatarica.*   9. S. fcapo dichotomo, foliis lanceolatis mucronatis, flo-
        ribus alternis diftantibus.  *Sp. pl.* 395.

                                       Tartarian

Tartarian Thrift.
*Nat.* of Ruffia.
*Cult.* 1731, by Mr. Philip Miller. *Mill. dict. edit.* 1.
Limonium 5.
*Fl.* June. H. ♃.

10. S. caule ramifque paniculatis triquetris, foliis obo- *pectinata.*
vatis petiolatis, fpicis fecundis.
Triangular-ftalk'd Thrift.
*Nat.* of the Canary Iflands. Mr. *Francis Maffon.*
*Introd.* 1780.
*Fl.* September and October. G. H. ♄.

11. S. c ule fruticofo: fuperne nudo ramofo, capitulis *fuffruti-*
feffilibus, foliis lanceolatis vaginantibus. *Sp. pl.* *cofa.*
396.
Narrow-leav'd fhrubby Thrift.
*Nat.* of Siberia.
*Introd.* 1781, by Meffs. Lee and Kennedy.
*Fl.* moft part of the Summer. G. H. ♄.

12. S. caule fruticofo foliofo, floribus folitariis, foliis *monopeta-*
lanceolatis vaginantibus. *Sp. pl.* 396. *la.*
Broad-leav'd fhrubby Thrift.
*Nat.* of Sicily.
*Cult.* 1732, by Mr. Philip Miller. *R. S. n.* 533.
*Fl.* July and Auguft. G. H. ♄.

13. S. caule herbaceo, foliis radicalibus alternatim pin- *finuata.*
nato-finuatis; caulinis ternis triquetris fubulatis
decurrentibus. *Sp. pl.* 396.
Scollop-leav'd Thrift.
*Nat.* of Sicily and the Levant.
*Cult.* 1731, by Mr. Philip Miller. *Mill. dict. edit.* 1.
Limonium 6.
*Fl.* moft part of the Summer. G. H. ♃.

*mucrona-* 14. S. caule crifpo, foliis ellipticis integris, fpicis fecun-
*ta.*          dis. *Linn. fuppl.* 187. *L'Heritier ftirp. nov.* 25.
              *t.* 13.
              Curl'd Thrift.
              *Nat.* of Barbary.
              *Introd.* 1784, by Mr. Francis Maffon.
              *Fl.* moft part of the Summer.                    G. H. ♃.


            L I N U M.  *Gen. pl.* 389.

       *Cal.* 5-phyllus. *Petala* 5. *Capf.* 5-valvis, 10-locula-
                ris. *Sem.* folitaria.

                    * *Foliis alternis.*

*ufitatiffi-* 1. L. calycibus capfulifque mucronatis, petalis crenatis,
*mum.*          foliis lanceolatis alternis, caule fubfolitario. *Sp.*
               *pl.* 397. *Curtis lond.*
              Common Flax.
              *Nat.* of Britain.
              *Fl.* June.                                        H. ☉.


*perenne.* 2. L. calycibus capfulifque obtufiufculis, foliis alternis
              lanceolatis integerrimis. *Sp. pl.* 397.
              Perennial Flax.
              *Nat.* of England.
              *Fl.* June——Auguft.                              H. ♃.


*narbo-*   3. L. calycibus acuminatis, foliis lanceolatis fparfis ftric-
*nenfe.*        tis fcabris acuminatis, caule tereti bafi ramofo. *Sp.*
               *pl.* 398.
              Narbonne Flax.
              *Nat.* of the South of France.
              *Cult.* 1759, by Mr. Ph. Miller. *Mill. dict. edit.* 7. *n.* 3.
              *Fl.* May——July.                                 G. H. ♃.


                                                              .4. L.

4. L. calycibus acuminatis, foliis ovato-lanceolatis *reflexum.*
acuminatis reflexis lævibus, filamentis connatis.
Linum fylveftre cæruleum folio acuto. *Bauh. prod.*
107. *pin.* 214.
Reflex'd-leav'd Flax.
*Nat.* of the South of Europe.
*Introd.* 1778, by Cafimir Gomez Ortega, M.D.
*Fl.* July. H. ♃.

5. L. calycibus acuminatis, foliis fparfis lineari-fetaceis *tenuifoli-*
retrorfum fcabris. *Sp. pl.* 398. *Jacqu. auftr.* 3. *um.*
*p.* 9. *t.* 215.
Narrow-leav'd Flax.
*Nat.* of England.
*Fl.* June and July. H. ♃.

6. L. calycibus fubulatis acutis, foliis lineari-lanceolatis *gallicum.*
alternis, paniculæ pedunculis bifloris, floribus fub-
feffilibus. *Syft. veget.* 302.
Annual Yellow Flax.
*Nat.* of the South of France.
*Introd.* 1777, by Monf. Thouin.
*Fl.* July. H. ⊙.

7. L. calycibus ovatis acutis muticis, foliis lanceolatis: *maritim-*
inferioribus oppofitis. *Sp. pl.* 400. *Jacqu. hort.* 2. *um.*
*p.* 72. *t.* 154.
Sea Flax.
*Nat.* of the South of Europe and the Levant.
*Cult.* 1596, by Mr. John Gerard. *Hort. Ger.*
*Fl.* July and Auguft. H. ♃.

8. L. calycibus rotundatis obtufis, foliis linearibus acu- *alpinum.*
tiufculis, caulibus declinatis. *Syft. veget.* 302. *Jacqu.*
*auftr.* 4. *p.* 11. *t.* 321.

Alpine

Alpine Flax.
*Nat.* of the Alps of Auftria.
*Introd.* 1775, by Monf. Thouin.
*Fl.* July and Auguft.                                 H. ♃.

*auftria-*    9. L. calycibus rotundatis obtufis, foliis linearibus acu-
*cum.*              tis rectiufculis. *Syft. veget.* 302. *Jacqu. auftr.* 5.
                   *p.* 8. *t.* 418.
             Auftrian Flax.
             *Nat.* of Auftria.
             *Introd.* 1775, by Monf. Thouin.
             *Fl.* June and July.                       H. ♂.

*fuffruti-*  10. L. foliis linearibus acutis fcabris, caulibus fuffruti-
*cofum.*           cofis. *Sp. pl.* 400.
             Upright Flax.
             *Nat.* of Spain.
             *Introd.* 1787, by Monf. Thouin.
             *Fl.* Auguft.                              G. H. ♄.

*arbore-*    11. L. foliis cuneiformibus, caulibus arborefcentibus.
*um.*              *Sp. pl.* 400.
             Tree Flax.
             *Nat.* of the Ifland of Candia.
             *Introd.* 1788, by John Sibthorp, M. D.
             *Fl.*                                      G. H. ♄.

                    * * *Foliis oppofitis.*

*africa-*    12. L. foliis oppofitis lineari-lanceolatis, floribus termi-
*num.*             nalibus pedunculàtis. *Sp. pl.* 401. *L'Herit. ftirp.*
                   *nov. tom.* 2. *tab.* 3.
             Shrubby Flax.
             *Nat.* of Africa.
             *Introd.* 1771, by Monf. Richard.
             *Fl.* June and July.                      G. H. ♄.
                                                       13. L.

13. L. foliis oppofitis ovato-lanceolatis, caule dichotomo, *catharti-* corollis acutis. *Sp. pl.* 401. *Curtis lond.* *cum.*
Purging Flax.
*Nat.* of Britain.
*Fl.* Auguft.　　　　　　　　　　　　　H. ☉.

14. L. foliis oppofitis, caule dichotomo, floribus tetran- *Radiola.* dris tetragynis. *Sp. pl.* 402.
Leaft Flax, or All-feed.
*Nat.* of Britain.
*Fl.* Auguft.　　　　　　　　　　　　　H. ☉.

## D R O S E R A. *Gen. pl.* 391.

*Cal.* 5-fidus. *Petala* 5. *Capf.* 1-locularis, apice 5-valvis.
*Sem.* plurima.

1. D. fcapis radicatis, foliis orbiculatis. *Sp. pl.* 402. *rotundi-*
Round-leav'd Sun-dew, or Rofa-folis. *folia.*
*Nat.* of Britain.
*Fl.* July and Auguft.　　　　　　　　　H. ☉.

2. D. fcapis radicatis, foliis oblongis. *Sp. pl.* 403. *longifolia.*
Long-leav'd Sun-dew.
*Nat.* of Britain.
*Fl.* June and July.　　　　　　　　　　H. ☉.

## G I S E K I A. *Linn. mant.* 554.

*Cal.* 5-phyllus. *Cor.* 0. *Capfulæ* 5, approximatæ,
fubrotundæ, 1-fpermæ.

1. GISEKIA. *Linn. mant.* 562. *pharna-*
Trailing Gifekia. *cioides.*
*Nat.* of the Eaft Indies.
*Introd.* 1783, by John Earl of Bute.
*Fl.* June.　　　　　　　　　　　　　　S. ☉.

　　　　　　　　CRAS-

CRASSULA. *Gen. pl.* 392.

*Cal.* 5-phyllus. *Petala* 5. *Squamæ* 5 nectariferæ ad
basin germinis. *Capf.* 5, polyspermæ.

*coccinea.* 1. C. foliis ovatis planis cartilagineo-ciliatis basi connato-
vaginantibus. *Syft. veget.* 304.
Scarlet-flower'd Craffula.
*Nat.* of the Cape of Good Hope.
*Introd.* 1714, by Profeffor Richard Bradley. *Bradl.*
*fucc.* 5. *p.* 7. *t.* 50.
*Fl.* June——Auguft. D. S. ♄.

*perfolia-* 2. C. foliis lanceolatis feffilibus connatis canaliculatis.
*ta.* Craffula perfoliata. *Sp. pl.* 404.
Perfoliate Craffula.
*Nat.* of the Cape of Good Hope.
*Cult.* 1725, by James Sherard, M.D. *Dill. elth.* 114.
*t.* 96. *f.* 113.
*Fl.* July and Auguft. D. S. ♄.

*ramofa.* 3. C. fruticofa, foliis fubulatis fupra planis connato-per-
foliatis lævibus patentiffimis, pedunculis elongatis,
floribus cymofis.
Craffula ramofa. *Thunb. nov. act. nat. cur.* 6. *p.* 341.
Craffula dichotoma. *Linn. fuppl.* 188.
Branching Craffula.
*Nat.* of the Cape of Good Hope. Mr. *Francis Maffon.*
*Introd.* 1774.
*Fl.* July and Auguft. D. S. ♄.

*expanfa.* 4. C. foliis femicylindrico-fubulatis fupra canaliculatis
patentibus, pedunculis axillaribus folitariis unifloris,
caulibus dichotomis.
Awl-leav'd Craffula.

*Nat.*

*Nat.* of the Cape of Good Hope. Mr. *Francis Maffon.*
*Introd.* 1774.
*Fl.* June and July.      D. S. ☉.

5. C. foliis femicylindraceis acutis fubtus gibbis lævibus    *mollis.*
     fuberectis, cymis terminalibus compofitis.
     Craffula mollis. *Thunberg nov. act. ac. nat. cur.* 6.
     *p.* 340. *Linn. fuppl.* 189.
     Fig Marygold-leav'd Craffula.
     *Nat.* of the Cape of Good Hope. Mr. *Francis Maffon.*
     *Introd.* 1774.
     *Fl.* Auguft.      D. S. ♄.

6. C. foliis fubulatis fubincurvis obfolete tetragonis pa-    *tetragona.*
     tentibus, caule erecto arborefcente radicante. *Syft.*
     *veget.* 305.
     Square-leav'd Craffula.
     *Nat.* of the Cape of Good Hope.
     *Introd.* 1714, by Profeffor Richard Bradley. *Bradl.*
     *fucc.* 5. *p.* 18. *t.* 41.
     *Fl.*      D. S. ♄.

7. C. foliis connatis oblongis remotis planis, caule fimplici,    *retroflexa.*
     cyma compofita, pedicellis retrofractis. *Thunberg*
     *nov. act. nat. cur.* 6. *p.* 338. *Linn. fuppl.* 188.
     Orange-flower'd Craffula.
     *Nat.* of the Cape of Good Hope. Mr. *Fr. Maffon.*
     *Introd.* 1788.
     *Fl.* June.      G. H. ☉.

8. C. caule herbaceo, foliis cordatis feffilibus, peduncu-    *lineolata.*
     lis fubterminalibus axillaribus approximatis umbel-
     liformibus.
     Channell'd Craffula.
     *Nat.* of the Cape of Good Hope. Mr. *Francis Maffon.*

C c 4      *Introd.*

*Introd.* 1774.

*Fl.* June——Auguft. D. S. ♂.

centau-
roides.

9. C. caule herbaceo dichotomo, foliis feffilibus oblongo-
ovatis cordatis planis, pedunculis axillaribus unifloris.

Craffula centauroides. *Sp. pl.* 404.

Centory-flower'd Craffula.

*Nat.* of the Cape of Good Hope.

*Introd.* 1774, by Mr. Francis Maffon.

*Fl.* May and June. D. S. ♂.

dichoto-
ma.

10. C. caule herbaceo dichotomo, foliis feffilibus ovato-
oblongis canaliculatis recurvatis, pedunculis axillaribus unifloris.

Craffula dichotoma. *Sp. pl.* 404.

Fork'd Craffula.

*Nat.* of the Cape of Good Hope.

*Introd.* 1774, by Mr. Francis Maffon.

*Fl.* June and July. D. S. ☉.

glomera-
ta.

11. C. caule herbaceo dichotomo fcabro, foliis lanceolatis, floribus ultimis fafciculatis. *Syft. veget.* 305.

α caule fcabrido.

Rough-ftalk'd clufter'd Craffula.

β caule glabrato.

Smooth-ftalk'd clufter'd Craffula.

*Nat.* of the Cape of Good Hope.

*Introd.* 1774, by Mr. Francis Maffon.

*Fl.* June and July. D. S. ☉.

pulchella.

12. C. caule herbaceo dichotomo, foliis ovato-oblongis
carnofis reflexis, floribus in dichotomiis pedunculatis : pedunculis turbinatis.

Reflex'd-leav'd Craffula.

*Nat.* of the Cape of Good Hope. Mr. *Fr. Maffon.*

*Introd.*

*Introd.* 1788.

*Fl.* July.                                                G. H. ⊙.

OBS. Affinis Craffulæ glomeratæ, a qua differt foliis brevioribus reflexis, pedunculis turbinatis germen inferum mentientibus, calycibus obtufis, corollis calyce fere longioribus.

13. C. foliis ovatis acutis quadrifariam imbricatis lævi-      *imbrica-*
bus, floribus axillaribus feffilibus.                            *ta.*
Imbricated Craffula.
*Nat.* of the Cape of Good Hope.
*Introd.* about 1760, by Mr. Clark.
*Fl.* June.                                                D. S. ♄.

14. C. foliis oppofitis obovatis fubcultratis obliquis con-      *cultrata.*
natis integerrimis.
Craffula cultrata.   *Sp. pl.* 405.
Sharp-leav'd Craffula.
*Nat.* of the Cape of Good Hope.
*Cult.* 1732, by James Sherard, M. D.   *Dill. elth.*
115. *t.* 97. *f.* 114.
*Fl.* July and Auguft.                                     D. S. ♄.

15. C. foliis oppofitis ovatis obliquis integerrimis acu-      *obliqua.*
tis diftinctis : margine fubcartilagineis.
Cotyledon ovata.   *Mill. dict.*
Oblique-leav'd Craffula.
*Nat.* of the Cape of Good Hope.
*Cult.* 1759, by Mr. Philip Miller.   *Mill. dict. edit.* 7.
Cotyledon 8.
*Fl.* April and May.                                        D. S. ♄.

16. C. foliis fubrotundis carnofis fupra punctatis, caule      *Cotyledon.*
arboreo.   *Syft. veget.* 305.   *Jacqu. mifcell.* 2. *p.*
29;. *tab.* 19.

                                                    Cotyledon

PENTANDRIA PENTAGYNIA. Craffula.

Cotyledon arborefcens. *Mill. dict.*
Tree Craffula.
*Nat.* of the Cape of Good Hope.
*Cult.* 1739. *Mill. dict. vol.* 2. Cotyledon. 1.
*Fl.* D. S. ♄ .

*alooides.* 17. C. foliis ovatis diftinctis acutis ciliatis, caule fimplici pilofiufculo, racemo compofito: ramis paniculæformibus.
Spiked-flower'd Craffula.
*Nat.* of the Cape of Good Hope. Mr. *Fr. Maffon.*
*Introd.* 1774.
*Fl.* June——Auguft. D. S. ♂ .
DESCR. *Caulis* vix fpithamæus. *Folia* carnofa, glabra, punctis minutis rubris impreffis adfperfa, uncialia. *Racemus* terminalis. *Pedunculi* diftantes,
patentes, fubdivifi in cymas trichotomas. *Calycis*
foliola glabra, lineam longa. *Petala* calyce duplo
longiora, alba, inferne virefcentia. *Nectaria* lutea.
*Filamenta* alba. *Antheræ* fubrotundæ, parvæ.

*capitella.* 18. C. foliis oblongo-lanceolatis acutis connatis ciliatis,
caule lævi, racemo elongato: floribus fafciculatis
fubfeffilibus.
Craffula capitella. *Thunb. nov. act. nat. cur.* 6. *p.* 339.
*Linn. fuppl.* 190.
Square-fpiked Craffula.
*Nat.* of the Cape of Good Hope. Mr. *Fr. Maffon.*
*Introd.* 1774.
*Fl.* July and Auguft. D. S. ♂ .

*ciliata.* 19. C. foliis oppofitis ovalibus planiufculis diftinctis ci-
liatis, corymbis terminalibus. *Sp. pl.* 405.
Ciliated Craffula.
*Nat.* of the Cape of Good Hope.
*Cult.*

*Cult.* 1732, by James Sherard, M.D. *Dill. elth.*
116. *t.* 98. *f.* 116.
*Fl.* July and Auguft.                      D. S. ♃.

20. C. foliis oppofitis patentibus connatis fcabris ciliatis, *fcabra.*
caule retrorfum fcabro.  *Sp. pl.* 405.
Rough-leav'd Craffula.
*Nat.* of the Cape of Good Hope.
*Cult.* 1732, by James Sherard, M.D. *Dill. elth.*
117. *t.* 99. *f.* 117.
*Fl.* June and July.                      D. S. ♄.

21. C. foliis alternis fubfpathulatis acutis integerrimis, *fparfa.*
racemo compofito.
Alternate-leav'd Craffula.
*Nat.* of the Cape of Good Hope.  Mr. *Fr. Maffon.*
*Introd.* 1774.
*Fl.* July.                      D. S. ♂.

22. C. foliis oblongis bafi attenuatis remote crenatis, *diffufa.*
pedunculis oppofitifoliis axillaribufque folitariis.
Diffufe Craffula.
*Nat.* of the Cape of Good Hope.  Mr. *Fr. Maffon.*
*Introd.* 1774.
*Fl.* July.                      D. S. ☉.

23. C. foliis petiolatis cordato-fubrotundis acutiufculis *fpathula-*
crenatis, corymbis paniculæformibus.                      *ta.*
Craffula fpathulata.  *Thunberg nov. act. nat. cur.* 6.
*p.* 330.
Crenated Craffula.
*Nat.* of the Cape of Good Hope.  Mr. *Fr. Maffon.*
*Introd.* 1774.
*Fl.* July and Auguft.                      D. S. ♄.

24. C.

*punctata.*  24. C. foliis oppoſitis ovatis punctatis ciliatis: inferio-
ribus oblongis.  *Sp. pl.* 406.
Dotted Craſſula.
*Nat.* of the Cape of Good Hope.
*Cult.* 1759, by Mr. Philip Miller.  *Mill. dict. edit.* 7.
*n.* 7.
*Fl.* April——Auguſt.                      D. S. ♄.

*margina-*  25. C. caule fruticoſo, foliis cordatis perfoliatis acumina-
*lis.*        tis planis patentibus intra marginem punctatis.
Marginated Craſſula.
*Nat.* of the Cape of Good Hope.  Mr. *Fr. Maſſon.*
*Introd.* 1774.
*Fl.* July and Auguſt.                     D. S. ♄.

*cordata.*  26. C. caule fruticoſo, foliis petiolatis cordatis obtuſis
integerrimis, cymis paniculæformibus.
Craſſula cordata.  *Thunb. nov. act. nat. cur.* 6. *p.* 330.
*Linn. ſuppl.* 189.
Heart-leav'd Craſſula.
*Nat.* of the Cape of Good Hope.  Mr. *Fr. Maſſon.*
*Introd.* 1774.
*Fl.* May——Auguſt.                         D. S. ♄.

*lactea.*  27. C. caule fruticoſo, foliis ovatis baſi attenuatis conna-
tis integerrimis intra marginem punctatis, cymis
paniculæformibus.
Snowy Craſſula.
*Nat.* of the Cape of Good Hope.  Mr. *Fr. Maſſon.*
*Introd.* 1774.
*Fl.* September and October.               D. S. ♄.

*orbicula-*  28. C. ſarmentis proliferis determinate folioſis, foliis pa-
*ris.*        tentiſſimis imbricatis.  *Syſt. veget.* 306.
Starry Craſſula.

                                                    *Nat.*

*Nat.* of the Cape of Good Hope.

*Cult.* 1732, by James Sherard, M. D.  *Dill. elth.*
119. *t.* 100. *f.* 118.

*Fl.* July.  D. S. ♃.

29. C. foliis fparfis femicylindraceis glabris, floribus late-  *rubens.*
ralibus folitariis fubfeffilibus, ramis villofis.

Craffula rubens. *Syft. veget.* 306. (exclufo fynonymo Oederi) *Allion. pedem.* 2. *p.* 123.

Sedum caule hirfuto, ramis fimplicibus, floribus in alis
feffilibus. *Hall. hift.* 960.

Sedum arvenfe flore rubente C. B.  *La Chenal in act.*
*helvet.* 7. *p.* 331. *tab.* 11.

Hardy Annual Craffula.

*Nat.* of Switzerland and Italy.

*Cult.* 1759, by Mr. Philip Miller.

*Fl.* May and June.  H. ☉.

DESCR. In noftra planta : *Caulis* teres, pubefcens,
decumbens, ramofus. *Rami* adfcendentes, villofi,
digitales. *Folia* fparfa, feffilia, glabra, craffitie
pennæ gallinæ, verfus apicem parum attenuata,
ibique rubicunda, inferiora fere uncialia, fuperiora
fenfim breviora. *Flores* in ramis inter folia, fparfi,
remoti, folitarii, breviffime pedunculati. *Calycis*
foliola carnofa, extus convexa, villofiufcula, lineam
longa. *Petala* calyce triplo longiora, alba, carinata: carina villofiufcula, rubicunda. *Nectaria* alba,
minima. *Filamenta* petalis paulo breviora. *Anthe-*
*ræ* fubrotundæ, parvæ, rubicundæ.

30. C. caule herbaceo, foliis patentibus, floribus verticil-  *verticil-*
latis ariftatis. *Syft. veget.* 306. (exclufo fynonymo  *laris.*
Magnolii.)

Whorl-flower'd Craffula.

*Nat.*

*Introd.*

*Introd.* 1788, by Mr. Zier.
*Fl.* July.                                                    H. ☉.

## MAHERNIA. *Linn. mant.* 8.

*Cal.* 5-dentatus.   *Petala* 5.   *Nectaria* 5, obcordata,
     filamentis suppofita.   *Capf.* 5-locularis.

*pinnata.*   1. M. foliis tripartito-pinnatifidis.   *Syft. veget.* 308.
             Hermannia pinnata.   *Sp. pl.* 943.
             Winged Mahernia.
             *Nat.* of the Cape of Good Hope.
             *Introd.* 1774, by Mr. Francis Maffon.
             *Fl.* June——Auguft.                         G. H. ♄.

## SIBBALDIA. *Gen. pl.* 393.

*Cal.* 10-fidus.   *Petala* 5, calyci inferta.   *Styli* e latere
               germinis.   *Sem.* 5.

*procum-*   1. S. foliolis tridentatis.   *Sp. pl.* 406.
*bens.*        Trailing Sibbaldia.
               *Nat.* of Britain.
               *Fl.* July and Auguft.                    H. ♃.

*POLYGY-*

# POLYGYNIA.

MYOSURUS. *Gen. pl.* 394.

*Cal.* 5-phyllus, bafi adnatus. *Nectaria* 5, fubulata, petaliformia. *Sem.* numerofa.

1. MYOSURUS. *Sp. pl.* 407. *Curtis lond.*      *minimus.*
Moufe-tail.
*Nat.* of Britain.
*Fl.* April and May.             H. ⊙.


ZANTHORHIZA. *L'Heritier ftirp. nov.*

*Cal.* o. *Pet.* 5. *Nectaria* 5, pedicellata. *Capf.* 1-fpermæ.

1. ZANTHORHIZA. *L'Heritier ftirp. nov. p.* 79. *t.* 38.   *apiifolia.*
Parfley-leav'd Zanthorhiza.
*Nat.* of North America.
*Introd.* about 1766, by Mr. John Bufh.
*Fl.* February——April.         H. ♄.

*Claſſis VI.*

# HEXANDRIA

## *MONOGYNIA.*

### BROMELIA. *Gen. pl.* 395.

*Cal.* 3-fidus, ſuperus. *Petala* 3. *Squama* nectarifera
ad baſin petali. *Bacca* 3-locularis.

*Ananas.*  1. B. foliis ciliato-ſpinoſis mucronatis, ſpica comoſa.
 *Sp. pl.* 408.

 α Ananas aculeatus, fructu ovato, carne albida. *Trew.*
 *ehret. p.* 1. *t.* 2.
 Queen Pine-apple.

pyrami-  β Ananas aculeatus, fructu pyramidato, carne aurea.
dalis.  *Tournef. inſt.* 653.
 Sugar-loaf Pine-apple.

lucida.  γ Ananas lucide virens folio vix ſerrato. *Dill. elth.* 25.
 *t.* 21. *f.* 23.
 King Pine-apple.

viridis.  δ Ananas aculeatus, fructu pyramidato ex viridi flaveſ-
 cente. *Mill. dict.*
 Green-fleſh'd Pine-apple.
 *Nat.* of South America.
 *Introd.* 1690, by Mr. Bentick. *Br. Muſ. Sloan. mſſ.*
 3370.
 *Fl.* March and April.  S. ♄.

*Pinguin.*  2. B. foliis ciliato-ſpinoſis mucronatis, racemo terminali.
 *Sp. pl.* 408.
 Broad-leav'd Wild Pine-apple.

 *Nat.*

*Nat.* of the Weſt Indies.
*Cult.* 1690, in the Royal Garden at Hampton-court.
*Catal. mſcr.*
*Fl.* March and April.                              S. ♄.


PITCAIRNIA. *L'Heritier ſert. angl.* (Hepetis.
*Swartz prodr.* 56.)

*Cal.* 3-phyllus, ſemiſuperus. *Petala* 3. *Squama* nečta-
rifera ad baſin petalorum. *Stigmata* 3, contorta.
*Capſ.* 3, introrſum dehiſcentes. *Sem.* alata.

1. P. foliis ciliato-ſpinoſis, pedunculis germinibuſque gla-   *bromeliæ-*
   berrimis.                                                    *folia.*
   Pitcairnia bromeliæfolia. *L'Herit. ſert. angl. tab.* 11.
   Scarlet Pitcairnia.
   *Nat.* of Jamaica.
   *Cult.* 1781, by Lord Aſhburton.
   *Fl.* June.                                        S. ♄.


2. P. foliis ciliato-ſpinoſis, pedunculis germinibuſque to-   *anguſti-*
   mentoſis.                                                   *folia.*
   Narrow-leav'd Pitcairnia.
   *Nat.* of the Iſland of Santa Cruz.
   *Introd.* 1777, by John Ryan, Eſq.
   *Fl.* December and January.                        S. ♄.


3. P. foliis integerrimis baſi ſubſpinoſis.                   *latifolia.*
   Broad-leav'd Pitcairnia.
   *Nat.* of the Weſt Indies.
   *Introd.* 1785, by Mr. Alex. Anderſon.
   *Fl.* Auguſt.                                       S. ♄.


D d                    TILLAN-

## TILLANDSIA. *Gen. pl.* 396.

*Cal.* 3-fidus, perfiftens.  *Cor.* 3-fida, campanulata.
*Capf.* 1-locularis.  *Sem.* pappofa.

*lingulata.*　1. T. foliis lanceolato-lingulatis integerrimis bafi ven-
　　　　　tricofis.  *Sp. pl.* 409.
　　　　Tongue-leav'd Tillandfia.
　　　　*Nat.* of Jamaica.
　　　　*Cult.* 1776, by Mr. James Gordon.
　　　　*Fl.*　　　　　　　　　　　　　　　　　S. ♄.


## TRADESCANTIA. *Gen. pl.* 398.

*Cal.* 3-phyllus.  *Petala* 3.  *Filamenta* villis articula-
　　　tis.  *Capf.* 3-locularis.

*virginia-*　1. T. erecta lævis, floribus congeftis.  *Sp. pl.* 411.
*na.*　　　Common Virginian Spider-wort.
　　　　*Nat.* of Virginia and Maryland.
　　　　*Introd.* before 1629, by Mr. John Tradefcant, Sen.
　　　　　*Park. parad.* 151. *f.* 4.
　　　　*Fl.* May——October.　　　　　　　　　H. ♃.

*malaba-*　2. T. erecta lævis, pedunculis folitariis longiffimis.  *Sp.*
*rica.*　　　*pl.* 412.
　　　　Grafs-leav'd Spider-wort.
　　　　*Nat.* of the Eaft Indies.
　　　　*Introd.* 1776, by John Fothergill, M.D.
　　　　*Fl.* July and Auguft.　　　　　　　　S. ♃.

*genicula-*　3. T. procumbens hirfuta.  *Sp. pl.* 412.
*ta.*　　　Knotted Spider-wort.
　　　　*Nat.* of the Weft Indies.
　　　　*Introd.* about 1783.
　　　　*Fl.* July.　　　　　　　　　　　　　S. ♃.
　　　　　　　　　　　　　　　　　　　　4. T.

4. T. repens lævis, fpathis diphyllis imbricatis. *Syfl.* *criflata.*
   *veget.* 315. *Jacqu. hort.* 2. *p.* 64. *t.* 137.
Commelina criftata. *Sp. pl.* 62.
Crefted Spider-wort.
*Nat.* of Ceylon.
*Introd.* 1770, by Monf. Richard.
*Fl.* July——September.          S. ☉.


5. T. acaulis lævis, foliis oblongo-lanceolatis canalicu- *difcolor.*
   latis carnofis difcoloribus.
Tradefcantia difcolor. *L'Herit. fert. angl. tab.* 12.
Tradefcantia fpathacea. *Swartz prodr.* 57.
Purple-leav'd Spider-wort.
*Nat.* of South America.
*Introd.* 1783, by Matthew Wallen, Efq.
*Fl.* moft part of the Summer.       S. ♃.


## PONTEDERIA. *Gen. pl.* 399.

*Cor.* 1-petala, 6-fida, bilabiata. *Stamina* 3 apici, 3
   tubo corollæ inferta. *Capf.* 3-locularis.


1. P. foliis cordatis, floribus fpicatis. *Sp. pl.* 412.     *cordata.*
Heart-leav'd Pontederia.
*Nat.* of Virginia.
*Cult.* 1759, by Mr. Ph. Miller. *Mill. dict. edit.* 7. *n.* 1.
*Fl.* July and Auguft.           H. ♃.


## HÆMANTHUS. *Gen. pl.* 400.

*Involucr.* 6-phyllum, multiflorum. *Cor.* 6-partita,
   fupera. *Bacca* 3-locularis.

1. H. foliis linguiformibus planis lævibus humi adpreffis *coccineus.*
bifariis, umbella coarctata faftigiata involucro bre-
viori, limbo patulo. *Linn. fil.*
D d 2         Hæmanthus

Hæmanthus coccineus.   *Sp. pl.* 412.
Scarlet Hæmanthus, or Blood-flower.
*Nat.* of the Cape of Good Hope.
*Cult.* 1731, by Mr. Philip Miller.   *Mill. dict. edit.* 1.
*Fl.* Auguſt——October.                    G. H. ♃.

*puniceus.*   2. H. foliis oblongis ellipticis acutis retuſis undulatis,
            umbella coarctata faſtigiata, limbo ſtaminibuſque
            erectis.   *Linn. fil.*
            Hæmanthus puniceus.   *Sp. pl.* 413.   (excluſis ſyno-
            nymis Sebæ, Swertii, Moriſoni et Rudbeckii.)
            Waved-leav'd Hæmanthus, or Blood-flower.
            *Nat.* of Africa.
            *Cult.* 1722, by James Sherard, M.D.   *Dill. elth.* 167.
            *t.* 140. *f.* 167.
            *Fl.* May——July.                    G. H. ♃.

*pubeſcens.*   3. H. foliis oblongo-lanceolatis undique hirſutis, umbel-
            la faſtigiato-rotundata, limbo ſtaminibuſque erectis.
            *Linn. fil.*
            Hæmanthus pubeſcens.   *Linn. ſuppl.* 193.
            Downy-leav'd Hermanthus.
            *Nat.* of the Cape of Good Hope.   Mr. *Fr. Maſſon.*
            *Introd.* 1774.
            *Fl.* Auguſt.                    G. H. ♃.

*ciliaris.*   4. H. foliis lanceolatis glabris ciliatis, involucro lato um-
            bella rotundata breviori, limbo reflexo.   *Linn. fil.*
            Hæmanthus ciliaris.   *Sp. pl.* 413.
            Amaryllis ciliaris.   *Linn. ſuppl.* 195.
            Fringed Hæmanthus.
            *Nat.* of the Cape of Good Hope.
            *Introd.* 1774, by Mr. Francis Maſſon.
            *Fl.*                    G. H. ♃.

                                        5. H.

J. F. Miller del.

Magornia

angustifolia.

M.ᶜ Kenzie. sc.

Tab. 4. Vol. 1 Pag. 447.

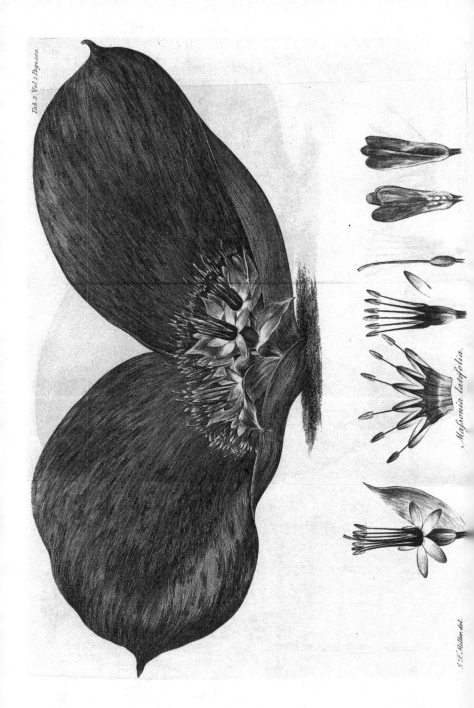

*Tab. 3 Vol. 1 Page 402.*

*Masonia latifolia.*

*S. F. Miller del.*

The material originally positioned here is too large for reproduction in this reissue. A PDF can be downloaded from the web address given on page iv of this book, by clicking on 'Resources Available'.

5. H. foliis bifariis oblongis planiufculis glabris, pedun-    *toxicari-*
   culis fpatha floreque longioribus. *Linn. fil.*    *us.*
   Amaryllis difticha. *Linn. fuppl.* 195.
   Fan-leav'd Hæmanthus.
   *Nat.* of the Cape of Good Hope. Mr. *Fr. Maffon.*
   *Introd.* 1774.
   *Fl.*                                       G. H. ♃.


6. H. foliis fetaceis, fcapo filiformi bafi fpirali flexuofo,    *fpiralis.*
   involucris fubulatis umbella 1-4-flora brevioribus.
   *Linn. fil.*
   Spiral-ftalk'd Hæmanthus.
   *Nat.* of the Cape of Good Hope.
   *Introd.* 1774, by Mr. Francis Maffon.
   *Fl.* September.                           G. H. ♃.


## M A S S O N I A.  *Linn. fuppl.* 27.

*Cor.* infera, limbo 6-partito.  *Filam.* collo tubi impo-
   fita.  *Capf.* 3-alata, 3-locularis, polyfperma.


1. M. foliis fubrotundis patentibus, laciniis corollæ pa-    *latifolia.*
   tulis.  TAB. 3.
   Maffonia latifolia.  *Linn. fuppl.* 193.
   Maffonia depreffa.  *Houtt. nat. hift.* 12. *p.* 424. *t.* 85.
   *f.* 1.
   Broad-leav'd Maffonia.
   *Nat.* of the Cape of Good Hope.  Mr. *Fr. Maffon,*
   *Introd.* 1775.
   *Fl.* March and April.                     G. H. ♃.


2. M. foliis oblongo-lanceolatis erectis, laciniis corollæ    *angufti-*
   reflexis.  TAB. 4.                                *folia,*
   Maffonia anguftifolia.  *Linn. fuppl.* 193.
   Narrow-leav'd Maffonia.

*Nat.* of the Cape of Good Hope. Mr. *Fr. Maſſon.*
*Introd.* 1775.
*Fl.* March and April.     G. H. ♃

## GALANTHUS. *Gen. pl.* 401.

*Petala* 3, concava. *Nectarium* ex petalis 3, parvis,
emarginatis. *Stigma* ſimplex.

*nivalis.*   1. GALANTHUS. *Sp. pl.* 413. *Jacqu. auſtr.* 4. *p.* 7.
*t.* 313.
Common Snow-drop.
*Nat.* of the South of Europe.
*Cult.* 1597. *Ger. herb.* 120. *f.* 1.
*Fl.* February and March.     H. ♃

## LEUCOJUM. *Gen. pl.* 402.

*Cor.* campaniformis, 6-partita, apicibus incraſſata.
*Stigma* ſimplex.

*vernum.*   1. L. ſpatha uniflora, ſtylo clavato. *Sp. pl.* 414. *Jacqu.*
*auſtr.* 4. *p.* 6. *t.* 312. *Curtis magaz.* 46.
Great Spring Snow-drop.
*Nat.* of Italy, Germany, and Switzerland.
*Cult* 1596, by Mr. John Gerard. *Hort. Ger.*
*Fl.* February and March.     H. ♃

*æſtivum.*   2. L. ſpatha multiflora, ſtylo clavato. *Sp. pl.* 414.
*Curtis lond. Jacqu. auſtr.* 3. *p.* 2. *t.* 203.
Summer Snow-drop.
*Nat.* of England.
*Fl.* April and May.     H. ♃

*autum-*
*nale.*   3. L. ſpatha multiflora, ſtylo filiformi. *Sp. pl.* 414.
Autumnal Snow-drop.
    *Nat.*

*Tab.5.Vol.1.Page.407.*

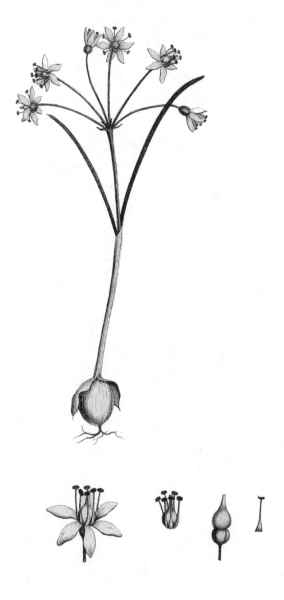

*Leucojum strumosum.*

*Nat.* of Portugal.
*Introd.* 1765, by Mr. Joſhua Brooks.
*Fl.* September.                                     H. ♃.

4. L. ſpatha diphylla multiflora : floribus erectis, ſtylo   *ſtrumo-*
baſi inflato globoſo. TAB. 5.                                 *ſum.*
Crinum tenellum. *Linn. ſuppl.* 194. (excluſo ſyno-
nymo.)
*Nat.* of the Cape of Good Hope. Mr. *Fr. Maſſon.*
*Introd.* 1774.
*Fl.* November.                                     G. H. ♃.
Descr. *Radix* bulboſa. *Folia* omnia radicalia, baſi
vagina incluſa, filiformia, parum compreſſa, glabra,
ſcapo longiora. *Umbella* terminalis, pauciflora.
*Pedunculi* filiformes, glabri, inæquales, unciales et
biunciales. *Spatha univerſalis* bivalvis : *valvulæ*
lanceolatæ, membranaceæ, altera ſemuncialis, altera ter brevior. *Spatha partialis* nulla. *Petala.* ſex,
alba, oblongo-lanceolata, patentia, tres lineas longa :
*tria interiora* obtuſiuſcula ; *exteriora* acuta, carinata : carina obtuſa, vireſcente. *Filamenta* ſex, receptaculo inſerta, ſubulata, æqualia, alba, petalis
breviora. *Antheræ* ſubrotundæ, parvæ, fuſcæ. *Germen* inferum, ſubglobosum, trigonum. *Stylus* baſi
in corpuſculum germine majus, inferne plicatum,
inflatus, dein ſubulatus obtuſe trigonus, longitudine ſtaminum. *Stigma* obſolete trifidum. *Capſula*
ſubglobosa, glabra, trigona, ſeu. potius quaſi e
tribus compoſita, trilocularis, trivalvis. *Semina*
plura.

D d 4                    TULBA-

## TULBAGIA. *Linn. mant.* 148.

*Cor.* infundibulif. limbo 6-fido.   *Nectarium* coronans faucem.   *Capf.* fupera.

*alliacea.*   1. T. nectario monophyllo fexdentato.
Tulbagia alliacea.   *Linn. fuppl.* 193.
Tulbagia capenfis.   *Linn. mant.* 223.   *Jacqu. hort.* 2. *p.* 52. *t.* 115.
Narciffus-leav'd Tulbagia.
*Nat.* of the Cape of Good Hope.
*Introd.* 1774, by Mr. Francis Maffon.
*Fl.* May.                                     G. H. ♃.

## NARCISSUS.   *Gen. pl.* 403.

*Petala* 6, æqualia: *Nectario* infundibuliformi, 1-phyllo.
*Stam.* intra nectarium.

*poeticus.*   1. N. fpatha uniflora, nectario rotato breviffimo fcariofo crenulato.   *Sp. pl.* 414.
α Narciffus albus, circulo purpureo.   *Bauh. pin.* 48.
Poetic Narciffus.
β Narciffus medio purpureus multiplex.   *Bauh. pin.* 54.
Double White Narciffus.
*Nat.* of France and Italy.
*Cult.* 1570.   *Lobel. adv.* 50.
*Fl.* April and May.                           H. ♃.

*Pfeudo-*   2. N. fpatha uniflora, nectario campanulato erecto crifpo
*Narcif-*        æquante petala ovata.   *Sp. pl.* 414.
*fus.*        α Narciffus fylveftris pallidus calice luteo.   *Bauh. pin.* 52.
Common Narciffus, or Daffodil.
β Narciffus fylveftris multiplex calice carens.   *Bauh. pin.* 54.
Common Double Narciffus, or Daffodil.
                                              *Nat.*

*Nat.* of England.
*Fl.* March and April.                    H. ♃.


3. N. ſpatha uniflora, nectario campanulato margine pa-    *bicolor.*
tulo criſpo æquante petala. *Sp. pl.* 415.
α Narciſſus albus, calice flavo, alter. *Bauh. pin.* 52.
Two-color'd Narciſſus.
β Narciſſus major. *Curtis magaz.* 51.
Great Narciſſus.
*Nat.* of the South of Europe.
*Cult.* 1629. *Park. parad.* 68. *n.* 1.
*Fl.* April and May.                    H. ♃.


4. N. ſpatha uniflora, nectario obconico erecto criſpo ſex-    *minor.*
fido æquante petala lanceolata. *Sp. pl.* 415. *Curtis
magaz.* 6.
Small Narciſſus.
*Nat.* of Spain.
*Cult.* 1629. *Park parad.* 105. *n.* 3.
*Fl.* March.                    H. ♃.


5. N. ſpatha ſubbiflora, floribus cernuis, petalis reflexis,    *trian-*
ſtaminibus tribus longioribus. *Curtis magaz.* 48.    *drus.*
Narciſſus triandrus. *Sp. pl.* 416.
Ruſh-leav'd Narciſſue.
*Nat.* of Portugal.
*Cult.* 1629. *Park. parad.* 93. *f.* 2.
*Fl.* April and May.                    H. ♃.


6. N. ſpatha ſubbiflora, nectario campanulato trifido    *orienta-*
emarginato petalis triplo breviore. *Linn. mant.* 62.    *lis.*
Oriental Narciſſus.
*Nat.* of the Levant.
*Cult.* 1758, by Mr. Philip Miller.
*Fl.* May.                    H. ♃.
                    7. N.

*odorus.*   7. N. ſpatha ſubbiflora, nectario campanulato ſexfido lævi
dimidio petalis breviore, foliis ſemicylindricis. *Syſt.
veget.* 317.
Sweet-ſcented Narciſſus.
*Nat.* of the South of Europe.
*Cult.* 1758, by Mr. Philip Miller.
*Fl.* April and May.                              H. ♃.

*Tazetta.*   8. N. ſpatha multiflora, nectario campanulato plicato
truncato triplo breviore petalis, foliis planis. *Syſt.
veget.* 317.
Polyanthus Narciſſus.
*Nat.* of Spain and Portugal.
*Cult.* 1597. *Ger. herb.* 110. *f.* 8.
*Fl.* March and April.                            H. ♃.

*Bulboco-*   9. N. ſpatha uniflora, nectario turbinato petalis majore,
*dium.*         genitalibus declinatis. *Sp. pl.* 417.
Hoop-petticoat Narciſſus.
*Nat.* of Portugal.
*Cùlt.* 1629. *Park. parad.* 107. *f.* 7.
*Fl.* April.                                      H. ♃.

*Jonquil-*   10. N. ſpatha multiflora, nectario campanulato brevi, fo-
*la.*           liis ſubulatis. *Syſt. veget.* 317. *Curtis magaz.* 15.
α flore ſimplici.
Single Jonquill.
β flore pleno.
Double Jonquill.
*Nat.* of Spain.
*Cult.* 1597. *Ger. herb.* 112.
*Fl.* April and May.                              H. ♃.

PANCRATIUM. *Gen. pl.* 404.
*Petala* 6 : *Nectario* 12-fido. *Stam.* nectario impoſita.

*mexica-*   1. P. ſpatha biflora. *Sp. pl.* 418.
*num.*
                                                  Mexican

Mexican Pancratium.
*Nat.* of Mexico.
*Cult.* 1732, by James Sherard, M. D. *Dill. elth.*
299, *t.* 222. *f.* 289.
*Fl.* Auguft.                                           S. ♃.

2. P. fpatha multiflora, foliis lanceolatis, corollæ laciniis     *caribæ-*
linearibus tubo longioribus.                                *um.*
Pancratium caribæum. *Sp. pl.* 418.
Caribean Pancratium.
*Nat.* of the Weft Indies.
*Introd.* 1730, by Mr. Philip Miller. *Mart. dec. 3.*
*p.* 27.
*Fl.* moft part of the Summer.                         S. ♃.

3. P. fpatha multiflora, petalis planis, foliis lingulatis.     *mariti-*
*Sp. pl.* 418.                                               *mum.*
Sea Pancratium, or Daffodil.
*Nat.* of the South of Europe.
*Cult.* 1629, by Mr. John Parkinfon. *Park. parad.*
108.
*Fl.* May.                                             H. ♃.

4. P. fpatha multiflora, foliis linearibus, ftaminibus nec-     *carolinia-*
tarii longitudine. *Sp. pl.* 418.                            *num.*
Carolina Pancratium.
*Nat.* of Jamaica and Carolina.
*Cult.* 1759, by Mr. Ph. Miller. *Mill. dict. edit. 7. n. 6.*
*Fl.* July and Auguft.                                 S. ♃.

5. P. fpatha multiflora, foliis enfiformibus, ftaminibus     *illyricum.*
nectario longioribus. *Sp. pl.* 418.
Illyrian Pancratium.
*Nat.* of the South of Europe.
*Cult.* 1615, by Mr. John Parkinfon. *Park. parad.* 96.
*Fl.*                                                  G. H. ♃.
                                                       6. P.

*litorale.*    6. P. fpatha multiflora, foliis lanceolato-linearibus bifariis, corollæ laciniis linearibus tubo brevioribus,
nectario fubintegro.

Pancratium littorale.    *Jacqu. hort.* 3. *p.* 41. *t.* 75.
*Syft. veget.* 318.

Pancratium foliis enfiformibus, fpatha multiflora, floribus magnis candidis fragrantibus. *Trew. ehret.* 6.
*t.* 27.

Tall Pancratium.

*Nat.* of the Weft Indies.

*Cult.* 1758, by Mr. Philip Miller.

*Fl.* moft part of the Summer.    S. ♃.

*verecun-*    7. P. fpatha multiflora, foliis linearibus, corollæ laciniis
*dum.*    lanceolatis tubo brevioribus, finubus laciniarum
nectarii ftaminiferis.

Narciffus-leav'd Pancratium.

*Nat.* of the Eaft Indies.

*Cult.* 1776, by Mrs. Theobald.

*Fl.* June——Auguft.    S. ♃.

DESCR. *Folia* fefquipedalia, femunciam lata. *Scapus*
erectus, compreffus, pedalis. *Spathæ* oblongo-lan-
leolatæ, acuminatæ, albidæ, marcefcentes: exteriores majores, fefquiunciales. *Flores* fuaveolentes,
pedicellati. *Pedicelli* trigoni, vix femunciales. *Corolla* monopetala: *Tubus* cylindrico-trigonus, virefcens, vix biuncialis, craffitie pennæ anferinæ;
*Limbus* campanulatus, fexpartitus: *laciniæ* lanceolatæ, acutæ, tubo paulo breviores, niveæ, extus medio
virefcentes. *Nectarium* campanulatum, corollæ laciniis brevius, fexfidum: laciniis bifidis. *Filamenta*
alba. *Antheræ* flavæ. *Stylus* declinatus, virefcens,
*Stigma* obfolete trilobum.

8. P.

8. P. ſpatha multiflora, foliis ovatis nervoſis petiolatis.    *amboi-*
    *Sp. pl.* 419.                                        *nenſe.*
    Broad-leav'd Pancratium.
    *Nat.* of Amboina.
    *Cult.* 1759, by Mr. Ph. Miller. *Mill. dict. edit.* 7. *n.* 5.
    *Fl.* April.                                   S. ♃.

## C R I N U M.    *Gen. pl.* 405.

*Cor.* infundibuliformis, ſemiſexfida: tubo filiformi;
    limbo patulo recurvato: laciniis ſubulatis canali-
    culatis. *Filamenta* fauci tubi inſerta, diſcreta. *Linn.*
    *fil.*

1. C. foliis oblongo-lanceolatis margine glaberrimis;    *america-*
    apice contracto-unguiculato, floribus pedicellatis,    *num.*
    tubo limbo breviore. *Linn. fil.*
    Crinum americanum α.    *Sp. pl.* 419.
    Great American Crinum.
    *Nat.* of South America.
    *Cult.* 1732, by James Sherard, M. D. *Dill. elth.* 194.
    *t.* 161. *f.* 195.
    *Fl.* July and Auguſt.                         S. ♃.

2. C. foliis lanceolatis cartilagineo-crenulatis apice pro-    *erubeſ-*
    ducto explanato, floribus ſeſſilibus, tubo limbo lon-    *cens.*
    giore. *Linn. fil.*
    Crinum americanum β.    *Sp. pl.* 419.
    Crinum foliis carinatis. *Mill. ic.* 73. *tab.* 110.
    Small American Crinum.
    *Nat.* of the Weſt Indies.
    *Cult.* 1756, by Mr. Philip Miller. *Mill. ic. loc. cit.*
    *Fl.* at various Seaſons.                      S. ♃.

A G A P A N-

## AGAPANTHUS.

*Cor.* infera, infundibuliformis, hexapetaloidea, regularis.

*umbella-*    1. AGAPANTHUS.
*tus.*
        Crinum africanum. *Sp. pl.* 419.
        Tulbaghia. *Heifteri Brunfvigia, pag.* x. *not. (b).*
        Mauhlia africana. *Dahl obf. bot.* 26.
        African blue Lily.
        *Nat.* of the Cape of Good Hope.
        *Cult.* 1692, in the Royal Garden at Hampton-court.
          *Pluk. phyt. t.* 195. *f.* 1.
        *Fl.* Auguft——January.         G. H. ♃.

## CYRTANTHUS.

    *Cor.* tubulofa, clavata, curva, 6-fida : laciniæ ovato-
    oblongæ. *Filam.* tubo inferta, apice conniventia.
    *Linn. fil.*

*anguftifo-*  1. C. foliis obtufe carinatis rectis, floribus cernuis. *Linn.*
*lius.*
        *fil.*
        Crinum anguftifolium. *Linn. fuppl.* 195.
        Narrow-leav'd Cyrtanthus.
        *Nat.* of the Cape of Good Hope. Mr. *Fr. Maffon.*
        *Introd.* 1774.
        *Fl.* July.         G. H. ♃.

*obliquus.*  2. C. foliis planis obliquis, floribus pendulis. *Linn. fil.*
        Crinum obliquum. *Linn. fuppl.* 195.
        Amaryllis umbrella. *L'Herit. fert. angl. tab.* 16.
        Oblique-leav'd Cyrtanthus.
        *Nat.* of the Cape of Good Hope. Mr. *Fr. Maffon.*
        *Introd.* 1774.
        *Fl.* July.         G. H. ♃.

                              AMARYL-

## AMARYLLIS. *Gen. pl.* 406.

*Cor.* hexapetaloidea, irregularis. *Filamenta* fauci tubi inserta, declinata, inæqualia proportione vel directione. *Linn. fil.*

1. A. spatha indivisa obtusa, flore sessili, corolla campanulata erecta basi breve tubulosa, staminibus erectis: alternis brevioribus. *Linn. fil.*     *lutea.*
Amaryllis lutea. *Sp. pl.* 420.
Yellow Amaryllis, or Autumnal Narcissus.
*Nat.* of the South of Europe.
*Cult.* 1596, by Mr. John Gerard. *Hort. Ger.*
*Fl.* September.      H. ♃.

2. A. spatha diphylla uniflora, corolla infundibuliformi æquali: laciniis revolutis, staminibus inclinatis: alternis brevioribus.     *Pumilio.*
Dwarf Amaryllis.
*Nat.* of the Cape of Good Hope. Mr. *Francis Masson.*
*Introd.* 1774.
*Fl.* November.      G. H. ♃.
DESCR. *Folium* radicale, lineare, inferne angustatum. *Scapus* teres, palmaris, virescens. *Flos* terminalis, solitarius. *Spatha* diphylla: *foliola* lineari-subulata, basi invicem amplexa, tubo corollæ longiora, virentia. *Corollæ tubus* infundibuliformis, uncialis, albidus, externe lineis sex elevatis notatus, interne lineis sex rubris cum prioribus alternantibus. *Limbi* laciniæ ovato-oblongæ, acutæ, reflexæ, tubo longiores, extus albidæ, intus lateritiæ. *Filamenta* filiformia, tubo infra faucem inserta, apice inflexa, albida: tria alterna breviora. *Antheræ* oblongæ, incumbentes, luteæ. *Germen* oblongum. *Stylus*
fili-

filiformis, ftaminibus longior, albidus. *Stigma* trifidum : laciniæ lineares, rubicundæ, apice albæ.

*Atamafco.* 3. A. fpatha bifida acuta, flore pedicellato, corolla campanulata fubæquali erecta bafi breve tubulofa, ftaminibus declinatis æqualibus. *Linn. fil.*
Amaryllis Atamafco. *Sp. pl.* 420.
Atamafco Lily.
*Nat.* of North America.
*Cult.* 1680, by Mr. Charles Hatton. *Morif. hift.* 2.
*p.* 366. *n.* 30.
*Fl.* May.         H. ♃.

*formofiffi-ma.* 4. A. fpatha indivifa, flore pedicellato, corolla bilabiata nutante profunde fexpartita, genitalibus declinatis. *Linn. fil.*
Amaryllis formofiffima. *Sp. pl.* 420. *Curtis magaz.* 47.
Jacobea Lily.
*Nat.* of South America.
*Cult.* 1658, in the Oxford Garden. *Hort. oxon. ed.* 2.
*p.* 116.
*Fl.* May and June.       S. ♃.

*reginæ.* 5. A. fpatha fubbiflora, pedicellis divaricatis, corollis campanulatis breve tubulofis nutantibus, fauce tubi hirfuta, foliis lanceolatis patulis. *Linn. fil.*
Amaryllis reginæ. *Sp. pl.* 421.
Amaryllis fpatha multiflora, corollis campanulatis æqualibus, marginibus undulatis. *Mill. ic. p.* 16.
*t.* 23. (errore fculptoris ; rectius 24.)
Mexican Lily.
*Nat.* of America.
*Cult.* 1725. *Mill. ic. loc. cit.*
*Fl.* May and June.       S. ♃.

6. A.

6. A. ſpatha ſubbiflora, corollis erectiuſculis baſi tubuloſis, *purpurea.*
   fauce tubi glabra, foliis lineari-lanceolatis. *Linn. fil.*
   Crinum ſpecioſum. *Linn. ſuppl.* 195.
   Purple-flower'd Amaryllis.
   *Nat.* of the Cape of Good Hope. Mr. *Francis Maſſon.*
   *Introd.* 1774.
   *Fl.* G. H. ♃.

7. A. ſpatha ſubbiflora, pedicellis erectis ſpatha breviori-   *equeſtris.*
   bus, tubo filiformi horizontali, limbo oblique patu-
   lo ſurſum curvo, fauce piloſa. *Linn. fil.*
   Amaryllis dubia. *Linn. am. ac.* 8. *p.* 254.
   Lilium americanum puniceo flore, Bella donna dic-
   tum. *Herm. par.* 194. *cum fig.*
   Lilium rubicundum. *Merian ſurin.* 22. *tab.* 22.
   Barbadoes Lily.
   *Nat.* of the Weſt Indies.
   *Introd.* 1778, by William Pitcairn, M.D.
   *Fl.* G. H. ♃.

8. A. ſpatha ſubbiflora, corollis baſi tubuloſis nutanti-   *reticula-*
   bus, fauce tubi glabra, ſcapo compreſſo, foliis oblon-   *ta.*
   gis baſi attenuatis. *Linn. fil.*
   Amaryllis reticulata. *L'Herit. ſert. angl. t.* 14.
   Flat-ſtalk'd Amaryllis.
   *Nat.* of Braſil.
   *Introd.* 1777, by Edward Whitaker Gray, M.D.
   *Fl.* April. S. ♃.

9. A. corollis erectiuſculis hexapetalis : laciniis planis,   *Belladon-*
   ſcapo compreſſo, foliis acute canaliculatis obtuſe   *na.*
   carinatis glaberrimis. *Linn. fil.*
   Amaryllis Belladonna. *Sp. pl.* 421. *J. Miller illuſtr.*
   Amaryllis ſpatha multiflora, corollis campanulatis
   æqualibus, genitalibus declinatis. *Ph. Mill. ic.*
   *p.* 15. *t.* 24. (errore ſculptoris ; rectius 23.)

E e Belladonna

Belladonna Lily.
*Nat.*
*Introd.* about 1712. *Mill. ic. loc. cit.*
*Fl.* July——September.                                   H. ♃.

*vittata.*   10. A. floribus pedicellatis, corollis cuneiformi-infundi-
       buliformibus, petalorum exteriorum rachibus inte-
       riorum margini adnatis, scapo tereti, stigmatibus
       sulcatis. *Linn. fil.*
       Amaryllis vittata. *L'Herit. sert. angl. tab.* 15.
       Superb Amaryllis.
       *Nat.*
       *Introd.* 1769, by Mr. William Malcolm.
       *Fl.* April and May.                              S. ♃.

*falcata.*   11. A. corollis pedunculatis erectis hexapetalis, scapo
       compresso longitudine umbellæ, foliis planis humi
       adpressis margine falcatis albo-cartilagineo-crena-
       tis. *Linn. fil.*
       Crinum falcatum. *Jacqu. hort.* 3. *p.* 34. *t.* 60. *Syst.*
       *veget.* 319.
       Sickled-leav'd Crinum.
       *Nat.* of the Cape of Good Hope.
       *Introd.* 1774, by Mr. Francis Masson.
       *Fl.*                                             G. H. ♃.

*ornata.*   12. A. floribus sessilibus, corollis basi tubulosis: tubo
       spathis limboque longiori curvo, limbi laciniis ob-
       longis aristatis: lacinia infima divaricata concava.
       *Linn. fil.*
       Cape Coast Lily.
       *Nat.* of Guinea.
       *Cult.* 1740, by Robert James Lord Petre. *Ehret.*
       *pict. t.* 5. *f.* 2?
       *Fl.* June and July.                              S. ♃.

                                                        13. A.

13. A. floribus pedicellatis, fpatha 12—20 flora, corollis *longifolia.*
     bafi tubulofis : tubo curvo brevi, limbi laciniis
     lanceolatis obtufis, foliis lato-fubulatis canalicu-
     latis apice flaccidis. *Linn. fil.*
Amaryllis longifolia. *Sp. pl.* 421.
Long-leav'd Amaryllis.
*Nat.* of the Cape of Good Hope.
*Introd.* 1773, by Mr. Francis Maffon.
*Fl.* July.                               G. H. ♃.

14. A. floribus pedicellatis, corollis bafi tubulofis : tubo *revoluta.*
     filiformi brevi curvo, foliis linearibus anguftis ca-
     naliculatis longis ab ortu flaccidis. *Linn. fil.*
Revolute Amaryllis.
*Nat.* of the Cape of Good Hope. Mr. *Fr. Maffon.*
*Introd.* 1774.
*Fl.* September.                          G. H. ♃.
DESCR. *Folia* bipedalia, femunciam lata, canalicula-
     ta. *Scapus* pedalis. *Umbella* 4—6 flora. *Flores*
     fuaveolentes. *Tubus* trigonus, virefcens, inflexus.
     *Laciniæ* biunciales, intus albæ, extus fecundum
     medium dilute rubentes, ad medium revolutæ.

15. A. floribus pedicellatis erectiufculis, corollis infundi- *aurea.*
     buliformi-clavatis fubhexapetalis : laciniis lineari-
     lanceolatis, genitalibus rectis, foliis linearibus
     erectis canaliculatis margine reflexo glabro. *Linn.*
     *fil.*
Golden Amaryllis.
*Nat.* of China.
*Introd.* 1777, by John Fothergill, M. D.
*Fl.* Auguft and September.               S. ♃.
DESCR. *Folia* linearia, carinata, margine parum re-
     flexa, fefquipedalia, vix unciam lata. *Scapus* pa-
     rum compreffus, vix bipedalis. *Spatha* diphylla,

lanceolata,

lanceolata, marcescens, triuncialis, 5—9 flora. *Pe-*
*dicelli* inæquales, longiores unciales, ad basin bractea suffulti membranacea, lanceolata, pedicellorum
longitudine. *Corolla* infundibuliformis, lutea. *Tu-*
*bus* trigonus, vix semuncialis. *Limbi* laciniæ li-
neari-lanceolatæ, apice incrassatæ, subuncinatæ,
duas cum semuncia longæ, obtuse carinatæ, carinis virescentibus. *Filamenta* fauci inserta, recta:
tria alterna corolla longiora, alterna breviora, albida. *Antheræ* lineari-oblongæ, rectæ, pallide
lutescentes. *Germen* ovatum, obtuse trigonum.
*Stylus* filiformis, rectus, staminibus longior, albidus. *Stigma* tripartibile, coccineum.

*orientalis.*   16. A. spatha multiflora, floribus pedicellatis sexpartitis
pedunculis sesquibrevioribus irregularibus, germi-
nibus cuneiformibus angulatis. *Linn. fil.*
Amaryllis orientalis. *Sp. pl.* 422.
Broad-leav'd African Amaryllis.
*Nat.* of the Cape of Good Hope.
*Introd.* 1767, by Mr. William Malcolm.
*Fl.*                                                    G. H. ♃.

*sarniensis.*   17. A. petalis linearibus planis, genitalibus rectiusculis
corolla longioribus, stigmatibus partitis revolutis.
*Linn. fil.*
Amaryllis sarniensis. *Sp. pl.* 421.
Guernsey Lily.
*Nat.* of Japan and the Island of Guernsey.
*Cult.* 1659, by General Lambert at Wimbleton.
*Douglas descr. of the Guernsey Lily, pag.* 11.
*Fl.* September and October.                 G. H. ♃.

*undulata.*   18. A. petalis linearibus canaliculatis undulatis, genitalibus

libus deflexis corolla brevioribus, ftigmate obfo-
leto. *Linn. fil.*
Amaryllis undulata. *Syft. veget.* 320. *Jacqu. hort.* 3.
*p.* 11. *t.* 13. *Meerburg ic.* 13. *J. Fr. Miller*
*ic.* 8.
Waved-flower'd African Amaryllis.
*Nat.* of the Cape of Good Hope.
*Introd.* about 1767, by John Blackburne, Efq.
*Fl.* April——June.                    G. H. ♃.

19. A. petalis lanceolatis undulatis, genitalibus deflexis   *radiata.*
divergentibus corolla duplo longioribus, ftigmate
obfoleto. *Linn. fil.*
Lilio-Narciffus V. *Trew Seligm. t.* 35.
Snowdrop-leav'd Amaryllis.
*Nat.*
*Cult.* 1758, by Mr. Philip Miller.
*Fl.* June.                          G. H. ♃.

BULBOCODIUM. *Gen. pl.* 407.

*Cor.* infundibulif. hexapetala: unguibus anguftis ftaminiferis. *Capf.* fupera.

1. B. foliis lanceolatis. *Sp. pl.* 422.                *vernum.*
Spring-flowering Bulbocodium.
*Nat.* of Spain.
*Cult.* 1731, by Mr. Philip Miller. *Mill. dict. edit.* 1.
Colchicum 7.
*Fl.* March.                         H. ♃.

ALLIUM. *Gen. pl.* 409.
*Cor.* 6-partita, patens. *Spatha* multiflora. *Umbella*
congefta. *Capf.* fupera.
   * *Folia caulina plana. Umbella capfulifera.*
1. A. caule planifolio umbellifero, umbella globofa, ftami-   *Ampelo-*
                    E e 3                    nibus   *prafum.*

nibus tricufpidatis, petalis carina fcabris.  *Sp. pl.*
423.
Great round-headed Garlick.
*Nat.* of England.
*Fl.* July and Auguft.                                    H. ♃.

*Porrum.*  2. A. caule planifolio umbellifero, ftaminibus tricufpida-
tis, radice tunicata.  *Sp. pl.* 423.
Common Leek.
*Nat.*
*Cult.* 1597.  *Ger. herb.* 138.
*Fl.* April and May.                                    H. ♂.

*lineare.*  3. A. caule planifolio umbellifero, umbella globofa, fta-
minibus tricufpidatis corolla duplo longioribus.  *Sp.*
*pl.* 423.
Linear-leav'd Garlick.
*Nat.* of Siberia.
*Cult.* 1768, by Mr. Philip Miller.  *Mill. dict. edit.* 8.
*Fl.* June.                                    H. ♃.

*Victoria-*  4. A. caule planifolio umbellifero, umbella rotundata,
*lis.*     ftaminibus lanceolatis corolla longioribus, foliis
ellipticis.  *Sp. pl.* 424.  *Jacqu. auftr.* 3. *p.* 9. *t.* 216.
Long-rooted Garlick.
*Nat.* of the Alps of Italy, Switzerland, and Auftria.
*Cult.* 1739, by Mr. Philip Miller.  *Rand. chel. n.* 11.
*Fl.* May.                                    H. ♃.

*fubhirfu-*  5. A. caule planifolio umbellifero, foliis inferioribus hir-
*tum.*     futis, ftaminibus fubulatis.  *Sp. pl.* 424.
Hairy Garlick, or Diofcorides's Moly.
*Nat.* of the South of Europe and the Levant.
*Cult.* 1596, by Mr. John Gerard.  *Hort. Ger.*
*Fl.* May.                                    H. ♃.
6. A.

6. A. caule planifolio umbellifero, ramulo bulbifero, fta- *magicum.*
   minibus fimplicibus. *Sp. pl.* 424.
   Homer's Garlick, or Moly.
   *Nat.*
   *Cult.* 1596, by Mr. John Gerard. *Hort. Ger.*
   *Fl.* June and July. H. ♃.

7. A. caule planifolio umbellifero, ftaminibus filiformi- *obliquum.*
   bus flore triplo longioribus, foliis obliquis. *Sp. pl.*
   424.
   Oblique-leav'd Garlick.
   *Nat.* of Siberia.
   *Cult.* before 1768, by Mr. Philip Miller. *Mill. dict.*
   *edit.* 8.
   *Fl.* June. H. ♃.

8. A. caule planifolio umbellifero, umbella faftigiata, *roseum.*
   petalis emarginatis, ftaminibus breviffimis fimplici-
   bus. *Syft. veget.* 321.
   Rofe Garlick.
   *Nat.* of the South of France.
   *Cult.* 1752, by Mr. Philip Miller. *Mill. dict. edit.* 6.
   *n.* 20.
   *Fl.* June. H. ♃.

   \* \* *Folia caulina plana. Umbella bulbifera.*
9. A. caule planifolio bulbifero, bulbo compofito, ftami- *sativum.*
   nibus tricufpidatis. *Sp. pl.* 425.
   Cultivated Garlick.
   *Nat.*
   *Cult.* 1551. *Turn. herb. part* 1. *fign.* B. iiij.
   *Fl.* June and July. H. ♃.

10. A. caule planifolio bulbifero, foliis crenulatis: vagi- *Scorodo-*
    nis ancipitibus, ftaminibus tricufpidatis. *Sp. pl.* 425. *prasum.*

Rocam-

Rocambole Garlick.
*Nat.* of Hungary, Denmark, and Sweden.
*Cult.* 1596, by Mr. John Gerard.  *Hort. Ger.*
*Fl.* July.                                    H. ♃.

*arenari-*     11. A. caule planifolio bulbifero, vaginis teretibus, fpa-
*um.*              tha mutica, ftaminibus tricufpidatis.  *Sp. pl.* 426.
               Sand Garlick.
               *Nat.* of Britain.
               *Fl.* June.                            H. ♃.

*carina-*      12. A. caule planifolio bulbifero, ftaminibus fubulatis.
*tum.*              *Sp. pl.* 426.
               Mountain Garlick.
               *Nat.* of England.
               *Fl.* May and June.                    H. ♃.


                    *** *Folia caulina teretia.*

*fphæroce-*    13. A. caule teretifolio umbellifero, foliis femiteretibus,
*phalon.*          ftaminibus tricufpidatis corolla longioribus.  *Sp. pl.*
               426.
               Small round-headed Garlick.
               *Nat.* of Italy, Siberia, and Switzerland.
               *Cult.* 1759.  *Mill. dict. edit.* - .*n.* 12.
               *Fl.* July.                            H. ♃.


*parviflo-*
*rum.*         14. A. caule fubteretifolio umbellifero, umbella globofa,
               ftaminibus fimplicibus corolla longioribus, fpatha
               fubulata.  *Sp. pl.* 427.
               Small-flower'd Garlick.
               *Nat.* of the South of Europe.
               *Introd.* 1781, by Monf. Thouin.
               *Fl.*                                  H. ♃.


                                        15. A.

15. A. caule fubteretifolio umbellifero, pedunculis exte- *defcen-*
rioribus brevioribus, ftaminibus tricufpidatis. *Sp.* *dens.*
*pl.* 427.
Purple-headed Garlick.
*Nat.* of Switzerland.
*Cult.* 1766, in Oxford Garden.
*Fl.* July.                                         H. ♃.

16. A. caule teretifolio umbellifero, floribus pendulis, *flavum.*
petalis ovatis, ftaminibus corolla longioribus. *Sp.*
*pl.* 428. *Jacqu. auftr.* 2. *p.* 25. *t.* 141.
Sulphur-color'd Garlick.
*Nat.* of Auftria.
*Cult.* 1768, by Mr. Philip Miller. *Mill. dict. edit.* 8.
*Fl.* June and July.                              H. ♃.

17. A. caule fubteretifolio umbellifero, floribus pendulis *pallens.*
biuncatis, ftaminibus fimplicibus corollam æquan-
tibus. *Sp. pl.* 427.
Pale-flower'd Garlick.
*Nat.* of the South of Europe.
*Introd.* 1779, by Abbé Pourret.
*Fl.* June and July.                              H. ♃.

18. A. caule fubteretifolio umbellifero, pedunculis capil- *panicula-*
laribus effufis, ftaminibus fubulatis, fpatha longif- *tum.*
fima. *Syft. veget.* 322.
Panicled Garlick.
*Nat.* of the South of Europe and the Levant.
*Introd.* 1780, by Sig. Giovanni Fabroni.
*Fl.*                                              H. ♃.

19. A. caule teretifolio bulbifero, ftaminibus tricufpida- *vineale.*
tis. *Sp. pl.* 428.

Crow

Crow Garlick.
*Nat.* of Britain.
*Fl.* June.                                                    H. ♃.

*oleraceum.*     20. A. caule teretifolio bulbifero, foliis fcabris femitere-
                    tibus fubtus fulcatis, ftaminibus fimplicibus. *Sp.*
                    *pl.* 429.
                 Purple-ftriped Garlick.
                 *Nat.* of England.
                 *Fl.* July.                                   H. ♃.

              **** *Folia radicalia. Scapus nudus.*

*nutans.*     21. A. fcapo nudo ancipiti, foliis linearibus planis, ftami-
                  nibus tricufpidatis.  *Sp. pl.* 429.
               Flat-ftalk'd Garlick.
               *Nat.* of Siberia.
               *Introd.* 1785, by William Pitcairn, M.D.
               *Fl.* July.                                      H. ♃.

*afcaloni-*   22. A. fcapo nudo tereti, foliis fubulatis, umbella globo-
*cum.*            fa, ftaminibus tricufpidatis.  *Sp. pl.* 429.
               Afcalonian Garlick, or Shallot.
               *Nat.* of Paleftine.
               *Cult.* 1633.  *Ger. emac.* 170. *f.* 4.
               *Fl.* June and July.                             H. ♃.

*fenefcens.*  23. A. fcapo nudo ancipiti, foliis linearibus fubtus con-
                  vexis lævibus, umbella fubrotunda, ftaminibus
                  fubulatis.  *Sp. pl.* 430.
               Narciffus-leav'd Garlick.
               *Nat.* of Germany and Siberia.
               *Cult.* 1596, by Mr. John Gerard.  *Hort. Ger.*
               *Fl.* June and July.                             H. ♃.

                                                    24. A.

24. A. ſcapo nudo ſubtriquetro, foliis linearibus planis *inodorum.*
ſubtus carinatis, umbella faſtigiata florifera, ſta-
minibus ſimplicibus.
Carolina Garlick.
*Nat.* of Carolina.
*Introd.* 1776, by the Dutcheſs Dowager of Portland.
*Fl.* March and April.   H. ♃.

25. A. ſcapo nudo ancipiti, foliis linearibus canaliculatis *angulo-*
ſubtus ſubangulatis, umbella faſtigiata. *Sp. pl.* *ſum.*
430. *Jacqu. auſtr.* 5. *p.* 11. *t.* 423.
Angular-ſtalk'd Garlick.
*Nat.* of Germany and Siberia.
*Cult.* 1739, by Mr. Philip Miller. *Rand. chel. n.* 6.
*Fl.* May.   H. ♃.

26. A. ſcapo nudo tereti, foliis linearibus, umbella erecta, *nigrum.*
petalis erectis, ſpatha mucronata bifida. *Sp. pl.*
430.
Allium multibulboſum. *Jacqu. auſtr.* 1. *p.* 9. *t.* 10.
Broad-leav'd Garlick.
*Nat.* of Auſtria.
*Cult.* 1759, by Mr. Philip Miller.
*Fl.* June and July.   H. ♃.

27. A. ſcapo nudo triquetro, foliis lanceolatis petiolatis, *urſinum.*
umbella faſtigiata. *Syſt. veget.* 323.
Ramſon Garlick.
*Nat.* of Britain.
*Fl.* April and May.   H. ♃.

28. A. ſcapo nudo foliiſque triquetris, ſtaminibus ſim- *trique-*
plicibus. *Syſt. veget.* 323.   *trum.*
Triangular Garlick.
*Nat.* of Spain.

*Cult.*

Cult. 1768, by Mr. Philip Miller.    Mill. dict. edit. 8.
Fl. May and June.                                    G. H. ♃.

*Cepa.*     29. A. scapo nudo inferne ventricoso longiore foliis te-
                retibus.   *Sp. pl.* 431.
            Common Onion.
            *Nat.*
            Fl. June and July.                          H. ♃.

*Moly.*     30. A. scapo nudo subcylindrico, foliis lanceolatis sessili-
                bus, umbella fastigiata.   *Sp. pl.* 432.
            Yellow Garlick, or Moly.
            *Nat.* of the South of Europe.
            *Cult.* 1604, by Edward Lord Zouch.   *Lobel. adv.* 2.
                p. 502.
            Fl. June.                                   H. ♃.

*tricoccum.*  31. A. scapo nudo semitereti, foliis lanceolato-oblongis
                planis glabris, umbella globosa, seminibus solitariis.
            Three-seeded Garlick.
            *Nat.* of North America.   Mr. *William Young.*
            *Introd.* 1770.
            Fl. July.                                   H. ♃.

*fistulosum.*  32. A. scapo nudo adæquante folia teretia ventricosa.
                *Syst. veget.* 323.
            Welsh Onion,
            *Nat.*
            *Cult.* 1629.   *Park. parad.* 511. f. 4.
            Fl. April and May.                          H. ♃.

*Schœno-*    33. A. scapo nudo adæquante folia teretia subulato-
*prasum.*        filiformia.   *Syst. veget.* 324.
            Common Cives.
            *Nat.* of Switzerland and Germany.
                                                        Cult.

*Cult.* 1597. *Ger. herb.* 139. *f.* 1.
*Fl.* May. H. ♃.

34. A. fcapo nudo tereti, foliis femicylindricis, ftamini- *fibiricum.*
bus fubulatis. *Syft. veget.* 324.
Allium Schœnoprafum β. *Sp. pl.* 433. *Murray nov.*
*comm. gotting.* 6. *p.* 33. *t.* 4.
Siberian Garlick.
*Nat.* of Siberia and North America.
*Introd.* 1777, by Chevalier Murray.
*Fl.* July and Auguft. H. ♃.

35. A. fcapo nudo tereti longiffimo, foliis linearibus ca- *gracile.*
naliculatis, ftaminibus fubulatis bafi connatis.
Jamaica Garlick.
*Nat.* of Jamaica.
*Introd.* 1787, by Hinton Eaft, Efq.
*Fl.* February. S. ♃.
DESCR. *Folia* narciffina, pedalia. *Scapus* tripedalis,
tenuis. *Petala* erecta alba, unguiculata : ungues
inferne cum ftaminibus connata in tubum viridem.
Forte fui generis planta.

### L I L I U M. *Gen. pl.* 410.

*Cor.* 6-petala, campanulata : linea longitudinali nectarifera. *Capf.* valvulis pilo cancellato connexis.

1. L. foliis fparfis, corollis campanulatis : intus glabris. *candi-*
*Sp. pl.* 433. *dum.*
α Lilium album flore erecto et vulgare. *Bauh. pin.* 76.
Common White Lily.
β Lilium album floribus dependentibus five peregri-
num. *Bauh. pin.* 76.
Nodding-flower'd White Lily.
*Nat.*

*Nat.* of the Levant.
*Cult.* 1597. *Ger. herb.* 146.
*Fl.* June and July.                                      H. ♃.

*bulbife-*
*rum.*
2. L. foliis fparfis, corollis campanulatis erectis : intus
fcabris. *Sp. pl.* 433. *Jacqu. auftr.* 3. *p.* 14. *t.* 226.
*Curtis magaz.* 36.
α Lilium purpurocroceum majus. *Bauh. pin.* 76.
Bulb-bearing, or Orange Lily.
β Lilium phœniceum. *Bauh. pin.* 77.
Purple Bulb-bearing Lily.
γ Lilium bulbiferum minus. *Bauh. pin.* 77.
Small Bulb-bearing Lily.
*Nat.* of Italy, Auftria, and Siberia.
*Cult.* 1596, by Mr. John Gerard. *Hort. Ger.*
*Fl.* June and July.                                      H. ♃.

*pomponi-*
*um.*
3. L. foliis fparfis fubulatis, floribus reflexis, corollis re-
volutis. *Sp. pl.* 434.
Pomponian Lily.
*Nat.* of Siberia and the Pyrenees.
*Cult.* 1629. *Park. parad.* 32. *f.* 3.
*Fl.* May and June.                                      H. ♃.

*chalcedo-*
*nicum.*
4. L. foliis fparfis lanceolatis, floribus reflexis, corollis
revolutis. *Sp. pl.* 434. *Curtis magaz.* 30.
Scarlet Martagon Lily.
*Nat.* of the Levant.
*Cult.* 1596, by Mr. John Gerard. *Hort. Ger.*
*Fl.* June.                                      H. ♃.

*fuperbum.*
5. L. foliis fparfis lanceolatis, floribus ramofo-pyramida-
tis reflexis, corollis revolutis. *Sp. pl.* 434.
Great yellow Martagon Lily.

*Nat.*

*Nat.* of North America.

*Cult.* 1738, by Peter Collinſon, Eſq. *Trew. ehret.* 2. *t.* 11.

*Fl.* June.                                                    H. ♃.

6. L. foliis verticillatis, floribus reflexis, corollis revolu-    *Marta-*
     tis. *Sp. pl.* 435. *Jacqu. auſtr.* 4. *p.* 27. *t.* 351.    *gon.*
     Purple Martagon Lily.

*Nat.* of Germany and Hungary.

*Cult.* 1596, by Mr. John Gerard. *Hort. Ger.*

*Fl.* June and July.                                      H. ♃.

7. L. foliis verticillatis, floribus reflexis, corollis revoluto-    *canadenſe.*
     campanulatis. *Syſt. veget.* 324.
     Canada Martagon Lily.

*Nat.* of North America.

*Cult.* 1629. *Park. parad.* 32. *f.* 2.

*Fl.* July and Auguſt.                                   H. ♃.

8. L. foliis verticillatis, flore erecto, corolla campanula-    *camſchat-*
     ta, petalis ſeſſilibus. *Sp. pl.* 435.    *cenſe.*
     Kamtſchatka Lily.

*Nat.* of Kamtſchatka.

*Introd.* 1783, by Mr. John Buſh.

*Fl.* May.                                                     H. ♃.

9. L. foliis verticillatis, floribus erectis, corolla campa-    *philadel-*
     nulata, petalis unguiculatis. *Sp. pl.* 435.    *phicum.*
     Philadelphian Martagon Lily.

*Nat.* of North America.

*Cult.* 1757, by Mr. Philip Miller. *Mill. ic.* 110. *t.* 165. *f.* 1.

*Fl.* July.                                                    H. ♃.

FRITIL-

## FRITILLARIA. *Gen. pl.* 411.

*Cor.* 6-petala, campanulata, fupra ungues cavitate nectarifera. *Stam.* longitudine corollæ.

*imperia-*    1. F. racemo comofo inferne nudo, foliis integerrimis.
*lis.*        *Sp. pl.* 435.
Crown Imperial.
*Nat.* of Perfia?
*Cult.* 1596, by Mr. John Gerard. *Hort. Ger.*
*Fl.* March and April.                H. ♃.

*perfica.*    2. F. racemo nudiufculo, foliis obliquis. *Sp. pl.* 436.
Perfian Fritillary, or Lily.
*Nat.* of Perfia?
*Cult.* 1596, by Mr. John Gerard. *Hort. Ger.*
*Fl.* April and May.                H. ♃.

*Melea-*    3. F. caule fubunifloro, foliis omnibus alternis. *Sp. pl.*
*gris.*        436. *Curtis lond. Jacqu. auftr.* 5. *p.* 45. *t. app.* 32.
Common Fritillary.
*Nat.* of England.
*Fl.* March——May.                H. ♃.

## EUCOMIS. *L'Herit. fert. angl.*

*Cor.* infera, 6-partita, perfiftens, patens. *Filam.* bafi in
nectarium adnatum connata.

*nana.*    1. E. fcapo clavato, foliis lato-lanceolatis acutis.
Fritillaria nana. *Burm. fl. cap.* 9. *Linn. mant.* 223.
Orchidea capenfis Tulipæ flore rofeo. *Pet. gaz.*
*t.* 85. *f.* 6.
Dwarf Eucomis.
*Nat.* of the Cape of Good Hope.
*Introd.* 1774, by Mr. Francis Maffon.
*Fl.* May.                G. H. ♃.

                        2. E.

2. E. ſcapo cylindrico, foliis linguiformibus obtuſis hu-  *regia.*
    mi adpreſſis.
Fritillaria regia. *Sp. pl.* 435.
Tongue-leav'd Eucomis, or Fritillary.
*Nat.* of the Cape of Good Hope.
*Cult.* 1709, by the Dutcheſs of Beaufort. *Br. Muſ.*
    *H. S.* 138. *fol.* 64.
*Fl.* March——May.             G. H. ♃.

3. E. ſcapo cylindrico, foliis ovato-oblongis undulatis pa-  *undulata.*
    tentibus, comæ foliis longitudine fere racemi.
Waved-leav'd Eucomis or Fritillary.
*Nat.* of the Cape of Good Hope.
*Introd.* about 1760, by Mr. Philip Miller.
*Fl.* March——May.            G. H. ♃.

4. E. ſcapo cylindrico, foliis oblongo-lanceolatis canali-  *punctata.*
    culatis patentibus, comæ foliis brevibus, racemis
    longiſſimis.
Eucomis punctata. *L'Herit. ſert. angl. tab.* 18.
Aſphodelus comoſus. *Houtt. nat. hiſt.* 12. *p.* 336.
    *tab.* 8 ?.
Spotted Eucomis.
*Nat.* of the Cape of Good Hope.
*Introd.* 1783, by Mr. John Græfer.
*Fl.* July.             G. H. ♃.

## U V U L A R I A. *Gen. pl.* 412.

*Cor.* 6-petala, erecta : *Nectarii* fovea baſeos petali.
    *Filamenta* breviſſima.

1. U. foliis amplexicaulibus. *Sp. pl.* 436.     *amplexi-*
Heart-leav'd Uvularia.               *folia.*
*Nat.* of Germany.

*Cult.* 1752, by Mr. Ph. Miller. *Mill. dict. edit.* 6. *n.* 2.
*Fl.* May.                                                    H. ♃.

*lanceola-*   2. U. foliis perfoliatis ovato-lanceolatis acuminatis.
*ta.*             Polygonatum ramofum flore luteo minus. *Corn.*
                  *canad.* 40. *tab.* 41.
                  Spear-leav'd Uvularia.
                  *Nat.* of North America.
                  *Introd.* 1785, by Mr. Archibald Menzies.
                  *Fl.* July.                                  H. ♃.

*perfolia-*   3. U. foliis perfoliatis ovatis.
*ta.*             Uvularia perfoliata. *Sp. pl.* 437.
                  Perfoliate Uvularia.
                  *Nat.* of North America.
                  *Introd.* 1734, by Peter Collinfon, Efq. *Coll. mfcr.*
                  *Fl.* May.                                   H. ♃.

### G L O R I O S A.   *Gen. pl.* 413.
*Cor.* 6-petala, undulata, reflexa.   *Stylus* obliquus.

*fuperba.*   1. G. foliis cirrhiferis. *Syft. veget.* 325.
                 Superb Lily.
                 *Nat.* of the Eaft Indies, and of Guinea.
                 *Introd.* 1690, by Mr. Bentick. *Br. Muf. Sloan. mff.*
                 3370.
                 *Fl.* July and Auguft.                        S. ♃.

### E R Y T H R O N I U M.   *Gen. pl.* 414.
*Cor.* 6-petala, campanulata: *Nectario* tuberculis 2,
         petalorum alternorum bafi adnatis.

*Dens ca-*   1. ERYTHRONIUM. *Sp. pl.* 437. *Jacqu. auftr.* 5.
*nis.*           *p.* 31. *tab. app.* 9. *Curtis magaz.* 5.
                 Dog's-tooth Violet.

                                                              *Nat.*

*Nat.* of Siberia, Italy, and Virginia.
*Cult.* 1596, by Mr. John Gerard. *Hort. Ger.*
*Fl.* March.                                    H. 4.


T U L I P A.   *Gen. pl.* 415.

*Cor.* 6-petala, campanulata.  *Stylus* nullus.

1. T. flore fubnutante, foliis lanceolatis.  *Sp. pl.* 438.   *fylveftris.*
Italian Yellow Tulip.
*Nat.* of the South of Europe.
*Cult.* 1597, by Mr. John Gerard.  *Ger. herb.* 116.
*f.* 1.
*Fl.* April and May.                             H. 4.


2. T. flore erecto, foliis ovato-lanceolatis.  *Sp. pl.* 438.   *gefneria-*
Common Tulip.                                                  *na.*
*Nat.* of the Levant.
*Cult.* 1577, by Mr. James Garret.  *Ger. herb.* 117.
*Fl.* April and May.                             H. 4.


3. T. caule multifloro polyphyllo, foliis linearibus.  *Sp.*   *breynia-*
*pl.* 438.                                                      *na.*
Cape Tulip.
*Nat.* of the Cape of Good Hope.
*Introd.* 1787, by Mr. Francis Maſſon.
*Fl.* July.                                      G. H. 4.


A L B U C A.   *Gen. pl.* 416.

*Cor.* 6-petala : interioribus conniventibus ; exterioribus patulis.  *Stylus* triqueter.

* *Staminibus tribus fertilibus.*

1. A. petalis interioribus apice glandulofis inflexis, foliis   *altiſſima.*
fubulatis canaliculato-convolutis.  *Dryander in act.*
*ftockh.* 1784. *p.* 292.  *Jacqu. ic.*

Tall Albuca.
*Nat.* of the Cape of Good Hope.
*Introd.* about 1780, by Meffrs. Kennedy and Lee.
*Fl.* April and May.                         G. H. ♃.

*major.*  2. A. petalis interioribus apice glandulofis inflexis, foliis
          lineari-lanceolatis planiufculis.  *Dryander in a&.*
          *ftockb.* 1784. *p.* 293.
          Albuca major.  *Sp. pl.* 438.
          Great Albuca.
          *Nat.* of the Cape of Good Hope.
          *Introd.* about 1767, by Mr. William Malcolm.
          *Fl.* May.                         G. H. ♃.

*minor.*  3. A. petalis interioribus apice glandulofis inflexis, foliis
          lineari-fubulatis canaliculatis.  *Dryander in a&.*
          *ftockb.* 1784. *p.* 294.
          Albuca minor.  *Sp. pl.* 438.
          Small Albuca.
          *Nat.* of the Cape of Good Hope.
          *Cult.* 1768, by Mr. Ph. Miller.  *Mill. di&. edit.* 8.
          *Fl.* May and June.                G. H. ♃.

*coar&ata.*  4. A. petalis interioribus apice fornicatis, foliis glabris,
           pedunculis longitudine bra&earum. *Dryander in a&.*
           *ftockb.* 1784. *p.* 295.
           Channel-leav'd Albuca.
           *Nat.* of the Cape of Good Hope.  Mr. *Francis Maffon.*
           *Introd.* 1774.
           *Fl.* May.                        G. H. ♃.

            * * *Staminibus omnibus fertilibus.*

*faftigia-*  5. A. petalis interioribus apice fornicatis, foliis glabris,
*ta.*        pedunculis longiffimis. *Dryander in a&. ftockb.* 1784.
             *p.* 296.

                                             Upright-

Upright-flower'd Albuca.

*Nat.* of the Cape of Good Hope. Mr. *Francis Maſſon.*

*Introd.* 1774.

*Fl.* May. G. H. ♃.

6. A. petalis interioribus apice fornicatis, foliis piloſo-    *viſcoſa.*
glanduloſis. *Dryander in act. ſtockh.* 1784. *p.* 297.

Albuca viſcoſa. *Linn. ſuppl.* 196. *Thunberg in act.*
*ſtockh.* 1786. *p.* 58.

Viſcous Albuca.

*Nat.* of the Cape of Good Hope.

*Introd.* about 1779, by *John Fothergill,* M. D.

*Fl.* May and June. G. H. ♃.

GETHYLLIS. *Linn. ſuppl.* 27.

*Cor.* 6-partita. *Cal.* 0. *Bacca* clavata, radicalis,
1-locularis.

1. G. foliis lineari-filiformibus ſpiralibus villoſis, limbi    *villoſa.*
laciniis ovato-oblongis. *Thunb. nov. gen.* 1. *p.* 14.
*n.* 3. *Linn. ſuppl.* 198.

Papiria villoſa. *Thunb. in act. lund.* 1. *p.* 111.

Hairy Gethyllis.

*Nat.* of the Cape of Good Hope.

*Introd.* 1787, by Mr. Francis Maſſon.

*Fl.* G. H. ♃.

2. G. foliis linearibus ſpiralibus ciliatis, limbi laciniis    *ciliaris.*
ovato-oblongis. *Thunb. nov. gen.* 1. *p.* 14. *n.* 2.
*Linn. ſuppl.* 198.

Papiria ciliaris. *Thunb. in act. lund.* 1. *p.* 111.

Fringed Gethyllis.

*Nat.* of the Cape of Good Hope.

*Introd.* 1788, by Mr. Francis Maſſon.

*Fl.* G. H. ♃.

3. G.

*ſpiralis.*    3. G. foliis linearibus ſpiralibus glabris, limbi laciniis
oblongis. *Thunb. nov. gen.* 1. *p.* 14. *n.* 1. *Linn.*
*ſuppl.* 198.
Gethyllis afra. *Sp. pl.* 633.
Papiria ſpiralis. *Thunb. in act. lund.* 1. *p.* 111.
Spiral Gethyllis.
*Nat.* of the Cape of Good Hope.
*Cult.* 1780, by John Fothergill, M.D.
*Fl.* July.                          G. H. ♃.

## HYPOXIS. *Gen. pl.* 417.

*Cor.* 6-partita, perſiſtens, ſupera. *Capſ.* baſi anguſtior.

*erecta.*    1. H. piloſa, capſulis ovatis. *Sp. pl.* 439.
Erect Hypoxis.
*Nat.* of North America.
*Introd.* 1784, by Mr. William Young.
*Fl.* June.                            H. ♃.

*decum-*
*bens.*    2. H. piloſa, capſulis clavatis. *Sp. pl.* 439.
Trailing Hypoxis.
*Nat.* of Jamaica.
*Cult.* 1755, by Mr. Ph. Miller. *Mill. ic.* 26. *t.* 39. *f.* 2.
*Fl.* moſt part of the Year.           S. ♃.

*plicata.*    3. H. ſcapo unifloro triquetro, foliis lanceolatis plicatis
villoſis. *Linn. ſuppl.* 197.
Fabricia plicata. *Thunberg in Fabric. it. norveg.* 29.
Plaited-leav'd Hypoxis.
*Nat.* of the Cape of Good Hope.
*Introd.* 1788, by Mr. Francis Maſſon.
*Fl.*                                G. H. ♃.

*ſtellata.*    4. H. ſcapo unifloro, foliis linearibus ſtriatis, petalis
maculatis. *Linn. ſuppl.* 197. *Syſt. veget.* 326.
                                      Fabricia

Fabricia ftellata. *Thunberg in Fabric. it. norveg.* 27.
Amaryllis capenfis. *Sp. pl.* 420. *Syft. veget.* 319.
Spotted-flower'd Hypoxis.
*Nat.* of the Cape of Good Hope.
*Introd.* 1788, by Mr. Francis Maffon.
*Fl.*                                                         G. H. ♃.

5. H. foliis linearibus, fcapis umbelliferis vel unifloris.        *aquatica.*
   *Linn. fuppl.* 197.
   Aquatic Hypoxis.
   *Nat.* of the Cape of Good Hope.
   *Introd.* 1787, by Mr. Francis Maffon.
   *Fl.*                                                      G. H. ♃.

6. H. foliis canaliculatis glabris ciliato-ferratis, fcapis        *ferrata.*
   unifloris. *Linn. fuppl.* 197.
   Fabricia ferrata. *Thunberg in Fabric. it. norveg.* 29.
   Channel-leav'd Hypoxis.
   *Nat.* of the Cape of Good Hope.
   *Introd.* 1788, by Mr. Francis Maffon.
   *Fl.* July.                                                G. H. ♃.

7. H. foliis lineari-enfiformibus villofis, ftigmate fimplici      *villofa.*
   trigono acuto. *Linn. fuppl.* 198.
   Fabricia villofa. *Thunberg in Fabric. it. norveg. p.* 31.
   Hairy Hypoxis.
   *Nat.* of the Cape of Good Hope. Mr. *Francis Maffon.*
   *Introd.* 1774.
   *Fl.* moft part of the Summer.                            G. H. ♃.


ORNITHOGALUM. *Gen. pl.* 418.
*Cor.* 6-petala, erecta, perfiftens, fupra medium patens.
        *Filamenta* alterna bafi dilatata.

1. O. fcapo diphyllo, pedunculo unifloro. *Linn. mant.* 62.    *uniflo-*
                    F f 4                         One-          *rum.*

One-flower'd Star of Bethlehem.
*Nat.* of Siberia.
*Introd.* 1781, by Baron Claes Alſtroemer.
*Fl.*                                              H. ♃.

*niveum.*   2. O. racemo paucifloro, petalis lanceolatis, foliis filiformibus canaliculatis, filamentis ſubulatis.
Snowy Star of Bethlehem.
*Nat.* of the Cape of Good Hope, Mr. *Fr. Maſſon.*
*Introd.* 1774.
*Fl.* Auguſt.                                    G. H. ♃.
DESCR. *Folia* glabra, vix digitalia. *Scapus* foliis
brevior. *Pedunculi* vix ſemunciales. *Bracteæ* oblongæ, acuminatæ, breviſſimæ. *Petala* quadrilinea-
ria, alba : tria exteriora carina vireſcente. *Filamenta* petalis dimidio breviora, nivea: alterna ſeſ-
quilatiora. *Antheræ* flavæ.

*umbella-*   3. O. corymbo paucifloro, pedunculis bracteis longiori-
*tum.*       bus, filamentis ſubulatis.
Ornithogalum umbellatum. *Sp. pl.* 441. *Jacqu.
auſtr.* 4. *p.* 22. *t.* 343.
Umbel'd Star of Bethlehem,
*Nat.* of England.
*Fl.* May and June.                            H. ♃.

*luteum.*   4. O. ſcapo anguloſo diphyllo, pedunculis umbellatis
ſimplicibus. *Sp. pl.* 439.
Yellow Star of Bethlehem.
*Nat.* of Britain.
*Fl.* March and April.                          H. ♃.

*minimum.*   5. O. ſcapo angulato diphyllo, pedunculis umbellatis ramoſis. *Sp. pl.* 440.
Small Star of Bethlehem,

                                                  *Nat.*

*Nat.* of Sweden.
*Cult.* 1739, by Mr. Ph. Miller. *Mill. dict. vol.* 2. *n.* 8.
*Fl.*                                                 H. ♃.

6. O. racemo longiffimo, petalis linearibus obtufis, fila-    *pyrenai-*
mentis lanceolatis æqualibus, ftylo longitudine fta-    *cum,*
minum.
Ornithogalum pyrenaicum. *Hudf. angl.* 143. *Jacqu.*
*auftr.* 2. *p.* 2. *tab.* 103.
Pyrenean Star of Bethlehem.
*Nat.* of England.
*Fl.* June and July.                                 H. ♃.

7. O. racemo longiffimo, petalis lanceolato-oblongis,    *ftachyo-*
filamentis late lanceolatis : alternis dimidio bre-    *des,*
vioribus.
Ornithogalum pyrenaicum. *Sp. pl.* 440. (exclufis
fynonymis Halleri et Clufii,)
Clofe-fpik'd Star of Bethlehem.
*Nat.* of the South of Europe.
*Introd.* about 1771.
*Fl.* April.                                         H. ♃.

8. O. racemo longiffimo, foliis lanceolato-enfiformibus.    *latifoli-*
*Sp. pl.* 440.                                       *um,*
Broad-leav'd Star of Bethlehem.
*Nat.* of Egypt and Arabia.
*Cult.* 1629. *Park. parad.* 137. *f.* 2.
*Fl.* June.                                          G. H. ♃.

9. O. racemo conico : floribus numerofis adfcendenti-    *pyrami-*
bus, petalis elliptico-oblongis planis, ftaminibus    *dale.*
lanceolatis æqualibus, ftylo breviffimo.
Ornithogalum pyramidale. *Sp. pl.* 441.
Pyramidal Star of Bethlehem.

*Nat.*

*Nat.* of Spain and Portugal.
*Cult.* 1739, by Mr. Ph. Miller. *Mill. dict. vol.* 2. *n.* 1.
*Fl.* June and July. H. ♃.

arabicum. 10. O. corymbo multifloro, filamentis fubulatis, corollà
late campanulata: petalis exterioribus obfole tridentatis.

Ornithogalum arabicum. *Sp. pl.* 441.
Great-flower'd Star of Bethlehem.
*Nat.* of Egypt and Madeira.
*Cult.* 1629, by Mr. John Parkinfon. *Park. parad.*
137. *f.* 1.
*Fl.* March and April. G. H. ♃.

thyrſoides. 11. O. corymbis multifloris racemiformibus, filamentis
alternis furcatis, foliis lanceolatis.

α floribus luteis, bracteis pedunculo brevioribus.
Ornithogalum dubium. *Houtt. nat. hiſt.* 12. *p.* 309.
*t.* 82. *f.* 3.
Yellow-flower'd Spear-leav'd Star of Bethlehem.
β floribus albis, bracteis longitudine pedunculi.
Ornithogalum thyrſoides. *Jacqu. hort.* 3. *p.* 17. *t.* 28.
Ornithogalum racemo conico laxo, pedunculis lon-
giſſimis, floribus erectis. *Mill. ic.* 128. *t.* 192.
White-flower'd Spear-leav'd Star of Bethlehem.
*Nat.* of the Cape of Good Hope.
*Cult.* 1757, by Mr. Philip Miller. *Mill. ic. loc. cit.*
*Fl.* June. G. H. ♃.

cauda- 12. O. racemo longiſſimo, foliis lanceolato-linearibus,
tum. corollis patentibus, ſtaminibus dilatatis: alternis
cuneiformibus.

Long-fpik'd Star of Bethlehem.
*Nat.* of the Cape of Good Hope. Mr. *Fr. Maſſon.*

*Introd.*

*Introd.* 1774.

*Fl.* February——Auguft. G. H. ♃.

Descr. Tota planta glabra. *Folia* fefquipedalia, fefquiunciam lata. *Scapus* teres, tripedalis. *Racemus* fefquipedalis : *pedunculi* fparfi, fub anthefi patentes, vix unciales. *Bractea* lanceolatæ, fubulatæ, longitudine pedunculorum, membranaceæ. *Petala* femuncialia, alba, vitta longitudinali viridi. *Filamenta* petalis breviora, alba. *Stylus* triqueter, albus, longitudine ftaminum.

13. O. floribus fecundis pendulis, nectario ftamineo cam-      *nutans.*
      paniformi. *Sp. pl.* 441. *Jacqu. auftr.* 4. *p.* 1. *t.* 301.
      Neapolitan Star of Bethlehem.
      *Nat.* of Italy.
      *Cult.* 1633. *Ger. emac.* 168. *f.* 9.
      *Fl.* April and May. G. H. ♃.

S C I L L A. *Gen. pl.* 419.

*Cor.* 6-petala, patens, decidua. *Filamenta* filiformia.

1. S. nudiflora, bracteis refractis. *Sp. pl.* 442.      *mariti-*
α Scilla vulgaris radice rubra. *Bauh. pin.* 73.      *ma.*
      Red-rooted officinal Squil.
β Scilla radice alba. *Bauh. pin.* 73.
      White-rooted officinal Squil.
      *Nat.* of Spain, Sicily, and Syria.
      *Cult.* 1648, in Oxford Garden. *Hort. oxon. edit.* 1.
      *p.* 48.
      *Fl.* April and May. G. H. ♃.

2. S. racemo conico oblongo. *Syft. veget.* 328.      *italica.*
      Italian Squil.
      *Nat.*
      *Cult.* 1629. *Park. parad.* 131. *f.* 6.
      *Fl.* April. H. ♃.

3. S.

*peruvia-*   3, S. corymbo conferto conico.   *Sp. pl.* 442.
*na,*   α flore cæruleo.
   Blue-flower'd Peruvian Squil.
  β flore albo.
   White-flower'd Peruvian Squil.
   *Nat.* of Portugal and Spain.
   *Cult.* 1629, by Mr. John Parkinſon.   *Park. parad.*
   125. *f.* 7.
   *Fl.* May.         H. ♃.

*amœna.*   4. S. floribus lateralibus alternis ſubnutantibus, ſcapo an-
   gulato. *Syſt. veget.* 328. *Jacqu. auſtr.* 3. *p.* 10. *t.* 218.
   Nodding Squil.
   *Nat.* of the Levant.
   *Introd.* about 1600, by Edward Lord Zouch.   *Lobel.*
   *adv.* 2. *p.* 486.
   *Fl.* March and April.      H. ♃.

*campanu-*   5. S. bulbo ſolido, racemo multifloro oblongo-ſubconico,
*lata.*   corollis campanulatis erectis, bracteis bipartitis
   pedunculo longioribus, foliis lanceolatis.
   Scilla hiſpanica.   *Mill. dict.*
   Hyacinthus ſtellaris, ſaturate cæruleus.   *Bauh. pin.* 46.
   Hyacinthus hiſpanicus ſtellato flore. *Cluſ. cur. poſt.* 20.
   Spaniſh Squil.
   *Nat.* of Spain and Portugal.
   *Cult.* 1759, by Mr. Ph. Miller. *Mill, dict. edit.* 7. *n.* 8.
   *Fl.* May and June.      H. ♃.

*bifolia.*   6. S. floribus racemoſis, foliis lanceolato-linearibus ſub-
   binis in ſcapo elevatis.
   Scilla bifolia.   *Sp. pl.* 443.   *Jacqu. auſtr.* 2. *p.* 11.
   *t.* 117.
  α Hyacinthus cæruleus mas minor.   *Fuchſ. hiſt.* 837.
   Two-leav'd Spring Squil.

              β Hyacin-

β Hyacinthus ftellatus, albo flore. *Cluf. hift.* 1. *p.* 184.
White-flower'd Spring Squil.
*Nat.* of France and Germany.
*Cult.* 1597, by Mr. John Gerard. *Ger. herb.* 98. *n.* 1.
*Fl.* February and March. H. 4.

7. S. bulbo tunicato, racemo paucifloro bracteato, corol-   *verna.*
lis campanulatis, foliis linearibus canaliculatis ra-
dicalibus pluribus.
Scilla verna. *Hudf. angl.* 142.
Small Squil.
*Nat.* of England.
*Fl.* June and July. H. 4.

8. S. racemo oblongo conico, petalis lineatis. *Syft. veget.* *lufitanica*
329.
Portugal Squil.
*Nat.* of Portugal.
*Introd.* 1777, by Edward Whitaker Gray, M. D.
*Fl.* May. H. 4.

9. S. racemo cylindraceo multifloro, petalis germine   *hyacinth-*
fefquilongioribus, pedunculis coloratis, foliis lanceo-   *oides.*
latis.
Scilla hyacinthoides. *Syft. veget.* 329. *Gouan illuftr.*
26.
Hyacinth Squil.
*Nat.* of Madeira.
*Introd.* 1778, by Mr. Francis Maffon.
*Fl.* Auguft. G. H. 4.

10. S. foliis filiformibus linearibus, floribus corymbofis,   *autumna-*
pedunculis nudis adfcendentibus longitudine floris.   *lis.*
*Sp. pl.* 443.
Autumnal Squil.

Nat.

*Nat.* of England.
*Fl.* Auguſt and September.                         H. ♃.


CYANELLA. *Gen. pl.* 420.

*Cor.* hexapetala : petalis 3-inferioribus propendentibus.
*Stamen* infimum declinatum, longius.

*lutea.*     1. C. foliis enſiformibus, ramis erectis. *Linn. ſuppl.* 201.
Yellow-flower'd Cyanella.
*Nat.* of the Cape of Good Hope.
*Introd.* 1788, by Mr. Francis Maſſon.
*Fl.* July.                                      G. H. ♃.

*capenſis.*   2. C. foliis undulatis, ramis patentiſſimis. *Syſt. veget.*
329. *Jacqu. hort.* 3. *p.* 21. *t.* 35.
Purple-flower'd Cyanella.
*Nat.* of the Cape of Good Hope.
*Cult.* 1768, by Mr. Philip Miller. *Mill. dict. edit.* 8.
*Fl.* July.                                      G. H. ♃.


ASPHODELUS. *Gen. pl.* 421.

*Cor.* 6-partita. *Nectarium* ex valvulis 6, germen te-
gentibus.

*luteus.*    1. A. caule folioſo, foliis triquetris ſtriatis. *Sp. pl.* 443.
*Jacqu. hort.* 1. *p.* 32. *t.* 77.
Yellow Aſphodel.
*Nat.* of Sicily.
*Cult.* 1596, by Mr. John Gerard. *Hort. Ger.*
*Fl.* May——July.                                H. ♃.

*ramoſus.*  2. A. caule nudo, foliis enſiformibus carinatis lævibus.
*Sp. pl.* 444.
Branchy Aſphodel, or King's Spear.

*Nat.*

*Nat.* of the South of Europe.
*Cult.* 1596, by Mr. John Gerard. *Hort. Ger.*
*Fl.* May. H. ♃.

3. A. caule nudo, foliis ſtriĉtis ſubulatis ſtriatis ſubfiſtu-    *fiſtuloſus.*
    loſis. *Sp. pl.* 444.
    Onion-leav'd Aſphodel.
    *Nat.* of the South of Europe.
    *Cult.* 1596, by Mr. John Gerard. *Hort. Ger.*
    *Fl.* June——September. G. H. ♃.


ANTHERICUM. *Gen. pl.* 422.

    *Cor.* 6-petala, patens. *Capſ.* ovata.

1. A. foliis planis glabris lineari-lanceolatis acutis, ſcapo    *floribun-*
    ſimplici, racemo multifloro cylindrico compaĉto,    *dum.*
    petalis patentibus, ſtaminibus glabris.
    Thick-ſpiked Anthericum.
    *Nat.* of the Cape of Good Hope. Mr. *Fr. Maſſon.*
    *Introd.* 1774.
    *Fl.* March and April. G. H. ♃.

2. A. foliis planis, ſcapo ramoſo, corollis revolutis. *Sp.*    *revolu-*
    *pl.* 445.    *tum.*
    Curl'd-flower'd Anthericum.
    *Nat.* of the Cape of Good Hope.
    *Cult.* 1731, by Mr. Philip Miller. *Mill. diĉt. edit.* 1.
    Phalangium 6.
    *Fl.* September——December. G. H. ♃.

3. A. foliis lineari-ſubulatis planis, ſcapo ramoſo, pedun-    *ramoſum.*
    culis ſolitariis, corollis planis, piſtillo reĉto.
    Anthericum ramoſum. *Sp. pl.* 445. *Jacqu. auſtr.* 2.
    *p.* 39. *t.* 161.
    Branchy Anthericum.
                                                                    *Nat.*

*Nat.* of the South of Europe.
*Cult.* 1597.  *Ger. herb.* 44. *f.* 1.
*Fl.* May and June.                                      H. ♃.

*elatum.*    4. A. foliis planis, fcapo ramofo, pedunculis aggregatis,
                corollis planis.
             Afphodelus foliis planis, caule ramofo, floribus fparfis.
                *Mill. ic.* 38. *t.* 56.
             Tall Anthericum.
             *Nat.* of the Cape of Good Hope.
             *Introd.* 1751, by Mr. Philip Miller.  *Mill. ic. loc. cit.*
             *Fl.* Auguſt and September.                  G. H. ♃.

*triflorum.*  5. A. foliis canaliculato-enſiformibus, fcapo ſimplici,
                bracteis remotis trifloris.
             Three-flówer'd Anthericum.
             *Nat.* of the Cape of Good Hope.
             *Introd.* 1782, by George Wench, Eſq.
             *Fl.* November.                               G. H. ♃.

*canalicu-*   6. A. foliis fubcarnofis pilofis enſiformi-triquetris latere
*latum.*         anguſtiori canaliculatis, fcapo ſimplici.
             Channel'd Anthericum.
             *Nat.* of the Cape of Good Hope.  Mr. *Fr. Maſſon.*
             *Introd.* 1774.
             *Fl.* April.                                  G. H. ♃.
             Descr. *Scapus* teres, pilofus.  *Racemus* multiflorus :
                *pedunculi* teretes, glabri, vix unciales.  *Bracteæ* lan-
                ceolatæ, acuminatæ, fub antheſi pedunculo breviores,
                glabræ.  *Petala* patula, alba, dorfo fordide virefcen-
                tia, femuncialia.  *Filamenta* fubulata, alba, petalis
                breviora, hifpidiufcula : tria alterna breviora, minus
                hifpida.  *Stylus* longitudine ſtaminum.

                                                          7. A.

7. A. foliis linearibus canaliculatis glabris margine car- *Albucoi-*
 tilagineis, fcapo fimplici.         *des.*

Striped-flower'd Anthericum.

*Nat.* of the Cape of Good Hope.   Mr. *Fr. Maſſon,*
*Introd.* 1788.

*Fl.* Auguſt.          G. H. ♃.

OBS. Petalis luteis, carina viridi, apice fornicatis, re-
 fert Albucam, fed differt petalis interioribus patulis
 et filamentis fupra bafin non complicatis.


8. A. foliis planis, fcapo fimpliciſſimo, corollis planis, *Liliago.*
 piſtillo declinato. *Sp. pl.* 445. *Jacqu. hort.* 1. *p.*
 36. *t.* 83.

Grafs-leav'd Anthericum.

*Nat.* of the South of Europe.

*Cult.* 1597.   *Ger. herb.* 44. *f.* 2.

*Fl.* May and June.        H. ♃.


9. A. foliis planis, fcapo fimpliciſſimo, corollis campa- *Liliaſ-*
 nulatis, ſtaminibus declinatis. *Sp. pl.* 445.    *trum.*

Savoy Anthericum, or Spider-wort.

*Nat.* of the Alps of Switzerland.

*Cult.* 1629.   *Park. parad.* 151. *f.* 1.

*Fl.* May and June.        H. ♃.


10. A. foliis carnofis teretibus, caule fruticofo. *Sp. pl.* *frutеf-*
 445.               *cens.*

Shrubby Anthericum.

*Nat.* of the Cape of Good Hope.

*Cult.* 1702, in Chelfea Garden.   *Pluk. amalth.* 168.

*Fl.* moſt part of the Summer.     G. H. ♄.


11. A. foliis carnofis fubulatis planiufculis. *Sp. pl.* 446. *Alooides.*

Aloe-leav'd Anthericum.

*Nat.* of the Cape of Good Hope.

       G g         *Cult.*

*Cult.* 1732, by James Sherard, M. D. *Dill. elth.*
312. *t.* 232. *f.* 299.

*Fl.* moſt part of the Summer.      G. H. ♃.

*Aſphodel-*    12. A. foliis carnoſis ſubulatis ſemiteretibus ſtrictis. *Sp.*
*oides.*          *pl.* 446. *Jacqu. hort.* 2. *p.* 85. *t.* 181.
Glaucous-leav'd Anthericum.
*Nat.* of the Cape of Good Hope.
*Cult.* 1759, by Mr. Ph. Miller. *Mill. dict. edit.* 7. *n.* 6.
*Fl.* moſt part of the Summer.      G. H. ♃.

*annuum.*    13. A. foliis carnoſis ſubulatis teretibus, ſcapo ſubrace-
moſo. *Sp. pl.* 446.
Annual Anthericum.
*Nat.* of the Cape of Good Hope.
*Cult.* 1748, by Mr. Philip Miller. *Mill. dict. edit.* 5.
Phalangium 4.
*Fl.* June and July.      H. ☉.

*hiſpidum.*    14. A. foliis carnoſis compreſſis hiſpidis. *Sp. pl.* 446.
Hairy-leav'd Anthericum.
*Nat.* of the Cape of Good Hope.
*Introd.* 1774, by Mr. Francis Maſſon.
*Fl.* May and June.      G. H. ♃.

*oſſifra-*    15. A. foliis enſiformibus, filamentis lanatis. *Sp. pl.*
*gum.*         447.
Lancaſhire Anthericum, or Aſphodel.
*Nat.* of Britain.
*Fl.* Auguſt.      H. ♃.

*calycu-*    16. A. foliis enſiformibus, perianthiis trilobis, filamentis
*latum.*        glabris, piſtillis trigynis. *Sp. pl.* 447.
Scotch Anthericum, or Aſphodel.
*Nat.* of Scotland.
*Fl.* July and Auguſt.      H. ♃.

17. A.

17. A. foliis filiformibus teretiufculis fcabridis, filamen- *filiforme.*
tis glabris, petalis lanceolatis.
Thread-leav'd Anthericum.
*Nat.* of the Cape of Good Hope. Mr. *Fr. Maffon.*
*Introd.* 1774.
*Fl.* April. G. H. ♃.
DESCR. *Scapus* filiformis, pedalis, foliis brevior, bafi
rufefcens, maculis pallidis. *Petala* tria exteriora
intus albida, extus viridia; tria interiora alba, ex-
tus linea viridi.


L E O N T I C E. *Gen. pl.* 423.

*Cor.* 6-petala. *Nectarium* 6-phyllum, unguibus corollæ infidens, limbo patens. *Cal.* 6-phyllus, deciduus.

1. L. foliis decompofitis : petiolo communi trifido. *Sp.* *Leonto-*
*pl.* 448. *petalum.*
Lion's-leaf.
*Nat.* of the Levant.
*Cult.* before 1597, by Lord Zouch. *Ger. herb.* 182.
*Fl.* G. H. ♃.

2. L. folio caulino triternato, florali biternato. *Sp. pl.* *Thalictr-*
448. *oides.*
Columbine-leav'd Leontice.
*Nat.* of North America.
*Introd.* 1784, by Mr. William Young.
*Fl.* May. H. ♃.


A S P A R A G U S. *Gen. pl.* 424.

*Cor.* 6-partita, erecta : petalis 3 interioribus apice
reflexis. *Bacca* 3-locularis, 2-fperma.

1. A. caule herbaceo tereti erecto, foliis fetaceis, fti- *officinalis.*
pulis paribus. *Syft. veget.* 332.

Common Aſparagus.
*Nat.* of England.
*Fl.* June and July.                                    H. ♃.

declina-
tus.
2. A. caule inermi tereti, ramis declinatis, foliis ſetaceis.
    *Syſt. veget.* 332.
Long-leav'd Aſparagus.
*Nat.* of the Cape of Good Hope.
*Introd.* 1787, by Mr. Francis Maſſon.
*Fl.*                                          G. H. ♃.

retrofrac-
tus.
3. A. aculeis ſolitariis, ramis teretibus reflexis retro-
    fractiſque, foliis faſciculatis ſetaceis. *Syſt. veget.*
    332.
Larch-leav'd Aſparagus.
*Nat.* of Africa.
*Cult.* 1759, by Mr. Ph. Miller. *Mill. dict. edit.* 7. *n.* 5.
*Fl.* Auguſt and September.              G. H. ♄.

aſiaticus.
4. A. aculeis ſolitariis, caule erecto, ramis filiformibus,
    foliis faſciculatis ſetaceis. *Syſt. veget.* 333.
Slender-ſtalk'd Aſparagus.
*Nat.* of Aſia.
*Cult.* 1768, by Mr. Philip Miller. *Mill. dict. edit.* 8.
*Fl.*                                          G. H. ♄.

albus.
5. A. aculeis ſolitariis, ramis angulatis flexuoſis, foliis
    faſciculatis triquetris muticis deciduis. *Syſt. veget.*
    333.
White Aſparagus.
*Nat.* of Spain and Portugal.
*Cult.* 1640. *Park. theat.* 455. *f.* 5.
*Fl.*                                          G. H. ♄.

acutifoli-
us.
6. A. caule inermi angulato fruticoſo, foliis aciformibus
                                              rigidulis

rigidulis perennantibus mucronatis æqualibus. *Syſt.*
*veget.* 333.

Acute-leav'd Afparagus.

*Nat.* of Spain, Portugal, and the Levant.

*Cult.* 1739, by Mr. Ph. Miller. *Mill. dict. vol.* 2. *n.* 5.

*Fl.*　　　　　　　　　　　　　　　　G. H. ♄.

7. A. caule inermi angulato fruticofo, foliis fubulatis　*aphyllus.*
　　ſtriatis inæqualibus divergentibus. *Syſt. veget.* 333.

Prickly Afparagus.

*Nat.* of the South of Europe.

*Cult.* 1640. *Park. theat.* 454. *f.* 4.

*Fl.*　　　　　　　　　　　　　　　　G. H. ♄.

8. A. fpinis quaternis, ramis aggregatis teretibus, foliis　*capenſis.*
　　fetaceis. *Syſt. veget.* 333.

Cape Afparagus.

*Nat.* of the Cape of Good Hope.

*Cult.* 1691, in the Royal Garden at Hampton-court.

　*Pluk. phyt. t.* 78. *f.* 3.

*Fl.*　　　　　　　　　　　　　　　　G. H. ♄.

9. A. foliis folitariis lineari-lanceolatis, caule flexuofo,　*farmento-*
　　aculeis recurvis. *Sp. pl.* 450.　　　　　　　　　*ſus.*

Linear-leav'd Afparagus.

*Nat.* of Ceylon.

*Cult.* 1714, by the Dutchefs of Beaufort. *Br. Muſ.*

　*H. S.* 131. *fol.* 26.

*Fl.* Auguſt.　　　　　　　　　　　　G. H. ♄.

## D R A C Æ N A. *Linn. mant.* 9.

*Cor.* 6-partita, erecta. *Filam.* medio fubcrasſiora.
　　*Bacca* 3-locularis, 1-fperma.

1. D. arborea, foliis fubcarnofis apice fpinofo. *Syſt. veget.*　*Draca.*
　333.

Afparagus Draco.   *Sp. pl.* 451.
Dragon Tree.
*Nat.* of the Eaft Indies.
*Cult.* 1640.   *Park. theat.* 1531.
*Fl.*                                D. S. ♄.

*ferrea.*     2. D. arborea, foliis lanceolatis acutis.   *Syft. veget.* 334.
Purple Dracæna.
*Nat.* of China.
*Introd.* 1771, by Benjamin Torin, Efq.
*Fl.* March and April.                     S. ♄.

*margina-*     3. D. fruticofa, foliis dentato-fpinofis, racemis axillaribus,
*ta.*             baccis polyfpermis.
Aloe purpurea.   *de Lamarck encycl.* 1. *p.* 85.
Aloe-leav'd Dracæna.
*Nat.* of the Ifland of Bourbon.
*Introd.* 1766, by Monf. Richard.
*Fl.* April.                            S. ♄.

*enfifolia.*     4. D. herbacea fubcaulefcens, foliis enfiformibus.   *Syft.*
            *veget.* 334.
Dianella nemorofa.   *de Lamarck encycl.* 2. *p.* 276.
Sword-leav'd Dracæna.
*Nat.* of the Eaft Indies.
*Cult.* 1759, by Mr. Philip Miller.
*Fl.* Auguft.                         S. ♃.

*borealis.*     5. D. herbacea fubcaulefcens, foliis ellipticis.   T A B. 6.
Oval-leav'd Dracæna.
*Nat.* of Newfoundland, Hudfon's Bay, and Canada.
*Introd.* 1778, by Daniel Charles Solander, LL.D.
*Fl.* June.                            H. ♃.

CONVAL-

Tab. 6. Vol. 1. Page 454

Ehret. del.

*Dracæna borealis.*

MᶜKenzie. sc.

## CONVALLARIA. *Gen. pl.* 425.

*Cor.* fexfida. *Bacca* maculofa, 3-locularis.

\* Lilium convallium T. *corollis campanulatis.*

1. C. fcapo nudo femitereti, foliis ellipticis.                    *majalis.*
Convallaria majalis.  *Sp. pl.* 451.  *Curtis lond.*
α flore fimplici albo.
Sweet-fcented Lily of the Valley.
β flore pleno.
Double Lily of the Valley.
γ flore rubro.
Red-flower'd Lily of the Valley.
*Nat.* of Britain.
*Fl.* June.                                             H. ♃.

2. C. fcapo nudo ancipiti, floribus racemofis fecundis,  *japonica.*
foliis linearibus fcapo triplo longioribus,  *Linn.*
*fuppl.* 204.  *Thunb. jap.* 139.
Grafs-leav'd Lily of the Valley.
*Nat.* of Japan.
*Introd.* 1784, by Mr. John Græfer.
*Fl.* June.                                          G. H. ♃.

\* \* Polygonata T. *corollis infundibuliformibus.*

3. C. foliis verticillatis.  *Sp. pl.* 451.                 *verticil-*
Whorl-leav'd Solomon's Seal.                            *lata.*
*Nat.* of the North of Europe.
*Cult.* 1656, by Mr. John Tradefcant, Jun.  *Trad. muf.*
155.
*Fl.* May.                                             H. ♃.

4. C. foliis alternis amplexicaulibus, caule ancipiti, pe-  *Polygo-*
dunculis axillaribus fubunifloris.  *Sp. pl.* 451.        *natum.*

G g 4                         α Polygo-

α Polygonatum latifolium flore majore odoro. *Bauh.*
*pin.* 303.
Common Solomon's Seal.

β Polygonatum Hellebori albi folio, caule purpuraſcente.
*Raj. ſyn.* 263.
Hellebore-leav'd Solomon's Seal.
*Nat.* of England.
*Fl.* May and June.      H. ♃.

*multiflo-*   5. C. foliis alternis amplexicaulibus, caule tereti, pe-
*ra.*      dunculis axillaribus multifloris. *Sp. pl.* 452.

α Polygonatum latifolium vulgare. *Bauh. pin.* 303.
Broad-leav'd Solomon's Seal.

β Polygonatum humile Anglicum. *Raj. ſyn.* 263.
Dwarf Broad-leav'd Solomon's Seal.
*Nat.* of England.
*Fl.* May and June.      H. ♃.

* * * Smilaces T. *corollis rotatis.*

*racemofa.*   6. C. foliis feſſilibus, racemo terminali compoſito. *Sp.*
*pl.* 452.
Cluſter-flower'd Solomon's Seal.
*Nat.* of Virginia and Canada.
*Cult.* 1656, by Mr. John Tradeſcant, Jun. *Trad.*
*muſ.* 155.
*Fl.* May and June.      H. ♃.

*ſtellata.*   7. C. foliis amplexicaulibus plurimis. *Sp. pl.* 452.
Star-flower'd Solomon's Seal.
*Nat.* of Virginia and Canada.
*Cult.* 1640, by Mr. John Parkinſon. *Park. theat.*
697. *f.* 7.
*Fl.* May and June.      H. ♃.

*bifolia.*   8. C. foliis cordatis, floribus tetrandris. *Syſt. veget.* 335.
Leaſt

Leaft Solomon's Seal.

*Nat.* of the North of Europe.

*Cult.* 1739, by Mr. Philip Miller. *Mill. dict. vol.* 2. Smilax 15.

*Fl.* May and June.                            H. ♃.


## POLIANTHES. *Gen. pl.* 426.

*Cor.* infundibuliformis, incurva, æqualis. *Filamenta* corollæ fauci inferta. *Germen* in fundo corollæ.

1. POLIANTHES. *Sp. pl.* 453.                    *tuberofa.*
Common Tuberofe.

*Nat.* of the Eaft Indies.

*Cult.* 1664. *Evel. kalend. hort.* 65.

*Fl.* Auguft and September.            G. H. ♃.


## HYACINTHUS. *Gen. pl.* 427.

*Cor.* campanulata : pori 3 melliferi *germinis.*

1. H. corollis campanulatis fexpartitis apice revolutis.    *non fcrip-*
   *Sp. pl.* 453. *Curtis lond.*                    *tus.*
   Common Hyacinth, or Harebells.

*Nat.* of Britain.

*Fl.* March——June.                        H. ♃.


2. H. corollis campanulatis fexpartitis, racemo cernuo.    *cernuus.*
   *Sp. pl.* 453.
   Bending Hyacinth.

*Nat.* of Spain.

*Cult.* 1759, by Mr. Ph. Miller. *Mill. dict. edit.* 7. *n.* 4.

*Fl.* April and May.                        H. ♃.


3. H. petalis exterioribus fubdiftinctis ; interioribus coa-    *ferotinus.*
   dunatis. *Sp. pl.* 453.

                                    Late-

Late-flowering Hyacinth.
*Nat.* of Spain and Barbary.
*Cult.* 1759. *Mill. dict. ed.* 7. *n.* 2.
*Fl.* June.                                        H. ♃.

*amethyf-*      4. H. corollis campanulatis femifexfidis bafi cylindricis.
*tinus.*                *Sp. pl.* 454.
                   Amethyft-colour'd Hyacinth.
                   *Nat.* of Spain and Italy.
                   *Cult.* 1759, by Mr. Ph. Miller. *Mill. dict. edit.* 7. *n.* 5.
                   *Fl.* April and May.                          H. ♃.

*revolutus.*    5. H. corollis campanulatis fexpartitis revolutis, foliis
                        oblongis undulatis. *Linn: fuppl.* 204.
                   Wave-leav'd Hyacinth.
                   *Nat.* of the Cape of Good Hope. Mr. *Fr. Maffon.*
                   *Introd.* 1774.
                   *Fl.* Auguft.                              G. H. ♃.

*orienta-*      6. H. corollis infundibuliformibus femifexfidis bafi ven-
*lis.*                  tricofis. *Sp. pl.* 454.
                   Garden Hyacinth.
                   *Nat.* of the Levant.
                   *Cult.* 1596, by Mr. John Gerard. *Hort. Ger.*
                   *Fl.* March and April.                        H. ♃.

*romanus.*      7. H. corollis campanulatis femifexfidis racemofis, fta-
                        minibus membranaceis. *Linn. mant.* 224.
                   Roman Grape Hyacinth.
                   *Nat.* of Italy.
                   *Introd.* 1786, by Mr. John Græfer.
                   *Fl.* May.                                  H. ♃.

*Mufcari.*      8. H. corollis ovatis : omnibus æqualibus. *Sp. pl.* 454.
                   Mufk Hyacinth.
                                                            *Nat.*

*Nat.* of the Levant.
*Cult.* 1597. *Ger. herb.* 105. *f.* 1.
*Fl.* April and May. H. ♃.

9. H. corollis fubovatis. *Sp. pl.* 454.      *monftro-*
   Feathered Hyacinth.      *fus.*
   *Nat.* of the South of Europe.
   *Cult.* 1629. *Park. parad.* 117. *f.* 5.
   *Fl.* May and June. H. ♃.

10. H. corollis angulato-cylindricis : fummis fterilibus   *comofus.*
     longius pedicellatis. *Sp. pl.* 455. *Jacqu. auftr.* 2.
     *p.* 16. *t.* 126.
   Purple Grape Hyacinth.
   *Nat.* of the South of Europe.
   *Cult.* 1596, by Mr. John Gerard. *Hort. Ger.*
   *Fl.* May. H. ♃.

11. H. corollis globofis uniformibus, foliis canaliculato-   *botryoides.*
     cylindricis ftrictis. *Sp. pl.* 455.
   Blue Grape Hyacinth.
   *Nat.* of Italy.
   *Cult.* 1596, by Mr. John Gerard. *Hort. Ger.*
   *Fl.* April. H. ♃.

12. H. corollis ovatis : fummis feffilibus, foliis laxis.   *racemo-*
     *Sp. pl.* 455. *Jacqu. auftr.* 2. *p.* 52. *t.* 187.    *fus.*
   Clufter'd Grape Hyacinth.
   *Nat.* of the South of Europe.
   *Cult.* 1596, by Mr. John Gerard. *Hort. Ger.*
   *Fl.* April. H. ♃.

LACHE-

## LACHENALIA. *Jacquin.*

*Cor.* 6-partita: Petala 3 exteriora difformia. *Capf.*
3-alata: loculamenta polyfperma. *Sem.* globofa,
receptaculo affixa. *Linn. fil.*

*Orchioi-*
*des.*

1. L. corollis campanulatis: petalis tribus inteiioribus
longioribus, floribus feffilibus, foliis lanceolatis fca-
po brevioribus.
Phormium Hyacinthoides. *Linn. fuppl.* 204.
Phormium orchioides. *Thunb. nov. gen.* 5. *p.* 95.
Hyacinthus Orchioides. *Sp. pl.* 455.
Spotted-leav'd Lachenalia.
*Nat.* of the Cape of Good Hope.
*Cult.* 1752, by Mr. Philip Miller. *Mill. dict. edit.* 6.
Hyacinthus Orchioides 2.
*Fl.* February——April. G. H. ♃.

*pallida.*

2. L. corollis campanulatis: petalis tribus interioribus
longioribus, floribus breviffime pedunculatis horizontalibus, foliis lineari-oblongis fcapo longioribus.
Pale-flower'd Lachenalia.
*Nat.* of the Cape of Good Hope.
*Introd.* 1782, by George Wench, Efq.
*Fl.* March and April. G. H. ♃.
*Descr. Petala exteriora* bafi in tubum extus gibbum,
cærulefcentem, connata, oblonga, obtufa, alba, infra apicem gibbo viridi notatæ; *interiora* obovata,
apice patentia, alba, carina verfus apicem pallide
viridi.

*contami-*
*nata.*

3. L. corollis campanulatis: petalis tribus fuperioribus
longioribus, floribus pedunculatis, foliis lineari-
fubulatis canaliculatis.
Mixed-colour'd Lachenalia.
*Nat.* of the Cape of Good Hope. Mr. *Fr. Maffon.*
*Introd.*

*Introd.* 1774.

*Fl.* February and March. G. H. ♃.

DESCR. *Folia* glabra, fupra maculis obfcure ruben-
tibus adfperfa, fcapo lóngiora. *Scapus* erectus,
femiteres, vix fpithameus, maculis fordide rubris.
*Racemus* multiflorus : *Pedunculi* erecti, filiformes,
albi, fefquilineares. *Bractea* ad bafin finguli
pedunculi, e lata bafi ovata, acuminata, concava,
patentiffima, albida, pedunculo brevior. *Petala*
exteriora bafi in tubum extus gibbum connata,
oblonga, obtufa, alba, fuperne rubentia, infra apicem gibbo retufo : fupremum (fcapo proximum) tri-
lineare ; duo inferiora breviora. *Petala interiora*
oblonga, minus obtufa, prope apicem carinata, carina rubra : duo fuperiora longitudine fupremi ex-
terioris ; infimum longitudine breviorum exterio-
rum.

4. L. corollis cylindraceis, petalis tribus interioribus    *tricolor.*
duplo longioribus emarginatis, floribus peduncula-
tis pendulis.

Phormium Aloides. *Linn. fuppl.* 205. *Thunb. nov.*
*gen.* 5. *p.* 94. *Syft. veget.* 336.

α foliis lineari-lanceolatis, petalis interioribus apice
rubris.

Lachenalia tricolor. *Jacqu. ic. Syft. veget.* 314.

Narrow leav'd three-colour'd Lachenalia.

β foliis oblongo-lanceolatis, corollis flavis bafi rubenti-
bus, apice viridibus.

Broad-leav'd three-colour'd Lachenalia.

*Nat.* of the Cape of Good Hope.

*Introd.* 1774, by Mr. Francis Maffon.

*Fl.* April. G. H. ♃.

5. L. corollis cylindraceis, petalis tribus interioribus    *pendula.*
longioribus integris, floribus pedunculatis penaulis.

Pendulous

Pendulous Lachenalia.
*Nat.* of the Cape of Good Hope. Mr. *Fr. Maffon.*
*Introd.* 1774.
*Fl.* March and April.                                 G. H. ♃.
DESCR. *Folia* oblongo-lanceolata, fucculenta, lævia,
fpithamæa. *Scapus* erectus, teres, craffitie pennæ
anferinæ, foliis paulo longior, fupra medium maculis
nonnullis rubefcentibus adfperfus. *Racemus* pauci-
florus: floribus diftantibus, cernuis: *Pedunculi* filiformes, rubentes, femunciales. *Bractea* ad bafin
finguli pedunculi, ovata, acuta, breviffima. *Petala
exteriora* bafi in tubum trigibbum connata, rubra,
apicibus virentibus: infimi apex incraffatus, intus
fornicatus, fupra depreffus. *Petala interiora* obtu-
fiffima, inferne flavicantia, linea dorfali rubente;
fuperne e purpureo rubicundæ, margine et medio
virentes.

*viridis.*    6. L. corollis cylindraceis: petalis tribus exterioribus
longiffimis fubulatis.
Hyacinthus viridis. *Sp. pl.* 454. *Jacqu. ic. mifcell.* 2.
*p.* 331.
Green-flower'd Lachenalia.
*Nat.* of the Cape of Good Hope.
*Introd.* 1774, by Mr. Francis Maffon.
*Fl.* Auguft.                                          G. H. ♃.


L A N A R I A.

*Cor.* fupera, lanata, filamentis longior: limbo 6-par-
tito, patulo. *Peric.* 3-loculare.

*plumofa.*    1. LANARIA.
Hyacinthus lanatus. *Sp. pl.* 455.
Woolly Lanaria.

                                                       *Nat.*

*Nat.* of the Cape of Good Hope.
*Introd.* 1787, by Mr. Francis Maſſon.
*Fl.*                                            G. H. ♃.


A L E T R I S. *Gen. pl.* 428.

*Cor.* infundibuliformis. *Stamina* inſerta laciniarum
baſi. *Capſ.* 3-locularis.

1. A. acaulis, foliis lanceolatis membranaceis, floribus *farinoſa.*
alternis. *Sp. pl.* 456.
American Aletris.
*Nat.* of North America.
*Cult.* 1768, by Mr. Ph. Miller. *Mill. dict. ed.* 8.
*Fl.* June.                                      H. ♃.


2. A. acaulis, foliis lanceolatis undulatis, ſpica ovata, *capenſis.*
floribus mutantibus. *Sp. pl.* 456.
Waved-leav'd Aletris.
*Nat.* of the Cape of Good Hope.
*Introd.* 1768, by Mr. William Malcolm.
*Fl.* November——April.                          G. H. ♃.


3. A. acaulis, foliis lanceolatis glaucis, floribus nutan- *glauca.*
tibus : limbo patente.
Glaucous Aletris.
*Nat.* of the Cape of Good Hope.
*Introd.* 1781, by George Wench, Eſq.
*Fl.* January.                                   G. H. ♃.
OBS. Differt ab Aletri capenſi foliis glaucis vix undulatis, quæ in capenſi maxime undulata, ſuperne
læte virentia. *Limbus corollæ* patens, laciniis longioribus quam in capenſi, obtuſis quidem, ſed non
rotundis ut in illa. *Flores* minores, anguſtiores.

4. A.

*Uvaria.*     4. A. acaulis, fcapo longiore foliis enfiformibus carinatis.

Aletris Uvaria. *Syft. veget.* 337.

Aloe Uvaria. *Sp. pl.* 460.

Great Orange-flower'd Aletris.

*Nat.* of the Cape of Good Hope.

*Cult.* 1707, in Chelfea Garden. *Br. Muf. Sloan. mfcr.* 3370.

*Fl.* Auguft and September.        G. H. ♃.

*pumila.*     5. A. acaulis, fcapo breviore foliis linearibus acute cà-rinatis.

Small Orange-flower'd Aletris.

*Nat.* of the Cape of Good Hope.     Mr. *Fr. Maffon.*

*Introd.* 1774.

*Fl.* September——November.        G. H. ♃.

*hyacin-*     6. A. acaulis, foliis lanceolatis carnofis, floribus gemi-
*thoides.*           natis. *Sp. pl.* 456.

*zeylani-*    α Aloe zeylanica pumila, foliis variegatis. *Comm. hort.* 2.
*ca.*           *p.* 41. *t.* 21.

Ceylon Aloe.

*guineen-*    β Aloe guineenfis, radice geniculata, foliis e viridi et
*fis.*           atro undulatim variegatis. *Comm. hort.* 2. *p.* 39.
          *t.* 20.

Aletris guineenfis. *Jacqu. hort.* 1. *p.* 36. *t.* 84.

Guinea Aloe.

*Nat.* α. of Ceylon ; β. of Guinea.

*Cult.* 1690, in the Royal Garden at Hampton-court.
*Catal. mfcr.*

*Fl.* June——November.        D. S. ♃.

*fragrans.*    7. A. caulefcens, foliis lanceolatis laxis. *Sp. pl.* 456.

Sweet-fcented Aletris.

*Nat.* of Africa.

*Cult.* 1768, by Mr. Ph. Miller.    *Mill. dict. ed.* 8.

*Fl.* February and March.        S. ♄.

YUCCA.

## YUCCA. *Gen. pl.* 429.

*Cor.* campanulato-patens. *Stylus* nullus. *Capf.* 3-locularis.

1. Y. foliis integerrimis. *Sp. pl.* 456.    *gloriofa.*
Superb Adam's Needle.
*Nat.* of America.
*Cult.* before 1596, by Mr. John Gerard. *Lobel. adv.* 2.
  *p.* 507.
*Fl.* July and Auguft.    H. ♄.

2. Y. foliis crenulatis ftrictis. *Sp. pl.* 457.    *aloifolia.*
Aloe-leav'd Adam's Needle.
*Nat.* of South America.
*Cult.* 1696, in the Royal Garden at Hampton-court.
  *Catal. mfcr.*
*Fl.* Auguft and September.    G. H. ♄.

3. Y. foliis crenatis nutantibus. *Sp. pl.* 457.    *draconis.*
Drooping-leav'd Adam's Needle.
*Nat.* of South Carolina.
*Cult.* 1732, by James Sherard, M. D. *Dill. elth.* 437.
  *t.* 234. *f.* 417.
*Fl.* October and November.    H. ♄.

4. Y. foliis ferrato-filiferis. *Syft. veget.* 337.    *filamento-*
Thready Adam's Needle.    *fa.*
*Nat.* of Virginia.
*Cult.* 1675. *Morif. hift.* 2. *p.* 419. *n.* 2.
*Fl.* September and October.    H. ♄.

H h    ALOE.

A L O E.    *Gen. pl.* 430.

*Cor.* erecta, ore patulo, fundo nectarifero.    *Filam.* receptaculo inserta.

*dichoto-*      1. A. ramis dichotomis, foliis ensiformibus serratis.
*ma.*                *Thunb. Aloe, n.* 1.    *Linn. suppl.* 206.
                Smooth-stem'd Tree Aloe.
                *Nat.* of the Cape of Good Hope.
                *Introd.* 1780, by Mr. William Forsyth.
                *Fl.*                                                    D. S. ♄.

*perfolia-*     2. A. foliis caulinis dentatis amplexicaulibus vaginanti-
*ta.*                bus, floribus corymbosis cernuis pedunculatis subcy-
                lindricis.    *Syst. veget.* 337.
*arbores-*      α Aloe foliis amplexicaulibus reflexis, margine dentatis,
*cens.*              floribus cylindricis, caule fruticoso.    *Mill. dict.*
                Narrow-leav'd Sword Aloe.
*africana.*     β Aloe foliis latioribus amplexicaulibus, margine et
                dorso spinosis, floribus spicatis, caule fruticoso.
                *Mill. dict.*
                Broad-leav'd Sword Aloe.
*barba-*        γ Aloe foliis dentatis erectis succulentibus subulatis, flo-
*densis.*            ribus luteis in thyrso dependentibus.    *Mill. dict.*
                Barbadoes Aloe.
*succotri-*     δ Aloe (*vera*) foliis longissimis et angustissimis, margi-
*na.*                nibus spinosis, floribus spicatis.    *Mill. dict.*
                Succotrine Aloe.
*purpuras-*     ε foliis purpurascentibus subtus inferne maculatis : ma-
*cens.*              culis parvis subrotundis.
                White-spined glaucous Aloe.
*glauca.*       ζ Aloe caule brevi, foliis amplexicaulibus bifariam ver-
                sis, spinis marginis erectis, floribus capitatis.    *Mill.*
                *dict.*
                Red-spined glaucous Aloe.

                                                                η foliis

η foliis lineatis, spinis rubris.                                    lineata.
  Red-spined striped Aloe.

ϑ Aloe foliis amplexicaulibus nigricantibus undique          ferox.
  spinosis. *Mill. dict.*
  Great Hedge-hog Aloe.

ι Aloe africana maculata spinosa minor. *Dill, elth.* 18.     sapona-
  *t.* 15. *f.* 16.                                              ria.
  Great Soap Aloe.

κ Aloe foliis latioribus amplexicaulibus maculatis mar-       obscura.
  gine spinosis, floribus spicatis. *Mill. dict.*
  Common Soap Aloe.

λ foliis maculatis marginibus et carinæ apice serrulatis.     serrulata.
  Hollow-leav'd perfoliate Aloe.

μ foliis planis suberectis marginibus et pagina inferiore     suberec-
  spinosis.                                                     ta.
  Upright perfoliate Aloe.

ν Aloe (*brevioribus*) foliis amplexicaulibus utraque         depressa.
  spinosis, floribus spicatis. *Mill. dict.*
  Short-leav'd perfoliate Aloe.

ξ Aloe foliis erectis subulatis radicatis undique inerme      humilis.
  spinosis. *Hort. cliff.* 131.
  Dwarf Hedge-hog Aloe.

ο Aloe africana mitræformis spinosa. *Dill. elth.* 21. *t.* 17.   mitræ-
  *f.* 19.                                                      formis.
  Great Mitre Aloe.

π foliis ovatis brevibus distantibus subtus tuberculatis.     brevifo-
  Small Mitre Aloe.                                             lia.
  *Nat.* of Africa.
  *Cult.* 1596, by Mr. John Gerard. *Hort. Ger.*
  *Fl.* most part of the Year.                    D. S. ♃ .


3. A. acaulis, foliis trigonis cuspidatis ciliatis, floribus   arachnoi-
   subspicatis erectis cylindricis. *Thunb. Aloe, n.* 7.        des.
   Aloe pumila δ. ε. *Sp. pl.* 460.

α Aloe africana humilis arachnoidea. *Comm. præl.* 78.        commu-
  *t.* 27.                                                     nis.

                         H h 2                      Common

Common Cobweb Aloe.

pumila. β Aloe africana minima atroviridis, fpinis herbaceis nu-
merofis ornata. *Boerh. ludgb.* 2. *p.* 131. *cum. fig.*
Small Cobweb Aloe.
*Nat.* of the Cape of Good Hope.
*Cult.* 1725. *Bradl. fucc.* 3. *p.* 12. *t.* 30.
*Fl.* moſt part of the Year. D. S. ♄.

*margari-* 4. A. acaulis, foliis trigonis cuſpidatis papillofis, floribus
*tifera.* racemofis cernuis cylindricis. *Thunb. Aloe, n.* 8.
Aloe pumila α. β. γ. *Sp. pl.* 460.

major. α Aloe africana margaritifera, folio undique verrucis
numerofiſſimis ornato. *Bradl. fucc.* 3. *p.* 1. *t.* 21.
Great Pearl Aloe.

minor. β Aloe africana margaritifera minor. *Dill. elth.* 19.
*t.* 16. *f.* 17.
Small Pearl Aloe.

minima. γ Aloe africana margaritifera minima. *Dill. elth.* 20.
*t.* 16. *f.* 18.
Leaſt Pearl Aloe.
*Nat.* of the Cape of Hope.
*Introd.* before 1725, by Profeſſor Bradley. *Bradl. fucc.*
*loc. cit.*
*Fl.* moſt part of the Year. D. S. ♄.

*verruco-* 5. A. acaulis, foliis enfiformibus acutis papillofis diſti-
*fa.* chis, floribus racemofis reflexis clavatis. *Thunb.*
*Aloe, n.* 9.
Aloe verrucofa. *Mill. dict.*
Aloe difticha β. *Sp. pl.* 459.
Warted Aloe.
*Nat.* of Africa.
*Cult.* 1731, by Mr. Philip Miller. *Mill. dict. edit.* 1.
*n.* 20.
*Fl.* moſt part of the Year. D. S. ♄.

6. A.

6. A. acaulis, foliis acinaciformibus papillofis, floribus     *carinata.*
racemofis cernuis curvatis.

Aloe carinata.   *Mill. dict.*

Aloe difticha γ.   *Sp. pl.* 459.

Aloe africana feffilis, foliis carinatis verrucofis.   *Dill.*
*elth.* 22. *t.* 18. *f.* 20.

Keel-leav'd Aloe.

*Nat.* of Africa.

*Cult.* 1732, by James Sherard, M. D.   *Dill. elth. loc.*
*cit.*

*Fl.* June and July.           D. S. ♄ .

7. A. fubacaulis, foliis acinaciformibus glabris pictis, flo-   *maculata.*
ribus racemofis cernuis curvatis. *Thunb. Aloe, n.* 10.

α foliis acutis.                             pulchra.

Aloe foliis linguiformibus variegatis, floribus pedunculatis cernuis, ore inæquali. *Mill. ic.* 195. *t.* 292.

Narrow-leav'd fpotted Aloe.

β foliis obtufis cum acumine.                 obliqua.

Broad-leav'd fpotted Aloe.

*Nat.* of the Cape of Good Hope.

*Cult.* 1759, by Mr. Ph. Miller.   *Mill. ic. loc. cit.*

*Fl.* July and Auguft.           D. S. ♄ .

8. A. fubacaulis, foliis linguæformibus denticulatis gla-   *lingua.*
bris diftichis, floribus racemofis erectis cylindricis.
*Thunb. Aloe, n.* 11.

Aloe linguæformis. *Linn. fuppl.* 206.

α foliis anguftioribus longioribus.               angufti-
Aloe linguiforme. *Mill. dict.*                  folia.

Aloe difticha α.   *Sp. pl.* 459.

Common Tongue Aloe.

β foliis latioribus brevioribus.                 craffifo-
Thick-leav'd Tongue Aloe.                   lia.

*Nat.* of the Cape of Good Hope.

*Cult.* 1731, by Mr. Ph. Miller. *Mill. dict. edit.* 1. *n.* 25.
*Fl.* moft part of the Year.                  D. S. ♄.

*plicatilis.*   9. A. fubacaulis, foliis linguæformibus lævibus diftichis,
floribus racemofis pendulis cylindricis.
Aloe difticha ε. plicatilis.   *Sp. pl.* 459.
Kumara difticha.   *Medic. Theodora, p.* 70. *tab.* 4.
Fan Aloe.
*Nat.* of Africa.
*Cult.* 1731, by Mr. Ph. Miller. *Mill. dict. edit.* 1. *n.* 12.
*Fl.* June and July.                  D. S. ♄.

*variega-*   10. A. fubacaulis, foliis trifariis pictis canaliculatis: angu-
*ta.*        lis cartilagineis, floribus racemofis cylindricis.
*Thunb. Aloe, n.* 12.
Aloe variegata.   *Sp. pl.* 459.
Partridge-breaft Aloe.
*Nat.* of the Cape of Good Hope.
*Cult.* 1720, by Mr. Thomas Fairchild.   *Blair's bot.
effays, tab.*
*Fl.* moft part of the Summer.                  D. S. ♄.

*vifcofa.*   11. A. fubcaulefcens, foliis imbricatis trifariis ovatis, flo-
ribus racemofis cernuis cylindricis.   *Thunb. Aloe,
n.* 13.
Aloe vifcofa.   *Sp. pl.* 460.
Upright triangular Aloe.
*Nat.* of Africa.
*Cult.* 1732, by James Sherard, M. D.   *Dill. elth.* 15.
*t.* 13. *f.* 13.
*Fl.* June and July.                  D. S. ♄.

*fpiralis.*   12. A. fubcaulefcens, foliis imbricatis octofariis ovatis, floribus racemofis recurvis.   *Thunb. Aloe, n.* 14.
Aloe fpiralis.   *Sp. pl.* 459.

α foliis

α foliis fpiraliter imbricatis.
Imbricated fpiral Aloe.

imbrica-
ta.

β foliis quinquefariam imbricatis.
Five-fided fpiral Aloe.

pentago-
na.

*Nat.* of Africa.

*Cult.* 1732, by James Sherard, M.D.   *Dill. elth.* 16.
t. 13. f. 14.

*Fl.* June and July.                                     D. S. ♄.

13. A. acaulis, foliis quinquefariis deltoideis.   *Thunb.*   *retufa.*
*Aloe, n.* 15.

Aloe retufa.   *Sp. pl.* 459.

Cufhion Aloe.

*Nat.* of the Cape of Good Hope.

*Cult.* 1720, by Mr. Thomas Fairchild.   *Blair's bot.*
*effays, tab.*

*Fl.* June.                                              D. S. ♄.

A  G  A  V  E.   *Gen. pl.* 431.

*Cor.* erecta.   *Filamenta* corolla longiora, erecta.

1. A. acaulis, foliis dentato-fpinofis.                 *america-*
α foliis margine concoloribus.                          *na.*
Agave americana.   *Sp. pl.* 461.
Common american Agave.

β foliis margine fufcis.                                 Karatto.
Agave Karatto.   *Mill. dict.*
Red-fpined American Agave.
*Nat.* of South America.
*Cult.* 1640.   *Park. theat.* 150.
*Fl.* Auguft——October.                         D. S. ♄.

2. A. acaulis, foliis dentatis.                         *vivipara.*
Agave vivipara.   *Sp. pl.* 461.

H h 4                          Vivipa-

Viviparous Agave.
*Nat.* of America.
*Cult.* 1731, by Mr. Philip Miller. *Mill. dict. edit.* 1.
Aloe 7.
*Fl.* Auguft——October.                    D. S. ♄.

*virginica.*  3. A. acaulis herbacea, foliis dentato-fpinofis.
Agave virginica. *Sp. pl.* 461. *Jacqu. ic. vol.* 2.
Virginian Agave.
*Nat.* of Carolina and Virginia.
*Introd.* 1765, by Mr. John Cree.
*Fl.* September.                           D. S. ♃.

*lurida.*  4. A. fubcaulefcens, foliis dentato-fpinofis.
α foliis latioribus.
Agave Vera Cruz. *Mill. dict.*
Broad-leav'd Vera Cruz Agave.
β foliis anguftioribus.
Agave rigida. *Mill. dict.*
Narrow-leav'd Vera Cruz Agave.
*Nat.* of South America.
*Cult.* 1731, by Mr. Ph. Miller. *Mill. dict. edit.* 1.
Aloe 4.
*Fl.*                                      D. S. ♄.

*tuberofa.*  5. A. caulefcens, foliis dentato-fpinofis.
α fpinis folitariis.
Agave tuberofa. *Mill. dict.*
Single-fpined tuberous-rooted Agave.
β fpinis duplicibus.
Double-fpined tuberous-rooted Agave.
*Nat.* of America.
*Cult.* 1739, by Mr. Ph. Miller. *Rand. chel.* Aloe 34.
*Fl.*                                      D. S. ♄.
Obs. Hæc cum præcedenti, dubiæ fpecies, ignota
fructificatione.

6. A.

6. A. caulefcens, foliis integerrimis.                    *fœtida.*
Agave fœtida. *Sp. pl.* 461.
Fœtid Agave.
*Nat.* of South America.
*Cult.* 1690, in the Royal Garden at Hampton-court.
  *Catal. mfcr.*
*Fl.*                                                D. S. ♄.


ALSTRŒMERIA. *Gen. pl.* 432.

*Cor.* 6-petala, fupera, irregularis. *Stam.* declinata.

1. A. caule erecto. *Sp. pl.* 461. *Jacqu. hort.* 1. *p.* 20.   *Pelegri-*
  *t.* 50.                                               *na.*
Spotted-flower'd Alftrœmeria.
*Nat.* of Peru.
*Introd.* 1753, by Meffrs. Kennedy and Lee.
*Fl.* June——September.                            G. H. ♃.


2. A. caule adfcendente. *Sp. pl.* 462.              *Ligtu.*
Striped-flower'd Alftrœmeria.
*Nat.* of Peru.
*Introd.* about 1776, by John Brown, Efq.
*Fl.* February and March.                          S. ♃.


HEMEROCALLIS. *Gen. pl.* 433.

*Cor.* campanulata: tubo cylindrico. *Stam.* declinata.

1. H. corollis flavis. *Sp. pl.* 462. *Jacqu. hort.* 2. *p.* 65.  *flava.*
  *t.* 139. *Curtis magaz.* 19.
Yellow Day Lily.
*Nat.* of Siberia and Hungary.
*Cult.* 1597, by Mr. John Gerard. *Ger. herb.* 90. *f.* 1.
*Fl.* June.                                        H. ♃.

                                                2. H.

*fulva.*  2. H. corollis fulvis.   *Sp. pl.* 462.
Copper-colour'd Day Lily.
*Nat.* of the Levant.
*Cult.* 1597, by Mr. John Gerard.   *Ger. herb.* 90. *f.* 2.
*Fl.* July and Auguſt.                                H. ♃.

ACORUS.   *Gen. pl.* 434.

*Spadix* cylindricus, teċtus floſculis.   *Cor.* 6-petalæ,
nudæ.   *Stylus* nullus.   *Capſ.* 3-locularis.

*Calamus.*  1. A. ſcapi mucrone longiſſimo foliaceo.
Acorus Calamus.   *Sp. pl.* 462.
Sweet Acorus, or Flag.
*Nat.* of England.
*Fl.* June and July.                             H. ♃.

*grami-*  2. A. ſcapi mucrone ſpadicem vix excedente.
*neus.*    Chineſe Sweet-graſs.
*Nat.* ————. Cultivated in China.
*Introd.* 1786, by Allan Cooper, Eſq.  Commander of
the Atlas Indiaman.
*Fl.* February.                               D. S. ♃.

ORONTIUM.   *Gen. pl.* 435.

*Spadix* cylindricus, teċtus floſculis.   *Cor.* 6-petalæ,
nudæ.   *Stylus* nullus.   *Folliculi* 1-ſpermi.

*aquati-*  1. O. foliis lanceolato-ovatis.   *Syſt. veget.* 340.
*cum.*    Aquatic Orontium.
*Nat.* of Virginia and Cañada.
*Introd.* 1775, by John Fothergill, M.D.
*Fl.* June.                                   H. ♃.

*japoni-*  2. O. foliis enſiformibus venoſis.   *Thunb. jap.* 144. *Ic.*
*cum.*    *Kœmpfer. t.* 12.  *Syſt. veget.* 340.
Japan

Japan Orontium.
*Nat.* of Japan.
*Introd.* 1783, by Mr. John Græfer.
*Fl.* January.                                        D. S. ♃.


J U N C U S.   *Gen. pl.* 437.

*Cal.* 6-phyllus.   *Cor.* o.   *Capf.* 1-locularis.

* *Culmis nudis.*

1. J. culmo fubnudo tereti mucronato, panicula termi-   *acutus.*
   nali, involucro diphyllo fpinofo. *Sp. pl.* 463.
   Sea Rufh.
   *Nat.* of England.
   *Fl.* July and Auguft.                              H. ♃.

2. J. culmo nudo ftricto, capitulo laterali. *Sp. pl.* 464.  *conglome-*
   Round-headed Rufh.                                        *ratus.*
   *Nat.* of Britain.
   *Fl.* June and July.                               H. ♃.

3. J. culmo nudo ftricto, panicula laterali. *Sp. pl.* 464.  *effufus.*
   Common foft Rufh.
   *Nat.* of Britain.
   *Fl.* May——Auguft.                                 H. ♃.

4. J. culmo nudo filiformi nutante, panicula laterali. *filiformis.*
   *Sp. pl.* 465.
   Leaft foft Rufh.
   *Nat.* of England.
   *Fl.* Auguft.                                       H. ♃.

5. J. culmo nudo, foliis floribufque tribus terminalibus.  *trifidus.*
   *Sp. pl.* 465.
   Three-flower'd Rufh,

                                                       *Nat.*

Nat. of Scotland.
Fl. July.　　　　　　　　　　　　　　　　　H. ♃.

*fquarro-*
*fus.*　　6. J. culmo nùdo, foliis fetaceis, capitulis glomeratis
　　　　　aphyllis.　*Sp. pl.* 465.
　　　　Mofs-rufh, or Goofe-corn.
　　　　Nat. of Britain.
　　　　Fl. June.　　　　　　　　　　　　　　H. ♃.

** * Culmis foliofis.*

*articula-*
*tus.*　　7. J. foliis nodofo-articulatis, petalis obtufis.　*Sp. pl.*
　　　　465.
　　　　Jointed Rufh.
　　　　Nat. of Britain.
　　　　Fl. July and Auguft.　　　　　　　　H. ♃.

*bulbofus.*　8. J. foliis linearibus canaliculatis, capfulis obtufis.　*Sp.*
　　　　*pl.* 466.
　　　　Bulbous Rufh.
　　　　Nat. of Britain.
　　　　Fl. Auguft.　　　　　　　　　　　　H. ♃.

*bufonius.*　9. J. culmo dichotomo, foliis angulatis, floribus folitariis
　　　　feffilibus.　*Sp. pl.* 466.
　　　　Toad-rufh.
　　　　Nat. of Britain.
　　　　Fl. July and Auguft.　　　　　　　H. ♃.

*pilofus.*　10. J. foliis planis pilofis, corymbo ramofo, floribus fo-
　　　　litariis.　*Hudf. angl.* 151.
　　　　Juncus pilofus.　*Sp. pl.* 468.　*Curtis lond.*
　　　　Hairy Rufh.
　　　　Nat. of Britain.
　　　　Fl. April and May.　　　　　　　　H. ♃.

11. J.

11. J. foliis planis pilofis, corymbo decompofito, floribus *fylvati-*
    fafciculatis feffilibus. *Hudf. angl.* 151. *Curtis lond.*  *cus.*
Wood Rufh.
*Nat.* of England.
*Fl.* May.                  H. ♃.

12. J. foliis planis fubpilofis, corymbis folio brevioribus,  *niveus,*
    floribus fafciculatis. *Sp. pl.* 468.
White-flower'd Rufh.
*Nat.* of the Alps of Switzerland.
*Introd.* 1770, by Monf. Richard.
*Fl.* June.                  H. ♃.

13. J. foliis planis fubpilofis, fpicis feffilibus pedunculatifque.  *campef-*
    *Sp. pl.* 468. *Curtis lond.*                *tris.*
Field Rufh.
*Nat.* of Britain.
*Fl.* April.                 H. ♃.

## A C H R A S. *Gen. pl.* 438.

*Cal.* 6-phyllus. *Cor.* ovata, 6-fida: fquamis totidem
    alternis interioribus. *Pomum* 10-loculare. *Sem.*
    folitaria, hilo marginali, apiceque unguiculato.

1. A. floribus folitariis, foliis cuneiformi-lanceolatis. *Sp.*  *mammofa.*
    *pl.* 469.
Mammee Sapota.
*Nat.* of South America.
*Cult.* 1739, by Mr. Philip Miller. *Mill. dict. vol.* 2.
    Sapota 2.
*Fl.*                    S. ♄.

2. A. floribus folitariis, foliis lanceolato-ovatis. *Syft.*  *Sapota.*
    *veget.* 342.
Common Sapota.
                         *Nat.*

Nat. of South America.
Cult. 1739, by Mr. Philip Miller.   Mill. dict. vol. 2.
   Sapota 1.
Fl.                                                    S. ♄.

*Salicifolia.*  3. A. floribus confertis, foliis lanceolato-ovatis.   Sp.
      pl. 470.
      Willow-leav'd Sapota.
      Nat. of South America.
      Cult. 1758, by Mr. Philip Miller.
      Fl.                                               S. ♄.

         P R I N O S.   Gen. pl. 441.

      Cal. 6-fidus.   Cor. 1-petala, rotata.   Bacca 6-sperma.

*verticil-*   1. P. foliis longitudinaliter serratis.   Sp. pl. 471.
*latus.*       Deciduous Winter-berry.
      Nat. of Virginia.
      Introd. 1736, by Peter Collinson, Esq. Coll. mss.
      Fl. July and August.                              H. ♄.

*glaber.*   2. P. foliis lanceolatis obtusiusculis glabris apice serratis.
      Prinos glaber.   Sp. pl. 471. (exclusis synonymis.)
      Ever-green Winter-berry.
      Nat. of Canada.
      Cult. 1759, by Mr. Ph. Miller. Mill. dict. edit. 7. n. 2.
      Fl. July and August.                              H. ♄.

*lucidus.*   3. P. foliis ellipticis acuminatis lævibus apice subserra-
         tis.
      Shining Winter-berry.
      Nat.
      Introd. about 1778, by Mr. James Gordon.
      Fl. July.                                         G. H. ♄.

               B U R S E R A.

## BURSERA. *Gen. pl.* 440.

*Cal.* 3-phyllus. *Cor.* 3-petala. *Capf.* carnofa, 3-valvis, 1-fperma.

1. BURSERA. *Sp. pl.* 471.        *gummife-*
 Jamaica Birch-tree.         *ra.*
 *Nat.* of the Weft Indies.
 *Cult.* 1690, in the Royal Garden at Hampton-court.
  *Catal. mfcr.*
 *Fl.*            S. ♄.

## BERBERIS. *Gen. pl.* 442.

*Cal.* 6-phyllus. *Petala* 6 : ad ungues glandulis 2.
  *Stylus* o. *Bacca* 2-fperma.

1. B. pedunculis racemofis. *Sp. pl.* 471.   *vulgaris.*
α fpinis triplicibus, ferraturis foliorum fetaceis elongatis,  rubra.
 baccis rubris.
 Common Berberry.
β fpinis multiplicibus, ferraturis foliorum fetaceis elon-  violacea.
 gatis, baccis violaceis.
 Purple-fruited Berberry.
γ fpinis triplicibus, ferraturis foliorum remotis breviffi-  canaden-
 mis.               fis.
 Berberis canadenfis. *Mill. dict.*
 Canada Berberry.
 *Nat.* α. β. of Britain ; γ. of Canada.
 *Fl.* April and May.      H. ♄.

2. B. pedunculis unifloris. *Sp. pl.* 472.   *cretica.*
 Cretan Berberry.
 *Nat.* of the Ifland of Candia.
 *Cult.* 1759, by Mr. Ph. Miller. *Mill. dict. edit.* 7. *n.* 3.
 *Fl.* April and May.      H. ♄.

CANA-

## CANARINA. *Linn. mant.* 148.

*Cal.* 6-phyllus. *Cor.* 6-fida, campanulata. *Stigmata* 6.
*Capſ.* infera, 6-locularis, polyſperma.

*Campa-*
*nula.*
1. CANARINA. *Syſt. veget.* 344.
Campanula canarienſis. *Sp. pl.* 238. *Syſt. veget.* 209.
Canary Bell-flower.
*Nat.* of the Canary Iſlands.
*Cult.* 1696, in the Royal Garden at Hampton-court.
*Pluk. alm.* 76. *t.* 276. *f.* 1.
*Fl.* January——March.                   G. H. ♃.

## FRANKENIA. *Gen. pl.* 445.

*Cal.* 5-fidus, infundibulif. *Petala* 5. *Stigma* 6-parti-
tum. *Capſ.* 1-locularis, 3-valvis.

*lævis.*
1. F. foliis linearibus confertis baſi ciliatis. *Syſt. veget.*
344.
Smooth Frankenia, or Sea Heath.
*Nat.* of England.
*Fl.* Auguſt.                             H. ♃.

*pulveru-*
*lenta.*
2. F. foliis obovatis retuſis ſubtus pulveratis. *Sp. pl.* 474.
Duſty Frankenia.
*Nat.* of England.
*Fl.* July.                               H. ☉.

## PEPLIS. *Gen. pl.* 446.

*Perianth.* campanulatum : ore 12-fido. *Petala* 6,
calyci inſerta. *Capſ.* 2-locularis.

*Portula.*
1. P. floribus apetalis. *Sp. pl.* 474. *Curtis lond.*
Water Peplis, or Purſlane.
*Nat.* of Britain.
*Fl.* September.                          H. ☉.

*DIGYNIA.*

# *D I G Y N I A.*

## O R Y Z A.  *Gen. pl.* 448.

*Cal.* Gluma 2-valvis, 1-flora. *Cor.* 2-valvis, fubæqualis, femini adnafcens.

1. ORYZA.  *Sp. pl.* 475.                                    *fativa.*
Common Rice.
*Nat.*
*Cult.* 1739, by Mr. Philip Miller. *Rand. chel.*
*Fl.* July.                                           S. ☉.

## A T R A P H A X I S.  *Gen. pl.* 449.

*Cal.* 2-phyllus. *Petala* 2, finuata. *Stigmata* capitata.
*Sem.* 1.

1. A. ramis fpinofis.  *Sp. pl.* 475.  *L'Herit. ftirp. nov.*   *fpinofa.*
27. *t.* 14.
Prickly-branch'd Atraphaxis.
*Nat.* of the Levant.
*Cult.* 1759, by Mr. Ph. Miller. *Mill. dict. edit.* 7. *n.* 1.
*Fl.* Auguft.                                      G. H. ♄.

2. A. inermis.  *Sp. pl.* 475.                              *undulata.*
Waved-leav'd Atraphaxis.
*Nat.* of the Cape of Good Hope.
*Cult.* 1732, by James Sherard, M. D.  *Dill. elth.*
36. *t.* 32.
*Fl.* June and July.                               G. H. ♄.

# *TRIGYNIA.*

**RUMEX.** *Gen. pl.* 451.

*Cal.* 3-phyllus.    *Petala* 3, conniventia.    *Sem.* 1, triquetrum.

\* *Hermaphroditi : valvulis floris granulo notatis.*

*Patientia.*    1. R. floribus hermaphroditis : valvulis integerrimis : unica granifera, foliis ovato-lanceolatis. *Syst. veget.* 346.
Patience Dock, or Rhubarb.
*Nat.* of Italy.
*Cult.* 1597, by Mr. John Gerard. *Ger. herb.* 313. *f.* 5.
*Fl.* June and July.                                    H. ♃.

*fanguineus.*    2. R. floribus hermaphroditis : valvulis integerrimis : unica granifera, foliis cordato-lanceolatis. *Sp. pl.* 476.
Bloody Dock.
*Nat.* of England.
*Fl.* June and July.                                    H. ♃.

*crispus.*    3. R. floribus hermaphroditis : valvulis integris graniferis, foliis lanceolatis undulatis acutis. *Sp. pl.* 476. *Curtis lond.*
Curl'd Dock.
*Nat.* of Britain.
*Fl.* June and July.                                    H. ♃.

*paludosus.*    4. R. floribus hermaphroditis : valvulis integris graniferis, foliis cordato-lanceolatis obtusiusculis, verticillis distinctis foliolo subjectis.
Rumex paludosus. *Hudf. angl.* 154.

Marsh

Marſh Dock.
*Nat.* of Britain.
*Fl.* Auguſt.        H. ♃.

5. R. floribus hermaphroditis : valvulis integerrimis
     graniferis, foliis lanceolatis glabris acutis integer-    *Hydrola-*
     rimis baſi attenuatis.                     *pathum.*
Rumex Hydrolapathum.   *Hudſ. angl.* 154.
Water Dock.
*Nat.* of England.
*Fl.* July.        H. ♃.

6. R. floribus hermaphroditis : valvulis dentatis : apice   *perſicari-*
     ſubulato : omnibus graniferis, foliis lanceolatis.     *oides.*
Rumex perſicarioides.   *Sp. pl.* 477.
Arſmart-leav'd Dock.
*Nat.* of Virginia.
*Introd.* 1773, by Chevalier Murray.
*Fl.* June and July.        H. ☉.

7. R. floribus hermaphroditis : valvulis trifido-ſetaceis :   *ægyptius.*
     unica granifera.   *Sp. pl.* 477.
Egyptian Dock.
*Nat.* of Egypt.
*Cult.* 1739, by Mr. Philip Miller.   *Rand. chel.* La-
     pathum 14.
*Fl.* June and July.        H. ☉.

8. R. floribus hermaphroditis : valvulis dentatis : apice   *dentatus.*
     lanceolato : omnibus graniferis, foliis lanceolatis.
Rumex dentatus.   *Linn. mant.* 226.
Dentated Dock.
*Nat.* of Egypt.
*Cult.* 1732, by James Sherard, M. D.   *Dill. elth.* 191.
     *t.* 158. *f.* 191.
*Fl.* July and Auguſt.        H. ☉.

*mariti-*    9. R. floribus hermaphroditis: valvulis dentatis gra-
*mus.*        niferis, foliis linearibus. *Sp. pl.* 478. *Curtis lond.*
              Golden Dock.
              *Nat.* of Britain.
              *Fl.* July and Auguft.                    H. ♃.

*acutus.*   10. R. floribus hermaphroditis: valvulis dentatis gra-
              niferis, foliis cordato-oblongis acuminatis. *Sp. pl.*
              478. *Curtis lond.*
              Sharp-pointed Dock.
              *Nat.* of Britain.
              *Fl.* June and July.                      H. ♃.

*obtuſifoli-* 11. R. floribus hermaphroditis: valvulis dentatis grani-
*us.*         feris, foliis cordato-oblongis obtuſiuſculis crenula-
              tis. *Sp. pl.* 478. *Curtis lond.*
              Blunt-leav'd Dock.
              *Nat.* of Britain.
              *Fl.* June and July.                      H. ♃.

*pulcher.*  12. R. floribus hermaphroditis: valvulis dentatis: ſub-
              unica granifera, foliis radicalibus panduriformi-
              bus. *Sp. pl.* 477.
              Fiddle Dock.
              *Nat.* of Britain.
              *Fl.* June and July.                      H. ♃.

        ** *Hermaphroditi: valvulis floris grano deſtitutis.*
*bucepha-*  13. R. floribus hermaphroditis: valvulis dentatis nudis,
*lophorus.*   pedicellis planis reflexis incraſſatis. *Syſt. veg.* 347.
              Baſil-leav'd Dock.
              *Nat.* of Italy.
              *Cult.* 1683, by Mr. James Sutherland. *Sutherl. hort.*
              *edin.* 7. *n.* 1.
              *Fl.* June.                               H. ☉.
                                                        14. R.

14. R. floribus hermaphroditis : valvulis lævibus, caule   *Lunaria.*
arboreo, foliis fubcordatis. *Sp. pl.* 479.
Tree Sorrel.
*Nat.* of the Canary Iflands.
*Cult.* 1698, by the Dutchefs of Beaufort. *Br. Muf.*
*Sloan. mff.* 3358.
*Fl.* June and July.                    G. H. ♄.

15. R. floribus hermaphroditis geminatis : valvularum   *veficari-*
alis omnibus maximis membranaceis reflexis, fo-   *us.*
liis indivifis. *Sp. pl.* 479.
Bladder Dock, or Sorrel.
*Nat.* of Africa.
*Cult.* 1656, by Mr. John Tradefcant, Jun. *Trad.*
*muf.* 75.
*Fl.* July and Auguft.                    H. ☉.

16. R. floribus hermaphroditis diftinctis : valvulæ alte-   *rofeus.*
rius ala maxima membranacea reticulata, foliis
erofis. *Syft. veget.* 347.
Rofe Dock.
*Nat.* of Egypt.
*Cult.* 1739, by Mr. Philip Miller. *Rand. chel.*
Acetofa 11.
*Fl.* July and Auguft.                    H. ☉.

17. R. floribus hermaphroditis diftinctis : valvulis cor-   *tingita-*
datis obtufis integerrimis, foliis haftato-ovatis.   *nus.*
*Syft. veget.* 347.
Tangier Dock.
*Nat.* of Barbary and Spain.
*Cult.* 1680, by Robert Morifon, M.D. *Morif. hift.*
2. *p.* 583. *n.* 8. *f.* 5. *t.* 28. *f.* 8.
*Fl.*                    H. ♃.

18. R.

*ſcutatus.*    18. R. floribus hermaphroditis, foliis cordato-haſtatis,
caule tereti. *Syſt. veget.* 347.
French Sorrel.
*Nat.* of France and Switzerland.
*Cult.* 1633. *Ger. emac.* 397. *f.* 4.
*Fl.* June and July.                                    H. ♃.

*digynus.*    19. R. floribus hermaphroditis digynis.   *Sp. pl.* 480.
Mountain Dock, or Sorrel.
*Nat.* of Britain.
*Fl.* June and July.                                    H. ♃.

**\* \* \*** *Floribus diclinis.*

*alpinus.*    20. R. floribus hermaphroditis ſterilibus femineiſque:
valvulis integerrimis nudis, foliis cordatis obtuſis
rugoſis. *Sp. pl.* 480.
Alpine Dock, or Monk's Rhubarb.
*Nat.* of France and Switzerland.
*Cult.* 1597. *Ger. herb.* 313. *f.* 6.
*Fl.* June and July.                                    H. ♃.

*ſpinoſus.*    21. R. floribus androgynis: calycibus femineis mono-
phyllis: valvulis exterioribus reflexo-uncinatis.
*Sp. pl.* 481.
Prickly-ſeeded Dock.
*Nat.* of Candia.
*Cult.* 1683. *Sutherland. hort. edin.* 49.
*Fl.* June and July.                                    H. ☉.

*tuberoſus.*    22. R. floribus dioicis, foliis lanceolato-ſagittatis, ramis
patentibus. *Syſt. veget.* 348.
Tuberous-rooted Dock.
*Nat.* of Italy.
*Cult.* 1748, by Mr. Philip Miller. *Mill. dict. edit.* 5.
Acetoſa 9.
*Fl.*                                                    H. ♃.
                                                        23. R.

23. R. floribus dioicis, foliis oblongis fagittatis. *Sp. pl.* *Acetofa.*
481.
Garden Sorrel.
*Nat.* of Britain.
*Fl.* June and July. H. ♃.

24. R. floribus dioicis, foliis lanceolato-haftatis. *Sp. pl.* *Acetofella.*
481. *Curtis lond.*
Sheep's Dock, or Sorrel.
*Nat.* of Britain.
*Fl.* May——July. H. ♃.

25. R. floribus dioicis, foliis omnibus petiolatis haftatis : *arifolius.*
auriculis fimplicibus divaricatis, caule erecto.
Rumex arifolius. *Linn. fuppl.* 212.
Rumex abyffinicus. *Jacqu. hort.* 3. *p.* 48. *t.* 93.
Halberd-leav'd Dock.
*Nat.* of Africa. *James Bruce,* Efq.
*Introd.* 1775.
*Fl.* December——April. G. H. ♄.

FLAGELLARIA. *Gen. pl.* 450.
*Cal.* 6-partitus. *Cor.* 0. *Bacca* 1-fperma.

1. FLAGELLARIA. *Sp. pl.* 475. *indica.*
Indian Flagellaria.
*Nat.* of the Eaft Indies and Guinea.
*Introd.* 1782, by the Earl of Tankerville and Dr.
Pitcairn.
*Fl.* S. ♄.

TRIGLOCHIN. *Gen. pl.* 453.
*Cal.* 3-phyllus. *Petala* 3, calyciformia. *Stylus* 0.
*Capf.* bafi dehifcens.

1. T. capfulis trilocularibus fublinearibus. *Sp. pl.* 482. *paluftre.*
I i 4 Marfh

Marſh Triglochin.
*Nat.* of Britain.
*Fl.* July and Auguſt.                          **H. ♂.**

*mariti-*      2. T. capſulis ſexlocularibus ovatis.  *Sp. pl.* 483.
*mum.*         Sea Triglochin.
               *Nat.* of Britain.
               *Fl.* May and June.                    **H. ♃.**


### MELANTHIUM. *Gen. pl.* 454.

*Cor.* 6-petala. *Filamenta* ex elongatis unguibus corollæ.

*virgini-*     1. M. floribus paniculatis, petalis unguiculatis extus hir-
*cum.*            ſutis.
               Melanthium virginicum.  *Sp. pl.* 483.
               Vìrginian Melanthium.
               *Nat.* of North America.
               *Cult.* 1768, by Mr. Philip Miller.  *Mill. dict. edit.* 8.
               *Fl.* June and July.                   **H. ♃.**

*lætum.*       2. M. racemo oblongo, petalis feſſilibus, foliis glabris
                  lanceolato-linearibus : caulinis remotis.
               Spear-leav'd Melanthium.
               *Nat.* of North America.
               *Introd.* 1770, by George W. Earl of Coventry.
               *Fl.* June.                            **H. ♃.**

*capenſe.*     3. M. petalis punctatis, foliis cucullatis.  *Sp. pl.* 483.
               Spotted-flower'd Melanthium.
               *Nat.* of the Cape of Good Hope.
               *Introd.* 1788, by Mr. Francis Maſſon.
               *Fl.*                                 **G. H. ♃.**

*viride.*      4. M. pedunculis unifloris cernuis.  *Linn. ſuppl.* 213.
               Green-flower'd Melanthium.

                                                     *Nat.*

*Nat.* of the Cape of Good Hope.
*Introd.* 1788, by Mr. Francis Maffon.
*Fl.*                                                                    G. H. ♃.

5. M. foliis triquetris glabris caule longioribus, floribus    *trique-*
    fpicatis. *Linn. fuppl.* 213.                                          *trum.*
    Rufh-leav'd Melanthium.
    *Nat.* of the Cape of Good Hope.
    *Introd.* 1788, by Mr. Francis Maffon.
    *Fl.* June.                                                        G. H. ♃.

6. M. corolla monopetala, foliis cucullatis lanceolatis.    *monopeta-*
    *Linn. fuppl.* 213. *Syft. veget.* 349.                          *lum.*
    Melanthium fpicatum. *Burm. fl. cap.* 11. *Houtt. nat.*
      *hift.* 12. *p.* 429. *tab.* 85. *f.* 2.
    Wurmbea capenfis. *Thunb. nov. gen.* 1. *p.* 18. *Syft.*
      *veget.* 348.
    One-petal'd Melanthium.
    *Nat.* of the Cape of Good Hope.
    *Introd.* 1788, by Mr. Francis Maffon.
    *Fl.*                                                                    G. H. ♃.

M E D E O L A.  *Gen. pl.* 455.

*Cal.* o.  *Cor.* 6-partita, revoluta.  *Bacca* 3-fperma.

1. M. foliis verticillatis, ramis inermibus. *Sp. pl.* 483.    *virginia-*
    Virginian Medeola.                                                    *na.*
    *Nat.* of Virginia.
    *Cult.* 1768, by Mr. Philip Miller. *Mill. dict. edit.* 8.
    *Fl.* June.                                                        H. ♃.

2. M. foliis alternis ovatis bafi fubcordatis obliquis.    *Afpara-*
    Medeola afparagoides. *Sp. pl.* 484. *Syft. veget.* 349.    *goides.*
    Dracæna Medeoloides. *Linn. fuppl.* 203. *Syft. veget.*
      334.
    Broad-leav'd Shrubby Medeola.
    *Nat.* of the Cape of Good Hope.
                                                                  *Cult.*

*Cult.* 1702, by the Dutchefs of Beaufort.   *Br. Muf.*
*H. S.* 138. *fol.* 16.
*Fl.* moft part of the Winter.                G. H. ♄.

*anguftifo-*   3. M. foliis alternis ovato-lanceolatis.
*lia.*            Medeola anguftifolia.   *Mill. dict.*
               Afparagus africanus fcandens Myrti folio anguftiore.
                  *Till. pif.* 17. *t.* 12. *f.* 2.
               Narrow-leav'd Shrubby Medeola.
               *Nat.* of the Cape of Good Hope.
               *Cult.* 1752, by Mr. Philip Miller.   *Mill. dict, edit.* 6.
                  Afparagus 12.
               *Fl.* December——March.                G. H. ♄.


            T R I L L I U M.   *Gen. pl.* 456.

         *Cal.* 3-phyllus.   *Cor.* 3-petala.   *Bacca* 3-locularis.

*cernuum.*   1. T. flore pedunculato cernuo.   *Sp. pl.* 484.
             Stalk-flower'd Trillium.
             *Nat.* of North America.
             *Cult.* 1759, by Mr. Ph. Miller.  *Mill. dict. edit.* 7. *n.* 1.
             *Fl.* April.                                  H. ♃.

*feffile.*   2. T. flore feffili erecto.   *Sp. pl.* 484.  *Curtis magaz.* 40.
             Seffile-flower'd Trillium.
             *Nat.* of North America.
             *Introd.* 1765, by Mr. John Cree.
             *Fl.* April and May.                          H. ♃.


            C O L C H I C U M.   *Gen. pl.* 457.

         *Spatha.*   *Cor.* 6-partita: tubo radicato.   *Capfulæ* 3,
                      connexæ, inflatæ.

*autum-*   1. C. foliis planis lanceolatis erectis.   *Sp. pl.* 485.
*nale.*                                                 *a* flore

*α* flore fimplici.
Common Meadow Saffron.
*β* flore pleno.
Double-flower'd Meadow Saffron.
*Nat.* of Britain.
*Fl.* September and October.     H. ♃.

2. C. foliis undulatis patentibus.   *Sp. pl.* 485.    *variega-*
Variegated Meadow Saffron.            *tum.*
*Nat.* of the Greek Iflands.
*Cult.* 1629.   *Park. parad.* 155. *f.* 5.
*Fl.* Auguft——October.     G. H. ♃.

### H E L O N I A S.   *Gen. pl.* 458.

*Cor.* 6-petala.   *Cal.* o.   *Capf.* 3-locularis.

1. H. foliis lanceolatis nervofis.   *Syf. veget.* 349.    *bullata.*
Spear-leav'd Helonias.
*Nat.* of Penfylvania.
*Cult.* 1758, by Mr. Ph. Miller. *Mill. ic.* 181. *t.* 272.
*Fl.* April and May.     H. ♃.

2. H. foliis caulinis fetaceis.   *Sp. pl.* 485.    *Afphodel-*
Grafs-leav'd Helonias.            *oides.*
*Nat.* of Penfylvania and Virginia.
*Introd.* 1765, by Mr. William Young.
*Fl.* May and June.     H. ♃.

# *T E T R A G Y N I A.*

## P E T I V E R I A.   *Gen. pl.* 459.

*Cal.* 4-phyllus.   *Cor.* o.   *Sem.* 1 : apice ariftis reflexis.

1. P. floribus hexandris.   *Sp. pl.* 486.    *alliacea.*
Common

Common Guinea Hen-weed.
*Nat.* of Jamaica.
*Cult.* 1758, by Mr. Philip Miller.
*Fl.* June and July.                    S. ♄.

*octandra.*    2. P. floribus octandris. *Sp. pl.* 486.
Dwarf Guinea Hen-weed.
*Nat.* of the West Indies.
*Cult.* 1739, by Mr. Philip Miller. *Rand. chel.*
*Fl.* June and July.                    S. ♄.


# *POLYGYNIA.*

## A L I S M A.   *Gen. pl.* 460.

*Cal.* 3-phyllus. *Petala* 3. *Sem.* plura.

*Plantago.*    1. A. foliis ovatis acutis, fructibus obtuse trigonis. *Sp.*
*pl.* 486. *Curtis lond.*
Greater Water Plantain.
*Nat.* of Britain.
*Fl.* June and July.                    H. ♃.

*Damaso-*    2. A. foliis cordato-oblongis, floribus hexagynis, capsulis
*nium.*         subulatis. *Sp. pl.* 486. *Curtis lond.*
Star-headed Water Plantain.
*Nat.* of England.
*Fl.* June——August.                    H. ♃.

*Ranun-*    3. A. foliis lineari-lanceolatis, fructibus globoso-squarro-
*culoides.*     sis. *Sp. pl.* 487.
Lesser Water Plantain.
*Nat.* of Britain.
*Fl.* August.                    H. ♃.

*Classis*

*Claßis VII.*

# HEPTANDRIA

## *MONOGYNIA.*

TRIENTALIS. *Gen. pl.* 461.

*Cal.* 7-phyllus. *Cor.* 7-partita, æqualis, plana. *Bacca*
exfucca.

1. T. foliis lanceolatis integerrimis. *Sp. pl.* 488.     *europæa.*
Common Trientalis, or Chick-weed Winter-green.
*Nat.* of Britain.
*Fl.* June.       H. ♃.

DISANDRA. *Linn. fuppl.* 32.

*Cal.* fub 7-partitus. *Cor.* rotata, fub 7-partita. *Capf.*
2-locularis, polyfperma.

1. DISANDRA. *Syft. veget.* 352.     *proftrata.*
Trailing Difandra.
*Nat.* of Madeira.
*Introd.* about 1771.
*Fl.* moft part of the Summer.     G. H. ♃.

ÆSCULUS. *Gen. pl.* 462.

*Cal.* 1-phyllus, 5-dentatus, ventricofus. *Cor.* 5-petala,
inæqualiter colorata, calyci inferta. *Capf.* 3-locu-
laris.

1. Æ. foliolis feptenis.     *Hippo-*
Æfculus Hippocaftanum. *Sp. pl.* 488.     *Cafta-*
                Common  *num.*

Common Horfe Chefnut.
*Nat.* of the North of Afia.
*Cult.* 1633, by Mr. John Tradefcant. *Ger. emac.* 1443.
*Fl.* April and May.        H. ♄.

*flava.*     2. Æ. foliolis quinis, corollæ laminis cordato-fubro
         tundis ; unguibus calyce duplo longioribus.
         Yellow-flower'd Horfe Chefnut.
         *Nat.* of North Carolina.
         *Cult.* 1764, by Mr. John Greening.
         *Fl.* May and June.        H. ♄.

*Pavia.*     3. Æ. foliis quinis, corollæ laminis obovatis ; unguibus
         longitudine calycis.
         Æfculus Pavia. *Sp. pl.* 488.
         Scarlet-flower'd Horfe Chefnut.
         *Nat.* of Carolina and Florida.
         *Cult.* 1712. *Philofoph. tranf. n.* 333. *p.* 424. *n.* 86.
         *Fl.* May and June.        H. ♄.

# *D I G Y N I A.*

## L I M E U M. *Gen. pl.* 463.

*Cal.* 5-phyllus. *Petala* 5, æqualia. *Capf.* globofa,
bilocularis.

*africa-*     1. LIMEUM. *Sp. pl.* 488.
*num.*       African Limeum.
         *Nat.* of the Cape of Good Hope.
         *Introd.* 1774, by Mr. Francis Maffon.
         *Fl.*                    G. H. ♃.

TETRAGY-

# TETRAGYNIA.

## SAURURUS. *Gen. pl.* 464.

*Cal.* amentum fquamis 1-floris. *Cor.* o. *Germina* 4.
*Baccæ* 4, monofpermæ.

1. S. caule foliofo polyftachyo. *Syft. veget.* 353.     *cernuus.*
Lizard's Tail.
*Nat.* of Virginia.
*Cult.* 1759, by Mr. Philip Miller. *Mill. dict. edit.* 7.
*Fl.* September.     H. ♃.

## APONOGETON. *Linn. fuppl.* 32.

*Cal.* amentum. *Cor.* o. *Capfulæ* 3-fpermæ.

1. A. fpica bifida, foliis lineari-oblongis natantibus, brac-   *diftachy-*
teis integris, floribus polyandris.     *on.*
Aponogeton diftachyon. *Linn. fuppl.* 215. *Thunb.*
*nov. gen.* 4. *p.* 74. *cum fig.*
Broad-leav'd Aponogeton.
*Nat.* of the Cape of Good Hope.
*Introd.* 1788, by Mr. Francis Maffon.
*Fl.* Moft part of the year.     G. H. ♃.
OBS. Linnæus in fupplemento hanc fpeciem cum
fequenti confundit, dum ftamina a fex ad duodecim
variare dicit.

2. A. fpica bifida, foliis lineari-lanceolatis erectis, bracteis   *anguftifo-*
bipartitis, floribus hexandris.     *lium.*
Narrow-leav'd Aponogeton.
*Nat.* of the Cape of Good Hope. Mr. *Fr. Maffon.*
*Introd.* 1788.
*Fl.* Moft part of the year.     G. H. ♃.

DESCR. *Folia* angustiora quam in præcedenti, utrinque attenuata. *Raches* pallide rubri. *Flores* pauci.
*Bracteæ* albæ, basi rubræ, ad basin fere bipartitæ
(vel, si mavis, binæ) : laciniæ lineari-oblongæ.
*Stamina* sex, bracteis ter vel quater breviora. *Styli*
tres.


# *HEPTAGYNIA.*


S E P T A S. *Gen. pl.* 465.

*Cal.* 7-partitus. *Petala* 7. *Germina* 7. *Capsulæ* 7,
polyspermæ.

*capensis.*  1. SEPTAS. *Sp. pl.* 489.
Round-leav'd Septas.
*Nat.* of the Cape of Good Hope.
*Introd.* 1774, by Mr. Francis Masson.
*Fl.* August and September.          G. H. ♃.


END OF THE FIRST VOLUME.

Printed in the United States
By Bookmasters